Antimicrobial Resistance and Hospital- and Community-Associated Infections

Antimicrobial Resistance and Hospital- and Community-Associated Infections

Guest Editor

Samantha Flores-Treviño

MDPI

Basel • Beijing • Wuhan • Barcelona • Belgrade • Novi Sad • Cluj • Manchester

Guest Editor
Samantha Flores-Treviño
Department of Infectious
Diseases
Autonomous University of
Nuevo Leon
Monterrey
Mexico

Editorial Office
MDPI AG
Grosspeteranlage 5
4052 Basel, Switzerland

This is a reprint of the Special Issue, published open access by the journal *Antibiotics* (ISSN 2079-6382), freely accessible at: https://www.mdpi.com/journal/antibiotics/special_issues/4A64Z4K650.

For citation purposes, cite each article independently as indicated on the article page online and as indicated below:

Lastname, A.A.; Lastname, B.B. Article Title. *Journal Name* **Year**, *Volume Number*, Page Range.

ISBN 978-3-7258-4247-6 (Hbk)
ISBN 978-3-7258-4248-3 (PDF)
https://doi.org/10.3390/books978-3-7258-4248-3

Cover image courtesy of Samantha Flores-Treviño

Contents

About the Editor

Samantha Flores-Treviño

Samantha Flores-Treviño, Ph.D., is a tenured Professor at the Department of Infectious Diseases of the Autonomous University of Nuevo Leon. She is a leading microbiologist and clinical researcher specializing in antimicrobial resistance, biofilms, and healthcare-associated infections. With over 60 publications in high-impact journals indexed in *Journal Citation Reports*, her work spans molecular diagnostics, resistant pathogens including ESKAPE microorganisms, and innovations in fecal microbiota transplantation. Dr. Flores-Treviño is actively involved in national multicenter studies and serves as a prominent voice in infectious disease prevention and translational microbiology in Mexico and Latin America.

antibiotics MDPI

Editorial

Antimicrobial Resistance and Hospital- and Community-Associated Infections

Samantha Flores-Treviño

Department of Infectious Diseases, Hospital Universitario Dr. José Eleuterio González, Universidad Autónoma de Nuevo León, Monterrey 64460, Mexico; samantha.florestr@uanl.edu.mx

Antimicrobial resistance (AMR) poses a significant global threat to human health, and was estimated to be associated with almost one million deaths in 2019 [1,2]. The World Health Organization (WHO) recently issued an updated *Bacterial Priority Pathogens List* including antimicrobial-resistant microorganisms [3]. Among Gram-negative bacteria resistant to last-resort antibiotics, studies on *Acinetobacter baumannii* and the Enterobacterales order are critical due to their ability to transfer resistance genes. In addition, studies on antimicrobial-resistant *Pseudomonas aeruginosa*, *Staphylococcus aureus* and *Enterococcus faecium* are of high priority due to the threat they pose in healthcare settings. Healthcare-associated infections (HAIs) caused by ESKAPE pathogens (vancomycin-resistant *Enterococcus* spp. [VRE], methicillin-resistant *S. aureus* [MRSA], carbapenem-resistant *Klebsiella pneumoniae*, multidrug-resistant [MDR] *A. baumannii*, MDR *P. aeruginosa*, and carbapenem-resistant *Enterobacter* spp.) are known to cause high morbidity and mortality in patients due to their acquired resistance to last-resort antibiotics [4–6]. Patients in intensive care units or those who are immunocompromised are most affected by HAIs caused by MDR pathogens, which can increase disease severity. The most frequent HAIs are bloodstream infections, ventilator-associated pneumonia and surgical site infections [7].

The contributions to this Special Issue regarding Gram-negative pathogens include the following: Luna-De-Alba et al. (2024) conducted an assessment of different antimicrobial combinations against MDR or extensive drug-resistant (XDR) *A. baumannii* strains. Most strains carried carbapenemase OXA-24/40, aminoglycoside-modifying enzymes, and *parC* gene mutations; overexpressed AdeIJK, AdeABC, and AdeFGH efflux pumps and CarO membrane porin; and under-expressed Omp33-36, OmpA, and CarO membrane porins, and showed low biofilm production. Interestingly, antimicrobial combinations such as levofloxacin-ampicillin/sulbactam and meropenem-colistin inhibited bacterial growth. In the contribution by Papa-Ezdra et al. (2024), MDR *P. aeruginosa* strains that caused an outbreak in an ICU were analyzed and found to all be clonally related and belong to ST309, an emerging high-risk clone in the Americas. The strains were resistant to ceftazidime, cefepime, amikacin, and ceftolozane-tazobactam, and harbored $bla_{\text{PER-1}}$ and *qnrVC* genes. Yamaki et al. (2024) assessed the clinical outcomes of antibiotic modifications in patients with infections caused by Gram-negative pathogens exhibiting resistance to extended-spectrum cephalosporins. Although 72% of patients had antibiotic regimen modifications, their clinical outcomes showed no differences, which highlights the importance of identifying patients at risk for resistant organisms early in admission.

In the contribution by Chotiprasitsakul et al. (2023), an assessment of the epidemiology of community-onset bloodstream infections in Thailand showed that 25% of infections were antimicrobial-resistant AMR, and one-third of Enterobacterales (*Escherichia coli* and *K. pneumoniae*) were not susceptible to ceftriaxone. Li et al. (2023) assessed the effectiveness of

 https://doi.org/10.3390/antibiotics14050514

multi-model strategies on HAIs caused by MDR pathogens in rehabilitation units in China, which decreased the burden of HAIs in general and of HAIs caused by MDR pathogens, in addition to the contamination rate of MDR pathogens in the environmental setting.

Kumar et al. (2024) overviewed the epidemiology of MDR sepsis and found that underlying comorbidities, old age, antibiotic overuse and inadequate empiric therapy contribute to recurrent sepsis, with high mortality rates. Effective sepsis treatment includes the use of antimicrobial combination therapy and the exploitation of local pathogen resistance patterns.

The contributions to this Special Issue regarding Gram-positive pathogens include the following: Balasiu and MacKenzie (2023) conducted an assessment of teicoplanin resistance in coagulase negative staphylococci (CoNS), which was challenging and influenced by technical factors. Their results emphasized the need for future studies focusing on the clinical efficacy of teicoplanin in relation to its susceptibility. Sohail et al. (2023) assessed MRSA isolates from Pakistan during the COVID-19 pandemic, of which 56% were HAIs and 44% were community-acquired. Most MRSA isolates detected were weak biofilm producers and adhesion genes (*clfB*, *icaAD*, *fib*, *sdrC*, *eno*, *fnbA*, *sdrE*, *icaBC*, *clfA*, *fnbB*, *sdrD*, and *cna*). In the contribution by Worku et al. (2023), the prevalence of MRSA was evaluated in surgical site infections in Ethiopia. Patients positive for *S. aureus* accounted for 21.6% of all patients, among which 24.5% had MRSA; moreover, the *mecA* gene was detected in 27.5% of isolates. Among the risk factors associated with MRSA infections were older age, prolonged hospitalization and previous antibiotic administration.

Clostridioides difficile is one of the most common pathogens in hospitalized patients receiving antimicrobial therapy and is the leading cause of hospitalization [8]. Salas-Treviño et al. (2025) showed that the co-colonization of *C. difficile* and other non-*difficile* Clostridia (*Clostridium ramosum* or *Clostridium innocuum*) in patients with antibiotic-associated diarrhea was correlated with treatment extension and failure.

The main conclusion drawn from the above contributions is that AMR is still an issue in most hospitals worldwide; thus, hospitals should strengthen their strategies for infection prevention, continue surveillance of antimicrobial resistance genes, and promote the implementation of antimicrobial stewardship programs. More prevention and control strategies are needed to reduce the burden of antimicrobial resistance in both hospitals and communities.

Funding: This research received no external funding.

Conflicts of Interest: The author declares no conflicts of interest.

List of Contributions:

1. Salas-Treviño, D.; Flores-Treviño, S.; Cisneros-Rendón, C.; Domínguez-Rivera, C.V.; Camacho-Ortiz, A. Co-Colonization of Non-*difficile* Clostridial Species in Antibiotic-Associated Diarrhea Caused by *Clostridioides difficile*. *Antibiotics* **2025**, *14*, 397. https://doi.org/10.3390/antibiotics14040397.
2. Luna-De-Alba, A.; Flores-Treviño, S.; Camacho-Ortiz, A.; Contreras-Cordero, J.F.; Bocanegra-Ibarias, P. Genetic Characterization of Multidrug-Resistant *Acinetobacter baumannii* and Synergy Assessment of Antimicrobial Combinations. *Antibiotics* **2024**, *13*, 1079. https://doi.org/10.3390/antibiotics13111079.
3. Yamaki, J.; Mikhail, M.; Beuttler, R.; Robinson, P.; Yücel, E.; Watanabe, A.H. Characterizing Antibiotic Regimen Modification Behavior, Patient Characteristics, and Outcomes for Patients with Gram-Negative Bacterial Infections, A Retrospective Single-Center Study. *Antibiotics* **2024**, *13*, 302. https://doi.org/10.3390/antibiotics13040302.
4. Papa-Ezdra, R.; Outeda, M.; Cordeiro, N.F.; Araújo, L.; Gadea, P.; Garcia-Fulgueiras, V.; Seija, V.; Bado, I.; Vignoli, R. Outbreak of *Pseudomonas aeruginosa* High-Risk Clone ST309 Serotype O11

Featuring blaPER-1 and qnrVC6. *Antibiotics* **2024**, *13*, 159. https://doi.org/10.3390/antibiotics13020159.

5. Chotiprasitsakul, D.; Trirattanapikul, A.; Namsiripongpun, W.; Chaihongsa, N.; Santanirand, P. From Epidemiology of Community-Onset Bloodstream Infections to the Development of Empirical Antimicrobial Treatment-Decision Algorithm in a Region with High Burden of Antimicrobial Resistance. *Antibiotics* **2023**, *12*, 1699. https://doi.org/10.3390/antibiotics12121699.
6. Worku, S.; Abebe, T.; Seyoum, B.; Alemu, A.; Shimelash, Y.; Yimer, M.; Abdissa, A.; Beyene, G.T.; Swedberg, G.; Mihret, A. Molecular Epidemiology of Methicillin-Resistant *Staphylococcus aureus* among Patients Diagnosed with Surgical Site Infection at Four Hospitals in Ethiopia. *Antibiotics* **2023**, *12*, 1681. https://doi.org/10.3390/antibiotics12121681.
7. Li, S.; Lin, J.; Tao, S.; Guo, L.; Huang, W.; Li, J.; Du, C.; Wang, Z.; Liu, L.; Chen, Y.; et al. Multi-Model Strategies for Prevention of Infection Caused by Certain Multi-Drug Resistant Organisms in A Rehabilitation Unit: A Semi-Experimental Study. *Antibiotics* **2023**, *12*, 1199. https://doi.org/10.3390/antibiotics12071199.
8. Balasiu, A.D.; MacKenzie, C.R. Teicoplanin-Resistant Coagulase-Negative Staphylococci: Do the Current Susceptibility Testing Methods Reliably Detect This Elusive Phenotype? *Antibiotics* **2023**, *12*, 611. https://doi.org/10.3390/antibiotics12030611.
9. Sohail, M.; Muzzammil, M.; Ahmad, M.; Rehman, S.; Garout, M.; Khojah, T.M.; Al-Eisa, K.M.; Breagesh, S.A.; Hamdan, R.M.A.; Alibrahim, H.I.; et al. Molecular Characterization of Community- and Hospital- Acquired Methicillin-Resistant *Staphylococcus aureus* Isolates during COVID-19 Pandemic. *Antibiotics* **2023**, *12*, 157. https://doi.org/10.3390/antibiotics12010157.
10. Kumar, N.R.; Balraj, T.A.; Kempegowda, S.N.; Prashant, A. Multidrug-Resistant Sepsis: A Critical Healthcare Challenge. *Antibiotics* **2024**, *13*, 46. https://doi.org/10.3390/antibiotics13010046.

References

1. Okeke, I.N.; de Kraker, M.E.A.; Van Boeckel, T.P.; Kumar, C.K.; Schmitt, H.; Gales, A.C.; Bertagnolio, S.; Sharland, M.; Laxminarayan, R. The scope of the antimicrobial resistance challenge. *Lancet* **2024**, *403*, 2426–2438. [CrossRef] [PubMed]
2. Muller, C. Antibiotics and Antimicrobials Resistance: Mechanisms and New Strategies to Fight Resistant Bacteria. *Antibiotics* **2022**, *11*, 400. [CrossRef] [PubMed]
3. WHO. *Bacterial Priority Pathogens List, 2024: Bacterial Pathogens of Public Health Importance to Guide Research, Development and Strategies to Prevent and Control Antimicrobial Resistance*; World Health Organization: Geneva, Switzerland, 2024.
4. Avershina, E.; Shapovalova, V.; Shipulin, G. Fighting Antibiotic Resistance in Hospital-Acquired Infections: Current State and Emerging Technologies in Disease Prevention, Diagnostics and Therapy. *Front. Microbiol.* **2021**, *12*, 707330. [CrossRef] [PubMed]
5. Wolford, H.; McCarthy, N.L.; Baggs, J.; Hatfield, K.M.; Maillis, A.; Olubajo, B.; Bishop, J.; Ferretti, M.; Craig, M.R.; Magill, S.S.; et al. Antimicrobial-Resistant Infections in Hospitalized Patients. *JAMA Netw. Open* **2025**, *8*, e2462059. [CrossRef] [PubMed]
6. Balasubramanian, R.; Van Boeckel, T.P.; Carmeli, Y.; Cosgrove, S.; Laxminarayan, R. Global incidence in hospital-associated infections resistant to antibiotics: An analysis of point prevalence surveys from 99 countries. *PLoS Med.* **2023**, *20*, e1004178. [CrossRef] [PubMed]
7. Abban, M.K.; Ayerakwa, E.A.; Mosi, L.; Isawumi, A. The burden of hospital acquired infections and antimicrobial resistance. *Heliyon* **2023**, *9*, e20561. [CrossRef] [PubMed]
8. Alexiou, S.; Diakou, A.; Kachrimanidou, M. The Role of Clostridioides difficile Within the One Health Framework: A Review. *Microorganisms* **2025**, *13*, 429. [CrossRef] [PubMed]

antibiotics

MDPI

Article

Genetic Characterization of Multidrug-Resistant *Acinetobacter baumannii* and Synergy Assessment of Antimicrobial Combinations

Aurora Luna-De-Alba [1], Samantha Flores-Treviño [2], Adrián Camacho-Ortiz [2], Juan Francisco Contreras-Cordero [1] and Paola Bocanegra-Ibarias [2,*]

1 Laboratory of Immunology and Virology, School of Biological Sciences, Autonomous University of Nuevo Leon, Monterrey 66455, Nuevo Leon, Mexico; maria.lunad@uanl.edu.mx (A.L.-D.-A.); contrerasjfco@gmail.com (J.F.C.-C.)

2 Department of Infectious Diseases, University Hospital Dr. José Eleuterio González, Autonomous University of Nuevo Leon, Monterrey 64460, Nuevo Leon, Mexico; samflorest@gmail.com (S.F.-T.); acamacho_md@yahoo.com (A.C.-O.)

* Correspondence: paola.bocanegraib@gmail.com

Abstract: Background/Objectives: *A. baumannii* is a prominent nosocomial pathogen due to its drug-resistant phenotype, representing a public health problem. In this study, the aim was to determine the effect of different antimicrobial combinations against selected multidrug-resistant (MDR) or extensive drug-resistant (XDR) isolates of *A. baumannii*. **Methods**: MDR or XDR *A. baumannii* isolates were characterized by assessing genes associated with drug resistance, efflux pumps, porin expression, and biofilm formation. The activities of antimicrobial combinations including tigecycline, ampicillin/sulbactam, meropenem, levofloxacin, and colistin were evaluated using checkerboard and time-to-kill assays on isolates with different susceptibility profiles and genetic characteristics. **Results**: Genetic characterization of MDR/XDR strains (n = 100) included analysis of OXA-24/40 gene carbapenemase (98%), genes encoding aminoglycoside-modifying enzymes (44%), and *parC* gene mutations (10%). AdeIJK, AdeABC, and AdeFGH efflux pumps were overexpressed in 17–34% of isolates. Omp33-36, OmpA, and CarO membrane porins were under-expressed in 50–76% of isolates; CarO was overexpressed in 22% of isolates. Isolates showed low biofilm production (11%). Synergistic activity was observed with levofloxacin-ampicillin/sulbactam and meropenem-colistin, which were able to inhibit bacterial growth. **Conclusions**: Genetic characteristics of *A. baumannii* were highly variable among the strains. Synergistic activity was observed with meropenem-colistin and levofloxacin-ampicillin/sulbactam combinations in the checkerboard method, but not in the time-to-kill assays. These discrepancies among both methods indicate that further studies are needed to determine the best therapeutic combination for treating infections by *A. baumannii*.

Keywords: genetic characterization; colistin; meropenem; efflux pumps; porins

1. Introduction

Acinetobacter baumannii is a Gram-negative coccobacillus associated with several hospital-acquired infections, occurring in critically ill patients, such as ventilator-associated pneumonia and bacteremia, with attributable mortality rates up to 35% [1]. This pathogen represents a worldwide public health problem due to its ability to survive on different surfaces of the hospital environment, and its ability to acquire and develop a diversity of antimicrobial resistance mechanisms against different antibiotics [2]. Whilst carbapenems are considered the first choice of treatment against *A. baumannii* infections, the relentlessly increasing prevalence of carbapenem-resistant *A. baumannii* (CRAB) strains signifies a threat to susceptible patients, increasing mortality up to 70% [1,3,4].

Given the rising rates of resistance to multiple antimicrobials and the lack of development of new molecules with efficacy against this pathogen, the antibiotic combination therapy has been considered as a strategy to effectively control *A. baumannii* infection [5]. The antibiotic combination therapy uses two or more drugs with different mechanisms of action to treat a bacterial infection, in order to improve therapeutic efficacy, delaying the development of drug resistance, reducing toxicity, and broadening the spectrum of antibacterial activity [6].

Although there is still no consensus on the optimal treatment of CRAB infections, colistin is most often used in combination with other antibiotics, such as carbapenems, fosfomycin, tigecycline, ampicillin/sulbactam, vancomycin, or rifampin [1]. However, resistance to colistin can occur in up to 30% of CRAB strains, complicating the treatment of CRAB infections [1,3].

Evaluating the in vitro activity of antimicrobial combinations against a bacterial pathogen is challenging due to the technically complex and time-consuming process. Some of the most used techniques to assess the in vitro activity of antimicrobial combinations are the checkerboard and time-to-kill assays. In this study, the aim was to determine the effect of different antimicrobial combinations against selected multidrug-resistant (MDR) or extensive drug-resistant (XDR) isolates of *A. baumannii.*

2. Results

2.1. Characteristics of MDR and XDR Isolates

During the two-year period, 263 *A. baumannii* clinical isolates were collected from 192 patients. Patients were predominantly hospitalized in the intensive care unit (55.7%, $n = 107$), in the COVID unit (24.5%, $n = 47$), in the internal medicine ward (15.1%, $n = 29$), and other medical wards (4.7%, $n = 9$).

Out of the 192 strains, 91.7% ($n = 176$) were either MDR (78.6%, $n = 151$) or XDR (13.1%, $n = 25$), of which 90.1% ($n = 173$) were CRAB isolates. Out of the 176 strains classified as either MDR or XDR, 100 isolates were obtained from respiratory tract specimens and were further selected for genetic characterization and synergy effect testing. These selected isolates presented high resistance to ceftazidime (100%), levofloxacin (100%), imipenem (98%), meropenem (98%), piperacillin/tazobactam (97%), cefepime (95%), and gentamicin (89%). Lower rates of resistance to tigecycline (65%), ampicillin/sulbactam (35%), and colistin (1%) were detected.

2.2. Genetic Characteristics of MDR and XDR Isolates

Regarding carbapenemases, 98% ($n = 98$) of the resistant strains carried the *OXA-24/40* gene (a class D carbapenemase) and 100% ($n = 100$) the *OXA-51* gene (a species-specific intrinsic carbapenemase). *KPC, VIM, IMP, NDM,* and *mcr* genes were not detected in any of the isolates. Genes encoding aminoglycoside-modifying enzymes were distributed heterogeneously among the isolates. At least one gene was detected in 44% of the strains, of which the most frequent was *aph*(3′)VIa (31%), followed by *ant*(2′)Ia (25%), *aph*(3′)IIa (12%), and *aac*(6)Ib (12%). In 53.2% (50/94) of gentamicin non-susceptible isolates, no genes encoding aminoglycoside-modifying enzymes were detected. Up to 10% of the isolates presented mutations in *parC* gene associated with fluoroquinolone resistance. No mutations were detected in *gyrA, pmrA,* and *pmrB* genes, associated with quinolone and polymyxin resistance.

In addition, efflux pump overexpression was observed in 34% of the isolates for AdeIJK (2.2–119.3-fold change), 29% for AdeABC (2.1–133.3-fold change), and 17% for AdeFGH (2.2–86.4-fold change). Membrane porins were under-expressed (<0.5 fold) in most isolates, 76% for Omp33-36, 54% for OmpA, and 50% for CarO, although 22% of isolates showed CarO overexpression (Figure 1). Regarding biofilm production, 11% ($n = 11$) of the isolates were biofilm producers, 6% presented high biofilm production, and 5% were low biofilm producers.

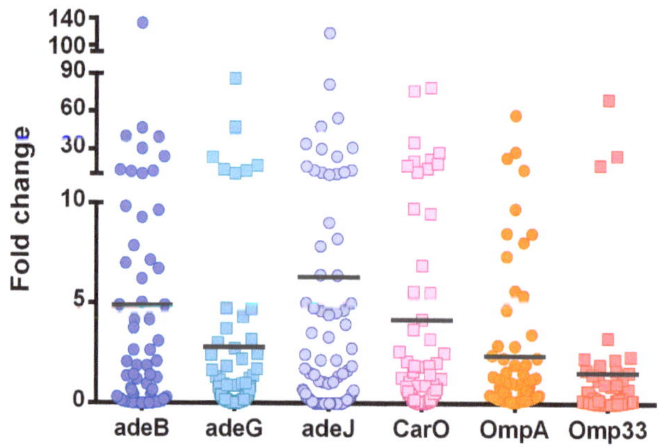

Figure 1. Expression levels of efflux pumps and membrane porins in MDR *A. baumannii* strains. The expression levels of AdeABC, AdeFGH, and AdeIJK pumps and CarO, OmpA, and Omp33-36 membrane porins are shown, compared to the reference strain (*A. baumannii* ATCC 17978), used as baseline. The line represents the mean of each fold change.

2.3. Activity of Antimicrobial Combinations by the Checkerboard Assay

Isolates were first classified into groups according to their antimicrobial resistance profile and genetic characteristics. However, the majority of isolates presented a unique pattern, and categorization was not possible. Consequently, 42 strains were randomly selected to evaluate the synergistic effects of antimicrobial combinations (Table S1). A heterogeneous behavior was observed among the isolates after exposure of combined tested antibiotics (Table 1). Although most of the isolates showed indifferent activity to several antimicrobial combinations (38.1–97.6%), some isolates presented additive activity to levofloxacin-SAM (28.6%), meropenem-colistin (19.0%), colistin-levofloxacin (14.3%), tigecycline-colistin (2.4%), and meropenem-levofloxacin (2.4%). Antagonistic activity was observed for meropenem-colistin (40.5%), tigecycline-levofloxacin (9.5%), meropenem-levofloxacin (4.8%), levofloxacin-SAM (2.4%), and tigecycline-meropenem (2.4%).

Table 1. Dual-therapy results for different antimicrobial combinations against MDR and XDR *A. baumannii* isolates.

Antimicrobial Combination *	No. (%) of Isolates with Combined Effect			
	Synergistic	Additive	Indifferent	Antagonistic
LEV + SAM	1 (2.4)	12 (28.6)	28 (66.7)	1 (2.4)
MEM + COL	1 (2.4)	8 (19.0)	16 (38.1)	17 (40.5)
TGC + MEM	0 (0.0)	0 (0.0)	41 (97.6)	1 (2.4)
TGC+ COL	0 (0.0)	1 (2.4)	41 (97.6)	0 (0.0)
TGC + LEV	0 (0.0)	0 (0.0)	38 (90.5)	4 (9.5)
MEM + LEV	0 (0.0)	1 (2.4)	39 (92.9)	2 (4.8)
COL + LEV	0 (0.0)	6 (14.3)	36 (85.7)	0 (0.0)

* SAM: ampicillin/sulbactam; MEM: meropenem; LEV: levofloxacin; TGC: tigecycline; COL: colistin.

Synergistic activity was observed with levofloxacin (16 μg/mL) and SAM (16/8 μg/mL) with FICI = 0.5 in one isolate (19-2211), and with meropenem (16 μg/mL) and colistin (1 μg/mL) with FICI = 0.3 in another isolate (20-0329), as shown in Table 2. The isolates in which the synergistic activity was observed showed different genetic characteristics. Isolate 19-2211 showed CarO overexpression and both OmpA and Omp33-36 under-expression. Isolate 20-0239 showed overexpression of adeFGH pump, OmpA, and Omp33-36. Neither isolate showed mutations associated with quinolone or polymyxin resistance, nor were they biofilm producers (Table S1).

Table 2. Comparison of concentrations used in monotherapy and dual therapy and antimicrobial activity.

Antimicrobial Combination *	Isolate	MIC of Individual Antibiotic (μg/mL)		MIC of Antibiotics in Combination (μg/mL)		FICI	Activity
		Antibiotic 1	Antibiotic 2	Antibiotic 1	Antibiotic 2		
LEV + SAM	19-2211	LEV (64)	SAM (16/8)	LEV (16)	SAM (16/8)	0.5	Synergistic
	20-0046	LEV (32)	SAM (16/8)	LEV (16)	SAM (32/16)	2.5	Indifferent
MEM + COL	20-0329	MEM (64)	COL (16)	MEM (16)	COL (1)	0.3	Synergistic
	19-0705	MEM (64)	COL (0.25)	MEM (256)	COL (0.25)	5.0	Antagonistic
TGC + MEM	19-0002	TGC (2)	MEM (64)	TGC (2)	MEM (8)	1.1	Indifferent
	19-0360	TGC (2)	MEM (64)	TGC (2)	MEM (8)	1.1	Indifferent
TGC+ COL	20-0008	TGC (2)	COL (1)	TGC (2)	COL (0.12)	1.1	Indifferent
	20-0327	TGC (2)	COL (0.5)	TGC (2)	COL (0.12)	1.2	Indifferent
TGC + LEV	19-0115	TGC (2)	LEV (32)	TGC (2)	LEV (2)	1.1	Indifferent
	20-0046	TGC (2)	LEV (2)	TGC (8)	LEV (8)	8.0	Antagonistic
MEM + LEV	20-0098	MEM (64)	LEV (4)	MEM (32)	LEV (16)	4.5	Antagonistic
	20-0406	MEM (64)	LEV (32)	MEM (32)	LEV (8)	0.8	Additive
COL + LEV	19-1092	COL (1)	LEV (16)	COL (0.25)	LEV (16)	1.3	Indifferent
	20-0048	COL (0.5)	LEV (16)	COL (0.25)	LEV (4)	0.8	Additive

* SAM: ampicillin/sulbactam; MEM: meropenem; LEV: levofloxacin; TGC: tigecycline; COL: colistin; FICI: fractional inhibitory concentration index.

2.4. Activity of Antimicrobial Combinations by the Time-to-Kill Method

The bacterial inhibitory effect of synergistic antimicrobial concentrations was evaluated using time-to-kill curves. According to the results, both antimicrobial combinations (levofloxacin-SAM and meropenem-colistin) were able to partially inhibit bacterial growth (Figure 2). Particularly, the combination of meropenem and colistin caused a decrease in bacterial growth during the first 4 h, unlike the effect shown in each antibiotic individually. However, this antimicrobial effect remained the same regardless of single or dual combination after 24 h, which increased after 48 h of incubation. Bacterial regrowth was observed after 8 h of incubation with meropenem plus colistin. Instead, an antagonistic effect was observed with the combination of SAM and levofloxacin, although higher bacterial inhibition was observed after using SAM alone.

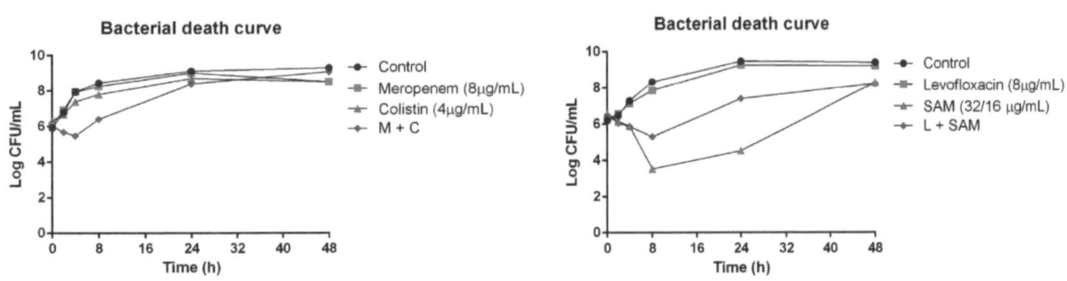

(a) (b)

Figure 2. Time-to-kill curve of MDR/XDR *A. baumannii* isolates under the combination of two antibiotics. The time-to-kill curve per hour is shown for two different MDR/XDR *A. baumannii* isolates: (**a**) isolate 20-0329, treated with the combination of meropenem (16 μg/mL) and colistin (1 μg/mL), and (**b**) isolate 19-2211, treated with the combination of levofloxacin (16 μg/mL) and SAM (16/8 μg/mL). C: colistin, L: levofloxacin; M: meropenem; MDR: multidrug-resistant; SAM: ampicillin/sulbactam; XDR: extensive drug-resistant.

3. Discussion

Over the years, *A. baumannii* has emerged as a prominent nosocomial pathogen due to its MDR phenotype, representing a public health problem. Increased morbidity and mortality can be associated with MDR and XDR phenotypes. Our study shows that 78.6% of the resistant isolates exhibited an MDR profile, while 13.1% exhibited an XDR profile, and 90.1% were CRAB. Past studies from Mexico also showed lower MDR values (44.3%) compared to ours, although XDR values were similar in our strains compared to previous reports of 11.4–56.6% [7,8]. These results suggest a greater capacity for adaptation and dissemination of XDR strains, highlighting the importance of *A. baumannii* in the hospital setting and the need to research alternative therapy solutions, such as antimicrobial combination therapy. In this study, we evaluated the prevalence of *A. baumannii* drug resistance phenotypes and characterized their genetic characteristics to assess antibiotic combinations with specific synergistic activity.

The genetic characterization of the strains, particularly carbapenem resistance, showed OXA-24/40 in 98% of the strains and the species-intrinsic OXA-51 gene in all the strains. Compared to previous results [9], our results show increased OXA-24/40 frequency (98% vs. 25.7%) and decreased OXA-58 detection (0% vs. 28.3%). Regarding aminoglycoside resistance, *aph*(3′)VIa was the most frequent gene in non-susceptible isolates, although lower than other studies (52% vs. 31%) [10]. Regarding fluoroquinolone resistance, only 10% of the isolates presented mutations in *parC* gene. Overexpression of efflux pumps (AdeIJK, AdeABC, and AdeFGH) associated with the MDR phenotype was also detected, similar to previous studies [11]. Efflux pump substrate affinities and expression levels can be associated with different resistance to multiple antibiotics, e.g., AdeIJK has a broader substrate spectrum than AdeABC pump [12]. Membrane porins were all under-expressed, Omp33-36 more than OmpA and CarO. Omp33-36 loss is more common in pneumonia isolates, as is the majority of isolates analyzed in our study [13]. A low frequency of efflux pump overexpression is associated with CRAB [8], which in our study was 98%.

Biofilm formation promotes antibiotic resistance in *A. baumannii,* as transmission of resistance mechanisms occurs among bacterial strains within biofilms. Previous studies show that the prevalence of antimicrobial resistance correlated with strong biofilm formation, as higher biofilm production was observed for XDR strains compared to MDR strains [14]. In this study, low biofilm production was detected, differing from previous studies (11% vs. 90.8%) [8]. These results suggest that our MDR *A. baumannii* possesses diverse resistance mechanisms, which help the bacteria to adapt and survive in different environments.

In vitro activity of drug combinations on *A. baumannii* strains with different genetic characteristics was assessed by checkerboard and time-to-kill assays. Selected drugs were chosen based on their specific mechanism of action, e.g., tigecycline (protein synthesis), SAM and meropenem (cell wall synthesis), levofloxacin (DNA replication), and colistin (cell membrane permeability). Synergistic activity was observed only with two antimicrobial combinations, meropenem-colistin and levofloxacin-SAM, in two different strains.

Meropenem-colistin combination showed not only synergistic and additive effects, but mainly antagonistic activity, which differs from previous results [15], in which antagonism was not observed in doripenem. Doripenem was used instead of meropenem due to its stability against carbapenemases. However, we did not use doripenem due to its lack of availability in our country. Furthermore, the synergistic activity of meropenem-colistin occurred in a strain resistant to both antibiotics. As available options to treat carbapenem- and colistin-resistant *A. baumannii* are limited and in most cases empirical, our results provide insight regarding the use of meropenem-colistin. However, while meropenem-colistin caused a decrease in bacterial growth during 4 h, bacterial regrowth was observed after 8 h, which increased after 48 h. In a previous study [16], the activity of colistin against *Klebsiella pneumoniae* in 24 h time-to-kill assays also showed initial killing followed by regrowth of strains at 24 h, suggesting a bacteriostatic effect rather than bactericidal, as might be our case. Thus, further studies are needed to assess a greater strain sample to investigate whether the synergistic activity of meropenem-colistin is related to the antimicrobial susceptibility profile or genetic characteristics. Levofloxacin plus SAM was the combination with higher additive effect, and synergy was observed in one isolate. A previous study showed 90% of synergistic activity on isolates resistant to both levofloxacin and SAM [17], which suggests it might be considered a good therapeutic option, although further studies are needed to assess strains with more diverse mechanisms of resistance. Tigecycline, in three different combinations, presented predominantly indifferent activity and no synergistic activity. However, previous studies reported synergy of tigecycline and levofloxacin in 16.7% of isolates [18], and of tigecycline plus colistin in 40.6% of isolates [15].

High variability was observed in the genetic characteristics of the population studied, which may explain the overall low synergistic activity observed. Synergistic activity was observed in *A. baumannii* strains with different genetic characteristics and different colistin, gentamycin, and SAM susceptibility. Synergistic activity may be influenced by the variability in the mechanisms involved in bacterial drug resistance [19]. In previous studies, higher synergism was observed in isolates with high MIC values [19–21]. Likewise, in our study, a MIC reduction was observed for meropenem (64 vs. 16 µg/mL), colistin (16 vs. 1 µg/mL), and levofloxacin (64 vs. 16 µg/mL) after using antibiotic combinations in dual therapy. According to our results, dual therapy reduced the concentration of antibiotics needed to inhibit bacterial growth compared to monotherapy not only in isolates with synergistic activity, but also in those with additive and indifferent effects. These results suggest dual therapy could offer an advantage for clinical treatment; however, more in vitro and in vivo studies are required.

One limitation of this study is that time-to-kill curves using optimal concentrations from the checkerboard assay did not confirm the previously observed synergistic activity. Several factors, such as bacteria and drug type, drug concentration, exposure time, and analysis method may account for the discrepancy between the checkerboard and time-to-kill assays, which can show variable results [22]. Therefore, the selection of the most appropriate method and the validation of the results with complementary methods are important. In addition, the analysis of antibiotic combinations and concentrations not included in time-to-kill assays might allow one to decipher the dose dependency of the observed synergistic or antagonistic activity. Furthermore, none of the in vitro synergistic assays are standardized, and their results may be controversial; therefore, data should be analyzed with caution and should be correlated with the clinical data of the patient.

4. Materials and Methods

4.1. Study Population

Consecutive *A. baumannii* isolates were collected during 2019 and 2020 from the routine microbiology laboratory of the Dr. José Eleuterio González University Hospital, a tertiary-care teaching hospital with 600 hospital beds located in Monterrey, Mexico. The hospital has a yearly average of 25,000 hospitalizations, and it receives patients transferred from other regional hospitals and from the northeastern states of Mexico. Only one isolate per patient and from respiratory tract specimens (bronchial lavage, bronchoalveolar lavage, endotracheal aspirate, and expectoration) were selected for the study.

4.2. Culture and Identification of Clinical Isolates

The strains were grown on blood agar plates (BD Bioxon, Mexico City, Mexico) and incubated at 37 °C for 24 h. *Acinetobacter calcoaceticus-baumannii* complex identification was performed using matrix-assisted laser desorption ionization-time of flight mass spectrometry (MALDI-TOF MS, Microflex LT system, Bruker Daltonics, Bremen, Germany) as described by the manufacturer. The identification of *A. baumannii* species was performed by *recA* and 16S-23S rRNA intergenic spacer genes amplification using the primers and conditions described previously [23].

4.3. *Antimicrobial Susceptibility Testing*

Antimicrobial susceptibility testing was conducted by disk diffusion according to the recommended methods in the M100 and breakpoints established in M02 protocols of the Clinical and Laboratory Standards Institute (CLSI, Wayne, PA, USA) [24]. The antibiotics tested were ampicillin/sulbactam (SAM), piperacillin/tazobactam, cefepime, ceftazidime, imipenem, meropenem, gentamicin (GEN), levofloxacin, and tigecycline (Thermo Fisher Scientific Oxoid Ltd., Basingstoke, UK). Colistin screening was performed using the colistin broth disk elution and confirmation was evaluated by broth microdilution as recommended by the CLSI.

The isolates were classified as non-MDR, MDR, or XDR according to previous recommendations [25]. Isolates non-susceptible (either intermediate or resistant) to three or more antibiotic categories were considered as MDR. Isolates non-susceptible to at least one agent from all but two or fewer antibiotic categories were considered as XDR. Only MDR or XDR isolates were selected for further analysis.

4.4. *Genetic Characterization*

4.4.1. Detection of Antimicrobial Resistance-Associated Genes

Either the presence or a mutation of different antimicrobial resistance-associated genes were evaluated by PCR and DNA sequencing. The carbapenemase-encoding genes analyzed were class A β-lactamases (*KPC*), metallo-β-lactamases (*IMP*, *VIM*, and *NDM*), and OXA-type (*OXA-23-like*, *OXA-24/40-like*, *OXA-51-like*, and *OXA-58-like*) using primers and conditions previously reported [26]. The genes encoding the following aminoglycoside-modifying enzymes were analyzed using the primers and conditions previously reported [27]: *aph*(3′)Ia, *aph*(3′)VIa, *aac*(3′)Ia, *aac*(3′)IIa, *acc*(6′)Ib, *aac*(6′)Ih, and *ant*(2′)Ia. Colistin resistance-associated mcr gene was also detected by PCR [28]. Mutations in *parC*, *gyrA*, *pmrA*, and *pmrB* genes were submitted for large-scale DNA sequencing (Macrogen, South Korea) using the primers and conditions suggested previously [29,30]. The sequences were analyzed on the BioEdit platform (Informer Technologies, Inc., Los Angeles, CA, USA). Mutations were searched using the reference strains *A. baumannii* (GenBank accession number X82165.1) for *gyrA* and (GenBank accession number X95819.1) for *parC*; *A. baumannii* strain AB67 (GenBank accession number MF673422.1) for *pmrA*, and *A. baumannii* strain S402 (GenBank accession number MK660501.1) for *pmrB*.

4.4.2. Assessment of Efflux Pump and Porin Expression

The expression of *adeB*, *adeG*, and *adeJ* (genes belonging to efflux pump systems AdeABC, AdeFGH, and AdeIJK, respectively) and *ompA*, *carO*, and *omp33* (genes harboring porins or outer membrane proteins) was determined by RT-qPCR [31–34]. Total RNA was extracted from a 4–5 h log phase culture of *A. baumannii* using the RNeasy mini kit (Qiagen, Venlo, The Netherlands) according to the manufacturer's instructions. RNA concentration and purity were determined using a NanoDrop spectrophotometer (ND-1000, Wilmington, NC, USA). Quantification of *adeB*, *adeJ*, *ompA*, *carO*, and *omp33* was performed using the SuperScript III platinum One-step qRT-PCR system (Invitrogen, Cergy Pontoise, France). *adeG* quantification was performed using the PowerUp SYBR Green Master Mix for qPCR (Applied biosystems, Foster, CA, USA). The RT-qPCR was performed using 20 ng of RNA and the primers and probes described in Supplementary Table S1 in a Bio-Rad CFX instrument (Bio-Rad, Hercules, CA, USA). *rpoB* gene was used as a housekeeping gene to normalize the expression of target genes [34]. The $2^{-\Delta\Delta CT}$ method was used to calculate the relative gene expression. Results were shown as the relative expression of the mRNA compared with that of *A. baumannii* ATCC 17978. Each experiment was performed in triplicate. A relative expression >2.0 and <0.5 were considered as overexpression and under-expression, respectively [35].

4.5. Assessment of Biofilm Formation

Semiquantitative determination of biofilm formation was performed by crystal violet staining as previously described, with some modifications such as no glucose supplementation to the broth [36]. The biofilm index (BI, ratio of optical density (OD) of biofilm cell to the OD of planktonic cells [ODbiofilm/ODplanktonic]) was used to normalize the amount of biofilm formed to the total cell content of each sample tested to avoid variations due to differences in bacterial growth. Biofilm production was classified according to the BI: non-producer (BI < 0.90), weak producer (BI = 0.90−1.20), and strong producer (BI > 1.20) as previously described [37]. *Staphylococcus aureus* ATCC 29213 (high biofilm producer) and *Escherichia coli* ATCC 25923 (low biofilm producer) were used as quality control strains.

4.6. Determination of Antibacterial Activity by Antimicrobial Combinations

Isolates were classified according to their susceptibility profile and genetic characteristics in order to evaluate the effect of different antibiotic combinations. Synergistic effects between tigecycline, SAM, meropenem, levofloxacin, and colistin were assessed for the selected isolates using the checkerboard microdilution method. The selection of antibiotics to be tested in combination was selected according to the mechanisms of action of each antibiotic and the pharmacological drug interactions between antibiotics. The combinations used were colistin-meropenem, colistin-levofloxacin, colistin-tigecycline, meropenem-levofloxacin, levofloxacin-SAM, tigecycline-levofloxacin, and tigecycline-meropenem [38].

4.6.1. Checkerboard Method

A bacterial inoculum of 0.5 McFarland was 1:150 diluted in Mueller–Hinton broth, and 100 µL was transferred to 96-well round-bottom plates (Corning Inc., Corning, NY, USA) containing serial dilutions of antibiotics. The antibiotics used were colistin (0.12–2 µg/mL), meropenem (8–256 µg/mL), levofloxacin (2–64 µg/mL), tigecycline (2–64 µg/mL), and SAM (8/4–256/128 µg/mL). The plate was then incubated for 24 h at 37 °C. The fractional inhibitory concentration index (FICI) was calculated from the sum of the fractions of inhibitory concentrations. The results were categorized as synergism (FICI $\leq$ 0.5), additive (FICI > 0.5–1), indifferent (1 < FICI $\leq$ 4), and antagonism (FICI $\geq$ 4) [39].

4.6.2. Time-to-Kill Method

Isolates, antibiotic combinations, and concentrations that showed best synergistic activity by the checkerboard test were further selected for analysis using time-to-kill assays. All assays were performed three times. A bacterial inoculum was tested against different antibiotics, individually or in combination with the previously selected concentrations. For the combinations which included levofloxacin, colistin, and sulbactam, an induction was previously carried out to express resistance to that specific antibiotic. An inoculum of 0.5 MacFarland from a 24 h bacterial culture was inoculated in a 15 mL tube containing the concentrations of levofloxacin-sulbactam and colistin-meropenem equivalent to the FICI demonstrating synergy by the checkerboard method. Cultures were incubated at 37 °C and 100 µL was obtained at 0, 2, 4, 8, 24, and 48 h after incubation, serially diluted in 0.9% saline, and transferred to trypticase soy plates to determine colony counts after incubation. Bactericidal activity was defined as a reduction of $\geq$3 $\log^{10}$ colony-forming unit (CFU)/mL compared to the initial inoculum after 24 h of exposure. A reduction of $\geq$2 $\log^{10}$ CFU/mL compared to the most active antimicrobial agent alone was considered as synergistic. An increase of $\geq$2 $\log^{10}$ CFU/mL compared to the most active antimicrobial agent alone was considered as antagonistic [40].

4.7. Statistical Analysis

Isolate classification was done using re-scaled distance cluster combination analysis in IBM SPSS Statistics 25 version. The graphs were created using IBM SPSS Statistics or GraphPad Prism (GraphPad Software Inc., San Diego, CA, USA) version 8.0.

5. Conclusions

Synergistic activity against MDR *A. baumannii* strains was observed with meropenem-colistin and levofloxacin-SAM combinations. However, the checkerboard and the time-to-kill assays showed discrepancies, indicating that further studies are needed to properly select the most appropriate method. Additionally, the variability of the genetic characteristics of *A. baumannii* strains might have influenced the low synergistic activity observed. In addition, the synergistic activity of meropenem-colistin occurred in a strain resistant to both antibiotics, and dual therapy reduced the concentration of antibiotics needed to inhibit bacterial growth compared to monotherapy. These results suggest dual therapy could offer an advantage for clinical treatment; however, more studies are needed to determine the best therapeutic combination for treating infections by *A. baumannii*.

Supplementary Materials: The following supporting information can be downloaded at: https://www.mdpi.com/article/10.3390/antibiotics13111079/s1, Table S1: Susceptibility profile and genetic characteristics of the *A. baumannii* isolates selected for the checkerboard assay.

Author Contributions: Conceptualization, A.L.-D.-A., J.F.C.-C. and P.B.-I.; methodology, A.L.-D.-A., J.F.C.-C. and P.B.-I.; software, A.L.-D.-A., J.F.C.-C., S.F.-T. and P.B.-I.; validation, S.F.-T. and P.B.-I.; formal analysis, A.L.-D.-A. and P.B.-I., investigation, A.L.-D.-A. and P.B.-I.; resources, J.F.C.-C., A.C.-O. and P.B.-I.; data curation, A.L.-D.-A. and P.B.-I.; writing—original draft preparation, A.L.-D.-A., S.F.-T. and P.B.-I.; writing—review and editing, S.F.-T. and P.B.-I.; visualization, J.F.C.-C., S.F.-T., P.B.-I. and A.C.-O.; supervision, S.F.-T., P.B.-I. and A.C.-O.; project administration, S.F.-T., P.B.-I. and A.C.-O.; funding acquisition, P.B.-I. All authors have read and agreed to the published version of the manuscript.

Funding: This research was funded by Fondo Sectorial de Investigacion para la Educacion of the Consejo Nacional de Ciencia y Tecnología (CONACyT) grant number A1-S-16970.

Institutional Review Board Statement: This study was conducted in accordance with the Declaration of Helsinki and approved by the Ethics Committee of the Dr. José Eleuterio González University Hospital (Approval number IF23-00003 on 23 March 2023).

Informed Consent Statement: Patient consent was waived due to anonymization of the clinical information.

Data Availability Statement: The original contributions presented in this study are included in the article and Supplementary Materials. Further inquiries can be directed to the corresponding author.

Conflicts of Interest: The authors declare no conflicts of interest.

References

1. Cavallo, I.; Oliva, A.; Pages, R.; Sivori, F.; Truglio, M.; Fabrizio, G.; Pasqua, M.; Pimpinelli, F.; Di Domenico, E.G. *Acinetobacter baumannii* in the critically ill: Complex infections get complicated. *Front. Microbiol.* **2023**, *14*, 1196774. [CrossRef] [PubMed]
2. Ayoub Moubareck, C.; Hammoudi Halat, D. Insights into *Acinetobacter baumannii*: A Review of Microbiological, Virulence, and Resistance Traits in a Threatening Nosocomial Pathogen. *Antibiotics* **2020**, *9*, 119. [CrossRef] [PubMed]
3. Iovleva, A.; Mustapha, M.M.; Griffith, M.P.; Komarow, L.; Luterbach, C.; Evans, D.R.; Cober, E.; Richter, S.S.; Rydell, K.; Arias, C.A.; et al. Carbapenem-Resistant *Acinetobacter baumannii* in U.S. Hospitals: Diversification of Circulating Lineages and Antimicrobial Resistance. *mBio* **2022**, *13*, e0275921. [CrossRef] [PubMed]
4. Piperaki, E.T.; Tzouvelekis, L.S.; Miriagou, V.; Daikos, G.L. Carbapenem-resistant *Acinetobacter baumannii*: In pursuit of an effective treatment. *Clin. Microbiol. Infect.* **2019**, *25*, 951–957. [CrossRef]
5. Murugaiyan, J.; Kumar, P.A.; Rao, G.S.; Iskandar, K.; Hawser, S.; Hays, J.P.; Mohsen, Y.; Adukkadukkam, S.; Awuah, W.A.; Jose, R.A.M.; et al. Progress in Alternative Strategies to Combat Antimicrobial Resistance: Focus on Antibiotics. *Antibiotics* **2022**, *11*, 200. [CrossRef]
6. Munita, J.M.; Arias, C.A. Mechanisms of Antibiotic Resistance. *Microbiol. Spectr.* **2016**, *4*, 1–24. [CrossRef]
7. Mancilla-Rojano, J.; Ochoa, S.A.; Reyes-Grajeda, J.P.; Flores, V.; Medina-Contreras, O.; Espinosa-Mazariego, K.; Parra-Ortega, I.; Rosa-Zamboni, D.; Castellanos-Cruz, M.D.C.; Arellano-Galindo, J.; et al. Molecular Epidemiology of *Acinetobacter calcoaceticus-Acinetobacter baumannii* Complex Isolated From Children at the Hospital Infantil de México Federico Gómez. *Front. Microbiol.* **2020**, *11*, 576673. [CrossRef]
8. Moreno-Manjón, J.; Castillo-Ramírez, S.; Jolley, K.A.; Maiden, M.C.J.; Gayosso-Vázquez, C.; Fernández-Vázquez, J.L.; Mateo-Estrada, V.; Giono-Cerezo, S.; Alcántar-Curiel, M.D. *Acinetobacter baumannii* IC2 and IC5 Isolates with Co-Existing bla(OXA-

143-like) and bla(OXA-72) and Exhibiting Strong Biofilm Formation in a Mexican Hospital. *Microorganisms* **2023**, *11*, 2316. [CrossRef]
9. Bocanegra-Ibarias, P.; Peña-López, C.; Camacho-Ortiz, A.; Llaca-Díaz, J.; Silva-Sánchez, J.; Barrios, H.; Garza-Ramos, U.; Rodríguez-Flores, A.M.; Garza-González, E. Genetic characterisation of drug resistance and clonal dynamics of *Acinetobacter baumannii* in a hospital setting in Mexico. *Int. J. Antimicrob. Agents* **2015**, *45*, 309–313. [CrossRef]
10. Nowak, P.; Paluchowska, P.M.; Budak, A. Co-occurrence of carbapenem and aminoglycoside resistance genes among multidrug-resistant clinical isolates of *Acinetobacter baumannii* from Cracow, Poland. *Med. Sci. Monit. Basic Res.* **2014**, *20*, 9–14. [CrossRef]
11. Rafiei, E.; Shahini Shams Abadi, M.; Zamanzad, B.; Gholipour, A. The frequency of efflux pump genes expression in *Acinetobacter baumannii* isolates from pulmonary secretions. *AMB Express* **2022**, *12*, 103. [CrossRef] [PubMed]
12. Sharma, S.; Kaushik, V.; Kulshrestha, M.; Tiwari, V. Different Efflux Pump Systems in *Acinetobacter baumannii* and Their Role in Multidrug Resistance. *Adv. Exp. Med. Biol.* **2023**, *1370*, 155–168. [CrossRef] [PubMed]
13. Nie, D.; Hu, Y.; Chen, Z.; Li, M.; Hou, Z.; Luo, X.; Mao, X.; Xue, X. Outer membrane protein A (OmpA) as a potential therapeutic target for *Acinetobacter baumannii* infection. *J. Biomed. Sci.* **2020**, *27*, 26. [CrossRef] [PubMed]
14. Khoshnood, S.; Sadeghifard, N.; Mahdian, N.; Heidary, M.; Mahdian, S.; Mohammadi, M.; Maleki, A.; Haddadi, M.H. Antimicrobial resistance and biofilm formation capacity among *Acinetobacter baumannii* strains isolated from patients with burns and ventilator-associated pneumonia. *J. Clin. Lab. Anal.* **2023**, *37*, e24814. [CrossRef] [PubMed]
15. Park, G.C.; Choi, J.A.; Jang, S.J.; Jeong, S.H.; Kim, C.M.; Choi, I.S.; Kang, S.H.; Park, G.; Moon, D.S. In Vitro Interactions of Antibiotic Combinations of Colistin, Tigecycline, and Doripenem Against Extensively Drug-Resistant and Multidrug-Resistant Acinetobacter baumannii. *Ann. Lab. Med.* **2016**, *36*, 124–130. [CrossRef]
16. Lagerbäck, P.; Khine, W.W.; Giske, C.G.; Tängdén, T. Evaluation of antibacterial activities of colistin, rifampicin and meropenem combinations against NDM-1-producing Klebsiella pneumoniae in 24 h in vitro time-kill experiments. *J. Antimicrob. Chemother.* **2016**, *71*, 2321–2325. [CrossRef]
17. Madadi-Goli, N.; Moniri, R.; Bagheri-Josheghani, S.; Dasteh-Goli, N. Sensitivity of levofloxacin in combination with ampicillin-sulbactam and tigecycline against multidrug-resistant Acinetobacter baumannii. *Iran. J. Microbiol.* **2017**, *9*, 19–25.
18. Principe, L.; D'Arezzo, S.; Capone, A.; Petrosillo, N.; Visca, P. In vitro activity of tigecycline in combination with various antimicrobials against multidrug resistant Acinetobacter baumannii. *Ann. Clin. Microbiol. Antimicrob.* **2009**, *8*, 18. [CrossRef]
19. Miyasaki, Y.; Morgan, M.A.; Chan, R.C.; Nichols, W.S.; Hujer, K.M.; Bonomo, R.A.; Murthy, A.R. In vitro activity of antibiotic combinations against multidrug-resistant strains of *Acinetobacter baumannii* and the effects of their antibiotic resistance determinants. *FEMS Microbiol. Lett.* **2012**, *328*, 26–31. [CrossRef]
20. Laishram, S.; Anandan, S.; Devi, B.Y.; Elakkiya, M.; Priyanka, B.; Bhuvaneshwari, T.; Peter, J.V.; Subramani, K.; Balaji, V. Determination of synergy between sulbactam, meropenem and colistin in carbapenem-resistant *Klebsiella pneumoniae* and *Acinetobacter baumannii* isolates and correlation with the molecular mechanism of resistance. *J. Chemother.* **2016**, *28*, 297–303. [CrossRef]
21. Nutman, A.; Lellouche, J.; Temkin, E.; Daikos, G.; Skiada, A.; Durante-Mangoni, E.; Dishon-Benattar, Y.; Bitterman, R.; Yahav, D.; Daitch, V.; et al. Colistin plus meropenem for carbapenem-resistant Gram-negative infections: In vitro synergism is not associated with better clinical outcomes. *Clin. Microbiol. Infect.* **2020**, *26*, 1185–1191. [CrossRef] [PubMed]
22. Ibrahim, S.; Al-Saryi, N.; Al-Kadmy, I.M.S.; Aziz, S.N. Multidrug-resistant *Acinetobacter baumannii* as an emerging concern in hospitals. *Mol. Biol. Rep.* **2021**, *48*, 6987–6998. [CrossRef] [PubMed]
23. Chen, T.L.; Siu, L.K.; Wu, R.C.; Shaio, M.F.; Huang, L.Y.; Fung, C.P.; Lee, C.M.; Cho, W.L. Comparison of one-tube multiplex PCR, automated ribotyping and intergenic spacer (ITS) sequencing for rapid identification of *Acinetobacter baumannii. Clin. Microbiol. Infect.* **2007**, *13*, 801–806. [CrossRef] [PubMed]
24. *M100*; Performance Standards for Antimicrobial Susceptibility Testing 33rd. Clinical and Laboratory Standards Institute: Wayne, PA, USA, 2023.
25. Magiorakos, A.P.; Srinivasan, A.; Carey, R.B.; Carmeli, Y.; Falagas, M.E.; Giske, C.G.; Harbarth, S.; Hindler, J.F.; Kahlmeter, G.; Olsson-Liljequist, B.; et al. Multidrug-resistant, extensively drug-resistant and pandrug-resistant bacteria: An international expert proposal for interim standard definitions for acquired resistance. *Clin. Microbiol. Infect.* **2012**, *18*, 268–281. [CrossRef] [PubMed]
26. Flores-Treviño, S.; Garza-González, E.; Mendoza-Olazarán, S.; Morfín-Otero, R.; Camacho-Ortiz, A.; Rodríguez-Noriega, E.; Martínez-Meléndez, A.; Bocanegra-Ibarias, P. Screening of biomarkers of drug resistance or virulence in ESCAPE pathogens by MALDI-TOF mass spectrometry. *Sci. Rep.* **2019**, *9*, 18945. [CrossRef]
27. Sheikhalizadeh, V.; Hasani, A.; Ahangarzadeh Rezaee, M.; Rahmati-Yamchi, M.; Hasani, A.; Ghotaslou, R.; Goli, H.R. Comprehensive study to investigate the role of various aminoglycoside resistance mechanisms in clinical isolates of *Acinetobacter baumannii. J. Infect. Chemother.* **2017**, *23*, 74–79. [CrossRef]
28. Bontron, S.; Poirel, L.; Nordmann, P. Real-time PCR for detection of plasmid-mediated polymyxin resistance (mcr-1) from cultured bacteria and stools. *J. Antimicrob. Chemother.* **2016**, *71*, 2318–2320. [CrossRef]
29. Lee, J.K.; Lee, Y.S.; Park, Y.K.; Kim, B.S. Mutations in the *gyrA* and *parC* genes in ciprofloxacin-resistant clinical isolates of *Acinetobacter baumannii* in Korea. *Microbiol. Immunol.* **2005**, *49*, 647–653. [CrossRef]
30. Sepahvand, S.; Doudi, M.; Davarpanah, M.A.; Bahador, A.; Ahmadi, M. Analyzing *pmrA* and *pmrB* genes in *Acinetobacter baumannii* resistant to colistin in Shahid Rajai Shiraz, Iran Hospital by PCR: First report in Iran. *Pak. J. Pharm. Sci.* **2016**, *29*, 1401–1406.

31. Coyne, S.; Rosenfeld, N.; Lambert, T.; Courvalin, P.; Périchon, B. Overexpression of resistance-nodulation-cell division pump AdeFGH confers multidrug resistance in *Acinetobacter baumannii*. *Antimicrob. Agents Chemother.* **2010**, *54*, 4389–4393. [CrossRef]
32. Lee, Y.; Yum, J.H.; Kim, C.K.; Yong, D.; Jeon, E.H.; Jeong, S.H.; Ahn, J.Y.; Lee, K. Role of OXA-23 and AdeABC efflux pump for acquiring carbapenem resistance in an *Acinetobacter baumannii* strain carrying the blaOXA-66 gene. *Ann. Clin. Lab. Sci.* **2010**, *40*, 43–48. [PubMed]
33. Rumbo, C.; Gato, E.; López, M.; Ruiz de Alegría, C.; Fernández-Cuenca, F.; Martínez-Martínez, L.; Vila, J.; Pachón, J.; Cisneros, J.M.; Rodríguez-Baño, J.; et al. Contribution of efflux pumps, porins, and β-lactamases to multidrug resistance in clinical isolates of *Acinetobacter baumannii*. *Antimicrob. Agents Chemother.* **2013**, *57*, 5247–5257. [CrossRef] [PubMed]
34. Kempf, M.; Abdissa, A.; Diatta, G.; Trape, J.F.; Angelakis, E.; Mediannikov, O.; La Scola, B.; Raoult, D. Detection of *Acinetobacter baumannii* in human head and body lice from Ethiopia and identification of new genotypes. *Int. J. Infect. Dis.* **2012**, *16*, e680–e683. [CrossRef] [PubMed]
35. AlQumaizi, K.I.; Kumar, S.; Anwer, R.; Mustafa, S. Differential Gene Expression of Efflux Pumps and Porins in Clinical Isolates of MDR Acinetobacter baumannii. *Life* **2022**, *12*, 419. [CrossRef] [PubMed]
36. Mendoza-Olazarán, S.; Camacho-Ortiz, A.; Martínez-Reséndez, M.F.; Llaca-Díaz, J.M.; Pérez-Rodríguez, E.; Garza-González, E. Influence of whole-body washing of critically ill patients with chlorhexidine on *Acinetobacter baumannii* isolates. *Am. J. Infect. Control* **2014**, *42*, 874–878. [CrossRef]
37. Bocanegra-Ibarias, P.; Garza-González, E.; Morfín-Otero, R.; Barrios, H.; Villarreal-Treviño, L.; Rodríguez-Noriega, E.; Garza-Ramos, U.; Petersen-Morfin, S.; Silva-Sanchez, J. Molecular and microbiological report of a hospital outbreak of NDM-1-carrying *Enterobacteriaceae* in Mexico. *PLoS ONE* **2017**, *12*, e0179651. [CrossRef]
38. Cokol-Cakmak, M.; Bakan, F.; Cetiner, S.; Cokol, M. Diagonal Method to Measure Synergy Among Any Number of Drugs. *J. Vis. Exp.* **2018**, *136*, 57713. [CrossRef]
39. Li, J.; Yang, X.; Chen, L.; Duan, X.; Jiang, Z. In Vitro Activity of Various Antibiotics in Combination with Tigecycline Against Acinetobacter baumannii: A Systematic Review and Meta-Analysis. *Microb. Drug Resist.* **2017**, *23*, 982–993. [CrossRef]
40. Bremmer, D.N.; Bauer, K.A.; Pouch, S.M.; Thomas, K.; Smith, D.; Goff, D.A.; Pancholi, P.; Balada-Llasat, J.M. Correlation of Checkerboard Synergy Testing with Time-Kill Analysis and Clinical Outcomes of Extensively Drug-Resistant *Acinetobacter baumannii* Respiratory Infections. *Antimicrob. Agents Chemother.* **2016**, *60*, 6892–6895. [CrossRef]

antibiotics

Article

Outbreak of *Pseudomonas aeruginosa* High-Risk Clone ST309 Serotype O11 Featuring bla_{PER-1} and *qnrVC6*

Romina Papa-Ezdra [1], Matilde Outeda [2], Nicolás F. Cordeiro [1], Lucía Araújo [1], Pilar Gadea [2], Virginia Garcia-Fulgueiras [1], Verónica Seija [2], Inés Bado [1,*] and Rafael Vignoli [1,*]

[1] Departamento de Bacteriología y Virología, Instituto de Higiene, Facultad de Medicina, Universidad de la República, Av. Alfredo Navarro 3051, CP 11600 Montevideo, Uruguay; rpapa@higiene.edu.uy (R.P.-E.); ncordeiro@higiene.edu.uy (N.F.C.); laraujo@higiene.edu.uy (L.A.); virginiagarcia@higiene.edu.uy (V.G.-F.)

[2] Departamento de Laboratorio Clínico, Área Microbiología, Hospital de Clínicas, Facultad de Medicina, Universidad de la República, Av. Italia s/n, CP 11600 Montevideo, Uruguay; matildeouteda@gmail.com (M.O.); pilargadea2009@gmail.com (P.G.); vseija1@gmail.com (V.S.)

* Correspondence: ibado@higiene.edu.uy (I.B.); rvignoli@higiene.edu.uy (R.V.)

Abstract: *Pseudomonas aeruginosa* is a leading cause of hospital-acquired infections worldwide. Biofilm production, antibiotic resistance, and a wide range of virulence factors contribute to their persistence in nosocomial environments. We describe an outbreak caused by a multidrug-resistant *P. aeruginosa* strain in an ICU. Antibiotic susceptibility was determined and bla_{PER-1} and *qnrVC* were amplified via PCR. Clonality was determined using PFGE and biofilm formation was studied with a static model. A combination of antibiotics was assessed on both planktonic cells and biofilms. WGS was performed on five isolates. All isolates were clonally related, resistant to ceftazidime, cefepime, amikacin, and ceftolozane-tazobactam, and harbored bla_{PER-1}; 11/19 possessed *qnrVC*. Meropenem and ciprofloxacin reduced the biofilm biomass; however, the response to antibiotic combinations with rifampicin was different between planktonic cells and biofilms. WGS revealed that the isolates belonged to ST309 and serotype O11. bla_{PER-1} and *qnrVC6* were associated with a tandem of IS*CR1* as part of a complex class one integron, with *aac(6′)-Il* and *ltrA* as gene cassettes. The structure was associated upstream and downstream with Tn*4662* and flanked by direct repeats, suggesting its horizontal mobilization capability as a composite transposon. ST309 is considered an emerging high-risk clone that should be monitored in the Americas.

Keywords: *Pseudomonas aeruginosa*; ESBL; bla_{PER-1}; transposon; ST309

1. Introduction

Antimicrobial resistance (AMR) is one of the main health challenges of the 21st century, which threatens to claim millions of lives annually and causes significant health costs in terms of gross domestic product (GDP), according to projections for the next 30 years [1]. Recent studies estimate that 4.5 million deaths were associated with and 1.27 million were attributable to bacterial AMR in 2019 [2].

Pseudomonas aeruginosa ranks among the six leading pathogens contributing to AMR-associated deaths and is one of the five pathogens most related with mortality and years of life lost, independently of AMR [2,3]. Although this Gram-negative rod is considered an opportunistic pathogen, *P. aeruginosa* is one of the main agents of hospital-acquired infections worldwide; moreover, it plays a key role in infections among immunocompromised patients, in patients with cystic fibrosis and burns, among others [4,5]. Lower respiratory tract and bloodstream infections followed by peritoneal and intra-abdominal infections and, to a lesser extent, urinary tract infections and infections of the skin and subcutaneous systems are the main syndromes associated with mortality caused by this microorganism [3].

The high associated mortality of *P. aeruginosa* can be attributed to several factors, including the frequent occurrence of diverse antibiotic resistance mechanisms, its ability to produce biofilm [6], its association with certain virulence factors [7], and its ability to persist in hospital and natural environments [8]. This is particularly noteworthy in high-risk clones (HRCs), a term often used to refer to multidrug-resistant or extensively drug-resistant *P. aeruginosa* clones with wide distribution, usually associated with epidemic outbreaks, and exemplified by ST235, ST111, ST233, ST244, ST357, ST308, ST175, ST277, ST654, and ST298 [7]. HRCs are often linked with the production of extended-spectrum β-lactamases (ESBLs) or carbapenemases, along with certain serotypes (especially O11 and O6) and potent exotoxins associated with type three secretion systems (T3SSs), such as ExoU or ExoS [7]. Recently, several authors have proposed the inclusion of the sequence type ST309 among the HRCs because it meets multiple characteristics described above and because of its wide dissemination [8–10].

Moreover, many isolates belonging to these HRCs are considered within the category of *P. aeruginosa* with Difficult-to-Treat Resistance (DTR-*P. aeruginosa*), a term recently proposed to denote isolates that exhibit non-susceptibility to all of the following antibiotics: piperacillin-tazobactam, ceftazidime, cefepime, aztreonam, meropenem, imipenem, ciprofloxacin, and levofloxacin [11].

As most of these antibiotics belong to the β-lactam family, the main mechanism of resistance, alongside low outer membrane permeability, lies in the production of specific β-lactamases such as ESBLs and carbapenemases. Among the ESBLs and carbapenemases reported in this microorganism, those acquired horizontally, such as CTX-M-2, PER-1 or variants of VIM, IMP, and GES, are noteworthy for their frequency and association with HRCs [7].

Class one integrons play a key role in disseminating carbapenemase- and ESBL-coding genes in *P. aeruginosa*, contributing to the development of multidrug resistance, by co-harboring resistance genes to other antibiotic groups, including aminoglycosides and fluoroquinolones [12]. These elements consist of two conserved segments, 5′-CS and 3′-CS, flanking a variable region where resistance genes are incorporated as gene cassettes. The 5′-CS contains the class one integrase-coding gene (*intI1*), promoters for the expression of *intI1* (Pint) and the gene cassettes (Pc), and the integrase recognition site (*attI1*) that is recognized by the integrase to mediate site-specific recombination, along with the cassette recognition site (*attC*), to incorporate and excise genes to the structure. In clinically relevant class one integrons, also known as 'sul1-type' integrons, the 3′-CS region typically consists of a truncated version of the quaternary ammonium compounds resistance gene *qacE1* (Δ*qacE1*) and the sulfonamide resistance gene *sul1* [13,14].

Concerning the structures from which class one integrons are derived, In4-like integrons are especially frequent in *P. aeruginosa* [12]. These structures include IS*6100* at the end of the 3′-conserved segment (3′-CS), which may be truncated or absent due to this insertion sequence. Moreover, additional resistance genes can appear through their association with the IS*CR1* element, followed by a partial duplication of the 3′-CS, constituting the so-called complex class one integrons [15]. On the other hand, although most class one integrons have lost their transposition functions, they can be mobilized when associated with Tn3-like transposons [12,14].

Previously, *P. aeruginosa* clinical isolates carrying the carbapenemase-coding gene $bla_{\text{VIM-2}}$ in class one integrons were reported from different settings in Uruguay [16]. Subsequently, its co-occurrence with $bla_{\text{PER-1}}$ was described, associated with novel resistance regions and transposition units [17]. In this study, we describe an outbreak caused by multidrug-resistant *P. aeruginosa* in an Intensive Care Unit (ICU), and characterize the molecular and microbiological features of the clone involved.

2. Results

2.1. Isolates

A total of 43 *P. aeruginosa* isolates were obtained from nineteen infected and four colonized patients (based on the clinical criteria established by the infectious diseases team), admitted to the Intensive Care Unit (ICU) of the Hospital de Clínicas of Montevideo between August 2021 and July 2022. Antibiotic susceptibility testing and pulsed-field gel electrophoresis (PFGE) were conducted on the first isolate obtained from each patient. As the four isolates from the colonized patients exhibited the same antibiotic susceptibility patterns and pulse types (as detailed below), they were excluded from subsequent microbiological studies. The 19 studied isolates were obtained from respiratory secretions (n = 13), blood culture (n = 3), bronchoalveolar lavage (n = 1), urine (n = 1), and a surgical wound (n = 1) (Table 1).

2.2. Pulsed-Field gel Electrophoresis

All 19 isolates were clonally related, exhibiting similarity coefficients >80%. Grouping analysis further identified the presence of two sub-clusters comprising nine and ten isolates, respectively, with similarity coefficients >90% (Figure 1). Also, the strains corresponding to colonization exhibited a compatible pattern with the clinical isolates, displaying similarity coefficients >80%.

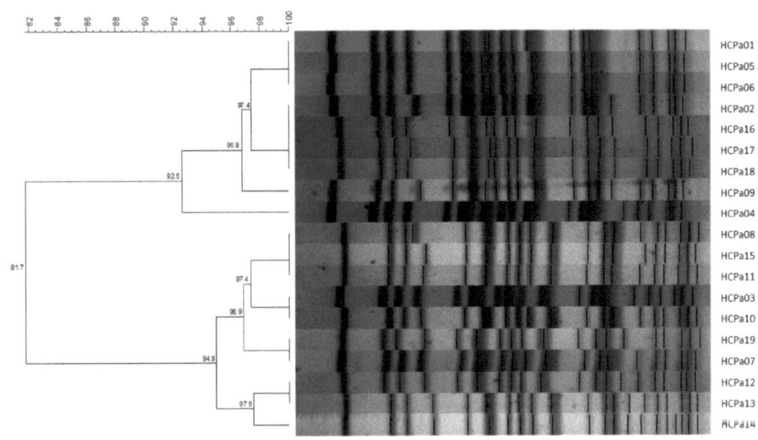

Figure 1. UPGMA dendrogram with Dice similarity coefficient; 1% opti., 2% tol. Similarity coefficients (%) are indicated in each node. Generated with BioNumerics v6.6 software.

Table 1. Microbiological characteristics of one representative isolate from each patient (n = 19).

Patient ID/Strain ‡	Date of 1st Isolate (dd/mm/yyyy)	Sample/s (in Chronological Order)	Genes (PCR)	Minimum Inhibitory Concentration (mg/L)								CZA (E-Test)	C/T (E-Test)	FDC (DD)
				PTZ	CAZ	FEP	IPM	MEM	GM	AK	CIP			
HCPa01	11/8/2021	Blood culture, catheter tip	*bla*$_{PER-1}$/*qnrVC*	32 (I)	≥64 (R)	≥32 (R)	≥16 (R)	8 (R)	4 (S)	≥64 (R)	0.25 (S)	16 (R)	>256 (R)	S
HCPa02	27/8/2021	Respiratory secretions, blood culture, bronchoalveolar lavage	*bla*$_{PER-1}$/*qnrVC*	32 (I)	≥64 (R)	≥32 (R)	8 (R)	≤0.25 (S)	4 (S)	≥64 (R)	0.5 (S)	24 (R)	>256 (R)	S
HCPa03	29/8/2021	Respiratory secretions, blood culture	*bla*$_{PER-1}$/*qnrVC*	32 (I)	≥64 (R)	≥32 (R)	8 (R)	1 (S)	4 (S)	≥64 (R)	1 (R)	24 (R)	>256 (R)	S
HCPa04	30/8/2021	Blood culture	*bla*$_{PER-1}$/*qnrVC*	8 (S)	≥64 (R)	≥32 (R)	2 (S)	0.5 (S)	4 (S)	≥64 (R)	0.5 (S)	8 (S)	>256 (R)	S
HCPa05	3/10/2021	Respiratory secretions, blood culture, catheter tip	*bla*$_{PER-1}$	16 (S)	≥64 (R)	≥32 (R)	≥16 (R)	0.5 (S)	4 (S)	≥64 (R)	0.25 (S)	16 (R)	>256 (R)	S
HCPa06	12/10/2021	Respiratory secretions	*bla*$_{PER-1}$/*qnrVC*	16 (S)	≥64 (R)	≥32 (R)	1 (S)	≤0.25 (S)	4 (S)	≥64 (R)	0.5 (S)	16 (R)	>256 (R)	S
HCPa07	20/10/2021	Respiratory secretions	*bla*$_{PER-1}$/*qnrVC*	8 (S)	≥64 (R)	≥32 (R)	2 (S)	1 (S)	8 (I)	≥64 (R)	0.25 (S)	8 (S)	128 (R)	S
HCPa08	10/12/2021	Respiratory secretions	*bla*$_{PER-1}$	16 (S)	≥64 (R)	≥32 (R)	2 (S)	0.5 (S)	4 (S)	≥64 (R)	0.25 (S)	8 (S)	>256 (R)	S
HCPa09	9/1/2022	Respiratory secretions	*bla*$_{PER-1}$	32 (I)	≥64 (R)	≥32 (R)	2 (S)	≤0.25 (S)	4 (S)	≥64 (R)	1 (R)	16 (R)	>256 (R)	S
HCPa10	11/1/2022	Blood culture, respiratory secretions	*bla*$_{PER-1}$/*qnrVC*	16 (S)	≥64 (R)	≥32 (R)	2 (S)	0.5 (S)	8 (I)	≥64 (R)	0.25 (S)	16 (R)	>256 (R)	S
HCPa11	21/1/2022	Respiratory secretions	*bla*$_{PER-1}$	16 (S)	≥64 (R)	≥32 (R)	2 (S)	0.5 (S)	4 (S)	≥64 (R)	0.25 (S)	16 (R)	128 (R)	S
HCPa12	21/1/2022	Urine, blood culture, respiratory secretions	*bla*$_{PER-1}$	16 (S)	≥64 (R)	≥32 (R)	2 (S)	2 (S)	4 (S)	≥64 (R)	≤0.06 (S)	16 (R)	>256 (R)	S
HCPa13	24/1/2022	Surgical wound	*bla*$_{PER-1}$	16 (S)	≥64 (R)	≥32 (R)	2 (S)	≤0.25 (S)	4 (S)	≥64 (R)	≤0.06 (S)	32 (R)	>256 (R)	S
HCPa14	12/2/2022	Respiratory secretions	*bla*$_{PER-1}$	32 (I)	≥64 (R)	≥32 (R)	1 (S)	≤0.25 (S)	≥16 (R)	≥64 (R)	≤0.06 (S)	16 (R)	>256 (R)	S
HCPa15	13/3/2022	Respiratory secretions, blood culture	*bla*$_{PER-1}$/*qnrVC*	16 (S)	≥64 (R)	≥32 (R)	2 (S)	1 (S)	4 (S)	≥64 (R)	0.25 (S)	8 (S)	>256 (R)	S
HCPa16	17/3/2022	Bronchoalveolar lavage	*bla*$_{PER-1}$/*qnrVC*	16 (S)	≥64 (R)	≥32 (R)	2 (S)	≤0.25 (S)	4 (S)	≥64 (R)	0.25 (S)	16 (R)	>256 (R)	S
HCPa17	27/3/2022	Respiratory secretions, surgical wound, bronchoalveolar lavage, urine	*bla*$_{PER-1}$	16 (S)	≥64 (R)	≥32 (R)	2 (S)	1 (S)	4 (S)	≥64 (R)	0.25 (S)	16 (R)	>256 (R)	S
HCPa18	7/4/2022	Respiratory secretions	*bla*$_{PER-1}$/*qnrVC*	16 (S)	≥64 (R)	≥32 (R)	2 (S)	≤0.25 (S)	8 (I)	≥64 (R)	0.125 (S)	32 (R)	>256 (R)	S
HCPa19	17/7/2022	Respiratory secretions, catheter tip	*bla*$_{PER-1}$/*qnrVC*	16 (S)	≥64 (R)	≥32 (R)	2 (S)	≤0.25 (S)	4 (S)	≥64 (R)	≥4 (R)	8 (S)	48 (R)	S

‡ Underlined strains were selected for whole genome sequencing. Abbreviations: PTZ, piperacillin-tazobactam; CAZ, ceftazidime; FEP, cefepime; IPM, imipenem; MEM, meropenem; GM, gentamicin; AK, amikacin; CIP, ciprofloxacin; CZA, ceftazidime-avibactam; C/T, ceftolozane-tazobactam; FDC, cefiderocol; DD, disk diffusion method.

2.3. Susceptibility Profiles and Mechanisms of Antibiotic Resistance

All isolates were resistant to ceftazidime, cefepime, and amikacin, while three were also resistant to ciprofloxacin. Additionally, four strains were resistant to imipenem and one of them was resistant to both imipenem and meropenem. Five strains were not susceptible to piperacillin-tazobactam and four to gentamicin (Table 1).

Additionally, isolates from colonized patients exhibited a similar susceptibility profile, characterized by resistance to ceftazidime, cefepime, and amikacin. They were susceptible to imipenem, meropenem, gentamicin, and ciprofloxacin, and either susceptible or intermediate to piperacillin-tazobactam.

Testing newer antibiotics, not widely available in our country, revealed resistance to ceftolozane-tazobactam and susceptibility to cefiderocol among all isolates. Regarding ceftazidime-avibactam, five isolates were susceptible (MIC = 8 mg/L) and the remaining fourteen were resistant (Table 1).

The double-disk synergy test between amoxicillin-clavulanic acid and ceftazidime resulted positive for all 19 isolates, and the presence of the ESBL-coding gene *bla*$_{PER-1}$ was identified using PCR in all of them. Additionally, the quinolone resistance determinant *qnrVC6* was detected in 11/19 isolates (Table 1). Isolates from the colonized patients also harbored the *bla*$_{PER-1}$ gene.

2.4. Effect of Antibiotics Combined with Rifampicin

The combined effects of either ciprofloxacin, meropenem, gentamicin, and amikacin with rifampicin was evaluated using the checkerboard assay on two strains, HCPa01 and HCPa12. A synergistic effect (FICI $\leq$ 0.5) was observed for all the combinations except for ciprofloxacin plus rifampicin. Meropenem 2 mg/L and rifampicin 4 mg/L resulted synergistic for strain HCPa01 (FICI = 0.5), meanwhile meropenem 0.06 and 0.5 mg/L were synergistic with rifampicin 2 and 4 mg/L (FICI = 0.375 and 0.28), respectively, for strain HCPa12. For both strains, gentamicin and rifampicin resulted synergistic at 0.5 and 4 mg/L (FICI = 0.375), meanwhile amikacin 8 mg/L exhibited synergy with rifampicin 4 mg/L (FICI = 0.375) in both strains. Furthermore, HCPa12 demonstrated synergy with amikacin 16 mg/L and rifampicin 0.5 mg/L (FICI = 0.28) (Table 2).

Table 2. Checkerboard assay results of the combined effect of rifampicin with either meropenem, gentamicin, amikacin, or ciprofloxacin for strains HCPa01 and HCPa12.

MIC †	HCPa01	HCPa12		MIC	HCPa01	HCPa12	MIC	HCPa01	HCPa12		MIC	HCPa01	HCPa12
RD_S	16	16		RD_S	16	16	RD_S	16	16		RD_S	16	16
MEM_S	8	2		GM_S	4	4	AK_S	64	64		CIP_S	0.25	0.06
RD_C	4	2	4	RD_C	4	4	RD_C	4	4	0.5	RD_C	8	8
MEM_C	2	0.5	0.06	GM_C	0.5	0.5	AK_C	8	8	16	CIP_C	0.06	0.03
FICI §	0.5	0.375	0.28	FICI	0.375	0.375	FICI	0.375	0.375	0.28	FICI	0.74	1

† MIC, Minimum inhibitory concentration expressed in mg/L for each single antibiotic (subfix S) and in combination (subfix C). § Fractional inhibitory concentration index. Antibiotic abbreviations: RD, rifampicin; MEM, meropenem; GM, gentamicin; AK, amikacin; CIP, ciprofloxacin.

2.5. Biofilm Characterization

HCPa01 and HCPa12 were categorized as strong biofilm producers at 24 h, exhibiting OD590 = 0.789 $\pm$ 0.11 and 0.674 $\pm$ 0.096, respectively (ODc = 0.126 $\pm$ 0.022).

To further characterize the strains, the activity of ciprofloxacin, meropenem, gentamicin, amikacin, and rifampicin was assessed against 24 h mature biofilms. Additionally, combinations that resulted synergistic in the checkerboard analysis (meropenem, gentamicin, and amikacin plus rifampicin) were evaluated too, along with a combination of rifampicin and ciprofloxacin. Rifampicin at concentrations of 8 and 16 mg/L led to a reduction in the biofilm biomass in HCPa01, compared with the control without antibiotics ($p < 0.05$), although this effect was not observed in HCPa12 (Figure 2a,b). Ciprofloxacin 0.25, 0.5, and 1 mg/L resulted in a reduction in biofilm biomass ($p < 0.05$) in both strains, as well as the combination of ciprofloxacin 0.5 mg/L and rifampicin 4 mg/L, although this effect was not statistically significant compared to ciprofloxacin alone (Figure 2c,d).

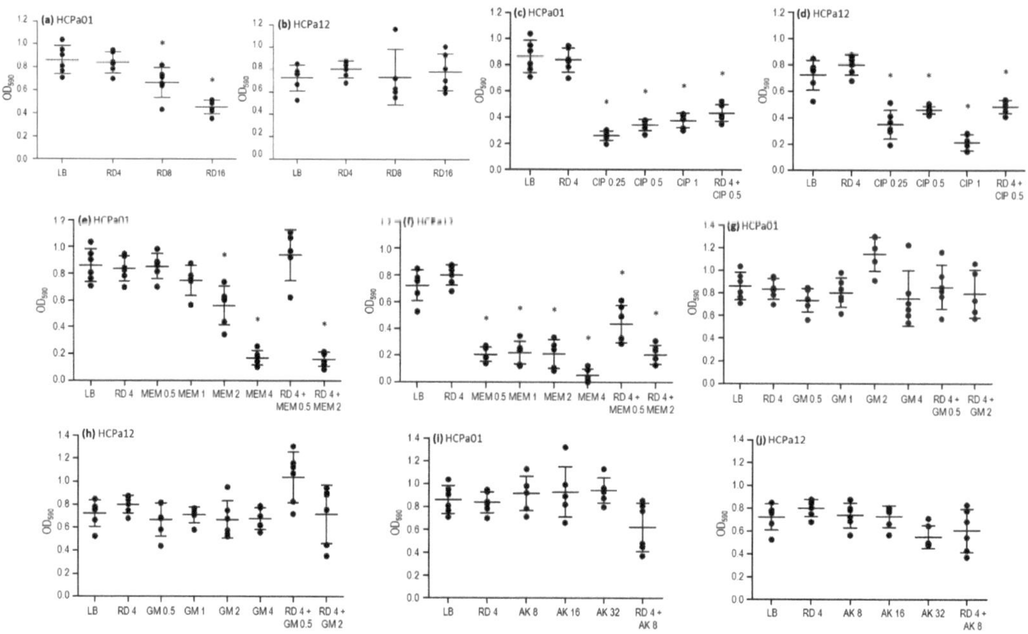

Figure 2. Activity of antibiotics alone and in combinations over 24 h on the mature biofilm of strains HCPa01 and HCPa12. Biofilm biomass is expressed as optical density measured at 590 nm (OD_{590}), antibiotic concentrations in mg/L (x axis). (**a**,**b**) Rifampicin (RD) alone; (**c**,**d**) ciprofloxacin (CIP) alone and combined with RD; (**e**,**f**) meropenem (MEM) alone and combined with RD; (**g**,**h**) gentamicin (GM) alone and combined with RD; (**i**,**j**) mikacin (AK) alone and combined with RD. The control without antibiotics is indicated as LB. Asterisks (*) indicate a significative decrease ($p < 0.05$) in biofilm biomass in comparison with the control without antibiotics.

Regarding meropenem, the biofilm biomass of strain HCPa01 decreased with 2 and 4 mg/L, while in HCPa12 0.5, 1, 2, and 4 mg/L produced a decrease in biomass compared to the control without antibiotics ($p < 0.05$). The combination of meropenem 2 mg/L plus rifampicin 4 mg/L also resulted in a reduction ($p < 0.05$) in biofilm biomass compared to the control without antibiotics and with rifampicin alone in both strains, as well as with meropenem alone in the case of HCPa01. The combination of meropenem 0.5 mg/L with rifampicin 4 mg/L showed a reduction in the biomass of the strain HCPa12 compared to the control without antibiotics and rifampicin alone, but it exhibited a biomass increase compared to meropenem 0.5 mg/L alone, although not statistically significant; meanwhile, such combination showed no effect against the biofilm of HCPa01 (Figure 2e,f).

Gentamicin 0.5, 1, 2, and 4 mg/L and amikacin 8, 16, and 32 mg/L showed no effect against the biofilms of both strains. Neither the combinations of gentamicin 0.5 or 2 mg/L, or amikacin 8 mg/L with rifampicin 4 mg/L evidenced a biofilm reduction. Conversely, an increase in the biomass of HCPa12 was observed when treated with 4 mg/L rifampicin and 0.5 mg/L gentamicin (Figure 2g–j).

2.6. Genetic Features

Whole genome sequencing (WGS) was performed on five isolates (HCPa01, HCPa02, HCPa10, HCPa12, and HCPa16). Sequence analyses revealed they all belonged to the sequence type ST309 and serotype O11.

Analysis using the Virulence Factor Database (VFDB) identified more than 200 virulence-related determinants in the five strains. Among the most relevant genes were those associated with type III secretion system effectors and regulators, including the toxin *exoU*, and adherence factors related with type IV pili and flagella biosynthesis, as well as their regulation and components. Additionally, genes were identified for alginate biosynthesis and regulation, elastase, rhamnolipid, and pyocyanin biosynthesis. Other genes related to the type IV and VI secretion systems, pyochelin, and pyoverdine synthesis and regulation were also found.

Regarding the antibiotic resistance determinants, AMRFinder revealed the presence of ten resistance determinants including *aac(6′)-Il* (*aacA7*), *aph(3′)-IIb* (aminoglycoside resistance), *bla*$_{OXA-1035}$ (OXA-50 family), *bla*$_{PDC-19a}$, *bla*$_{PER-1}$ (β-lactam resistance), *catB7* (phenicol resistance), *fosA* (fosfomycin resistance), *qacE*Δ*1* (quaternary ammonium resistance), and *sul1* (sulfonamide resistance) in all five strains studied, meanwhile the quinolone resistance determinant *qnrVC6* was detected in all strains except HCPa12. Finally, four isolates presented the *crpP1* locus variant, while HCPa01 had the *crpP* variant.

2.7. *qnrVC6 and bla*$_{PER-1}$ *Genetic Environment*

The genetic environments of both *qnrVC6* and *bla*$_{PER-1}$ were studied in detail in the strain HCPa10, and consist of a 43.3 kb structure. Both genes are embedded into a complex class one integron comprising the class one integron *intI1–aac(6′)-Il–ltrA–qacE*Δ*1-sul1*, followed by a tandem structure consisting of IS*CR1*–*qnrVC6*–IS*CR1*–*bla*$_{PER-1}$, concluding in a second copy of *qacE*Δ*1-sul1*. Upstream of this structure there is a reverse-oriented transposon belonging to the Tn*3* family, named Tn*4662*, which comprises transposase and resolvase genes (*tnpA* and *tnpR*, respectively), the resolvase system formed by three *res* sites (I, II and III), a toxin/antitoxin gene pair (*relE* and *relB*), and four additional ORFs, all flanked by the inverted repeats IRL and IRR. Adjacent to the Tn*4662* and immediately upstream of *intI1*, there is a fragment of Tn*As1* comprising the truncated transposase gene (Δ*tnpA*), the resolvase gene (*tnpR*), and the *res* site RIII. Finally, downstream of the complex class one integron there is the insertion sequence IS*6100*, with a second copy of Tn4662 positioned 6.4 kb apart and directly oriented. Upstream of the first copy of Tn*4662* and downstream the second, there are 5 bp direct repeats (DRs) (5′-TACTC), flanking the composite transposon designated as *Tn7723*. Moreover, upstream of the first DR there are 979 bp of an MFS transporter-coding gene, and downstream the second DR there are the remaining 228 bp of the same gene (Figure 3).

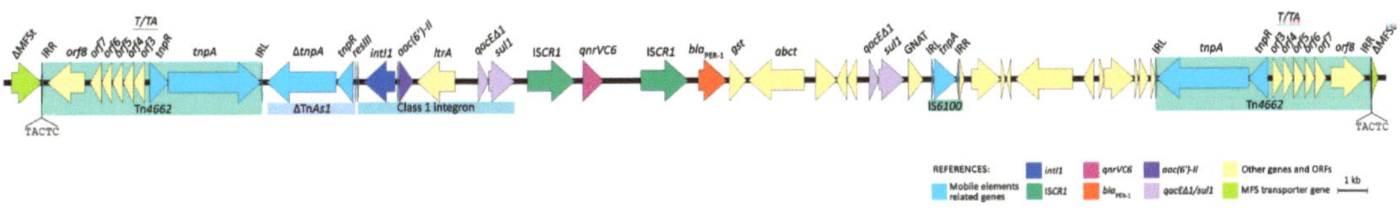

Figure 3. Genetic environment of *qnrVC6* and *bla*$_{PER-1}$ in *P. aeruginosa* HCPa10, featuring a composite transposon (Tn*7723*) delimited by two copies of Tn4662 and flanked by direct repeats (5′-TACTC). Genes and ORFs are represented by arrows and colored according to their function, as indicated in the reference. Linear map generated in EasyFig v2.1.

The BLAST analysis of the structure in the GenBank database revealed that the overall structure of the genetic environment of *qnrVC6–bla*$_{PER-1}$ is similar to others previously described in three plasmids and one chromosome of *Pseudomonas aeruginosa*, one *Pseudomonadaceae* plasmid, and one *Acinetobacter johnsonii* plasmid (Figure 4). The platform is mostly composed of a class one integron, with different gene cassettes among the isolates, followed by the module IS*CR1*–*qnrVC6*–IS*CR1*–*bla*$_{PER-1}$–*gst*–*abct*–*qacE*Δ*1-sul1*, and generally associated upstream with Tn*3*-family derived resolvase and transposase genes (complete or partial). The integron identified in HCPa10 represents a new class one integron (5′-CS–*aac(6′)-Il–ltrA*–3′-CS), and is also novel to the platform, since the aforementioned isolates were associated with different gene cassettes, including *aac(6′)-Ib4–aadA4* (in CP113227), *bla*$_{VIM-2}$ (in OP329419), *aac(6′)-Ib4–bla*$_{IMP-45}$*–bla*$_{OXA-1}$*–catB3* (in MF344570, CP061377 and CP104871), or *arr-3* (in CP121777).

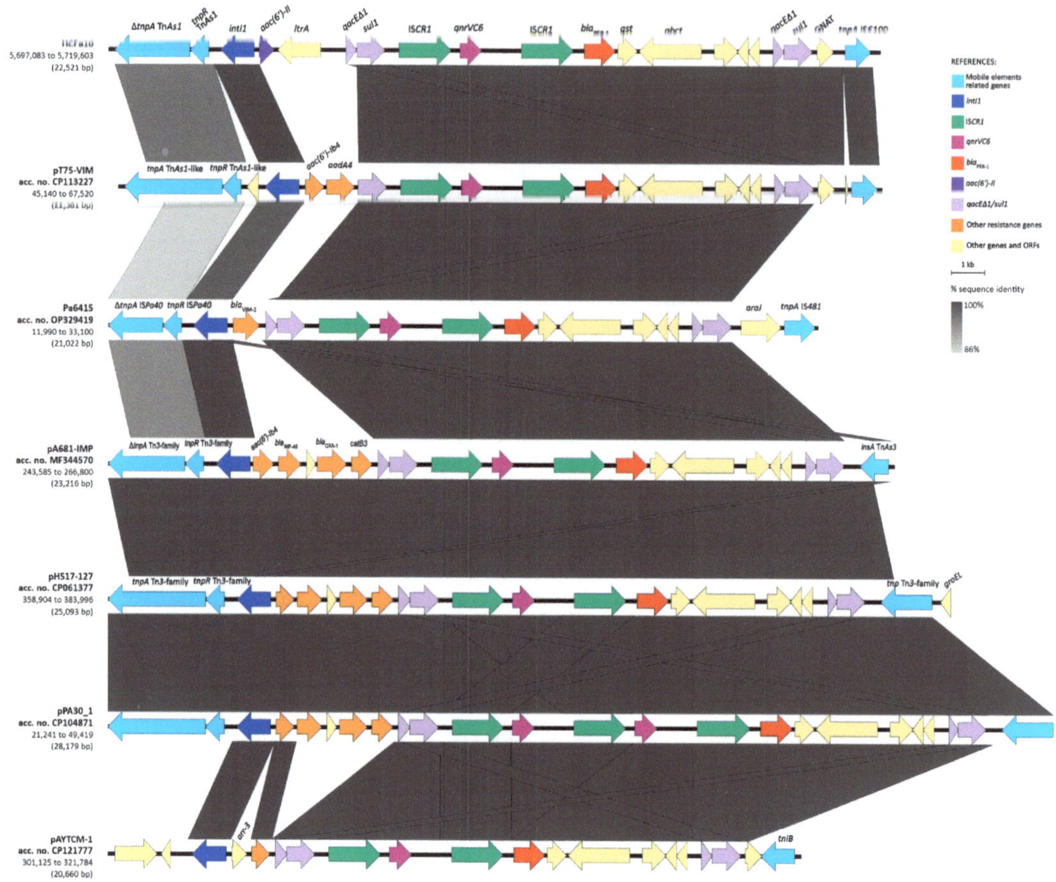

Figure 4. Sequence comparison (BLAST) of the resistance region harboring *qnrVC6* and *bla*$_{PER-1}$ with other genetic platforms from GenBank. Genes and ORFs are represented by arrows and colored according to their function and homologous segments are represented in shades of gray according to sequence identity, both as indicated in the reference. Linear map generated in EasyFig v2.1.

3. Discussion

P. aeruginosa is listed among the leading bacterial pathogens responsible for infection-related deaths in recent years, and is a significant contributor to the burden of AMR [2,3]. Lower respiratory infections and bloodstream infections stand out as the main causes of death attributed to this pathogen [3]. This microorganism is particularly frequent as a cause of nosocomial infections in ICUs [18].

In our study, *P. aeruginosa* was predominantly isolated from respiratory secretions, followed by blood samples from patients admitted to an ICU. All isolates were resistant to ceftazidime, cefepime, and amikacin. It is noteworthy that both ceftazidime and cefepime are among the most used antibiotics in this setting, while meropenem is rarely administered. Moreover, newer antibiotics such as ceftolozane/tazobactam and cefiderocol are not yet available in Uruguay, or their use is restricted to particular cases, as is the case of ceftazidime/avibactam. It has been suggested that the use of older antipseudomonal β-lactams such as ceftazidime, cefepime, and piperacillin/tazobactam may exert selective pressure, contributing to the emergence of resistance to newer agents like ceftolozane/tazobactam [19].

The high-level resistance to both ceftazidime and cefepime observed in all isolates can be attributed to the presence of the ESBL-coding gene $bla_{PER\text{-}1}$. Moreover, the expression of PER-1 may also explain the resistance to ceftolozane/tazobactam and ceftazidime/avibactam [20,21]. Interestingly, although $bla_{PER\text{-}1}$ is most frequently detected in *P. aeruginosa* isolates from Europe and the Middle East, to our knowledge, it has only been reported from Uruguay and Chile in the Americas [17,22,23].

On the other hand, although the quinolone resistance gene *qnrVC6* was found in half of the isolates, most of them were susceptible to ciprofloxacin, which may be expected under the assumption that generally, *qnr* genes confer low level quinolone resistance [24]. Also, the isolates analyzed via WGS harbored either *crpP* or *crpP1* locus variants, However, neither of these have been associated with fluoroquinolone resistance [25]. The resistance to amikacin can be attributed to the presence of the aminoglycoside N-acetyltransferase (6′) type 1 gene *aac(6′)-II* (*aacA7*) [26]. Discrepancies in antibiotic susceptibility observed among strains isolated from different patients may be explained by the development of adaptative resistance, mainly triggered by antibiotic selective pressure, and differences in gene regulation and expression [12].

In this study, we describe an outbreak of *P. aeruginosa* in an ICU involving 23 patients. This microorganism is well documented as a cause of outbreaks within such healthcare settings, mainly due to its capacity to colonize and survive in different surfaces [10]. The biofilm formation capacity is a key strategy for environmental colonization, which is also associated with higher antibiotic resistance and persistent infections, as well as to the clonal success of high-risk clones [27].

Both strains selected for biofilm formation analysis demonstrated strong biofilm-producing capabilities and exhibited variations in antibiotic susceptibility when comparing biofilm and planktonic growth stages. Notably, both ciprofloxacin and meropenem demonstrated the ability to reduce the biofilm biomass when assessed individually; meanwhile, the combination with rifampicin did not yield additional effects. Aminoglycosides, when evaluated individually or combined with rifampicin, did not exhibit a significant impact on biofilm biomass. Conversely, in planktonic cells, the combination of either meropenem, gentamicin, or amikacin showed a synergistic effect. Although combination therapy with rifampicin has proven effective in treating staphylococcal biofilms [28], and even in vitro against carbapenemase-producing *Escherichia coli* and *Klebsiella pneumoniae* [29], limited evidence exists regarding its efficacy against *P. aeruginosa* biofilms. Previous studies have shown positive in vitro outcomes when combining carbapenems and rifampicin at sub-MIC concentrations [29], and a rifampicin-driven potentiation of aminoglycoside activity [30]. However, in both cases, the assays were conducted with planktonic *P. aeruginosa*. Further studies are needed to further understand the role of rifampicin against *P. aeruginosa* in both planktonic and biofilm forms.

In this work, both $bla_{PER\text{-}1}$ and *qnrVC6* were located in a genetic platform consisting of a tandem with two copies of the IS*CR1* element. This arrangement was associated upstream to a class one integron harboring *aac(6′)-II* and *ltrA* as gene cassettes, and downstream with a second copy of *qacEΔ1/sul1*, constituting a complex class one integron. Similar configurations were identified in the database, differing only in the gene cassettes carried by the class one integrons. These variations included either aminoglycoside, chloramphenicol, or β-lactam resistance genes such as *aac(6′)-Ib4*, *aadA4*, *catB3*, $bla_{OXA\text{-}1}$, $bla_{IMP\text{-}45}$, or $bla_{VIM\text{-}2}$, the latter reported recently by our group in a clinical *P. aeruginosa* obtained from the same setting [17]. Upstream of the class one integron, most platforms were associated with different Tn3-family elements, typically featuring a resolvase gene and a complete or truncated transposase gene. Meanwhile, the elements found downstream comprised diverse genes derived from various transposon families such as IS6 (IS*6100* in our case), Tn*3*, or IS*481*. Notably, the presence of IS*6100* is characteristic of In4-like class one integrons, which usually lack part of all of the 3′-CS region, including the IRt [12], this latter being absent in the platform described here.

Interestingly, the whole region is bounded by two oppositely oriented transposons known as Tn*4662*. These transposons are flanked upstream and downstream by 5 pb direct repeats, interrupting an MFS transporter gene. This suggests that the entire structure might have undergone mobilization as a composite transposon, denoted here as Tn*7723*. This mobilization pattern as has been previously demonstrated for other Tn3-family derived structures [31].

As suggested by data obtained from the WGS of five representative strains, the outbreak described here was caused by *P. aeruginosa* belonging to sequence type ST309 and serotype O11. ST309 has recently been proposed as a high-risk clone given its wide distribution among different continents including Asia, Europe, Oceania, and the Americas [8,32]. This lineage was initially identified in low proportions as a minor clone in Greece and Korea, with the latter being associated with $bla_{\mathrm{VIM\text{-}2}}$; subsequent reports have documented clonal spread. These include a massive and persistent colonization of a dental care unit waterline in a University Hospital in France [33], isolates from children with bacteremia in Mexico [10], extensively drug-resistant clinical isolates from the United States [9], the Philippines [34], and Brazil [8], and intestinal colonization and environmental samples from a long-term care facility in France [32]. ST309 clones are reported to carry several acquired resistance genes in class one integrons, including ESBLs ($bla_{\mathrm{GES\text{-}19,\,\text{-}20,\,\text{-}26}}$), carbapenemases ($bla_{\mathrm{VIM\text{-}2}}$ and $bla_{\mathrm{IMP\text{-}15}}$), aminoglycoside-modifying enzymes (*aadA1*, *aacA4*, *aac(6′)-II* and *aac(6′)-33*), and fluoroquinolone resistance genes (*qnrVC1*), among others [8]. In our case, the ESBL detected among all isolates was $bla_{\mathrm{PER\text{-}1}}$, accompanied by *qnrVC6* within a complex class one integron, comprising novel components to the resistome of the ST309 lineage.

Regarding the serotype O11, it stands out as one of the most frequent serotypes among *P. aeruginosa* isolates worldwide, alongside O1 and O6 [35]. It has been frequently reported among high-risk clones such as ST235, ST357, ST308, and ST298 [7]. Serotype O11 was related with high prevalence among critically ill patients [35], worse clinical outcomes, extended hospital stays, and more virulent phenotypes [36]. As for virulence, the exotoxin ExoU, secreted by the type III secretion system (T3SS), is a key virulence factor of *P. aeruginosa* pathogenicity [18] and has been frequently reported in O11 isolates [35,36]. Both the gene *exoU* and T3SS-coding genes were detected in our isolates.

The clone here described exhibits concerning characteristics, including antibiotic resistance genes associated with transferable elements, several virulence factors, a resistance to new antibiotics, and the ability to form biofilms. This outbreak persisted for at least one year in the ICU, suggesting the presence of an environmental source as documented in previous studies, where biofilms could play an essential role [27,32,33]. Additionally, an enhanced ability to develop biofilms has been proposed as an underlying factor behind the success of high-risk clones [7].

In summary, we describe a large hospital outbreak caused by a successful high-risk clone of *P. aeruginosa* ST309. This clone carries resistance genes to broad-spectrum cephalosporins, amikacin, and new β-lactam/β-lactamase inhibitor combinations such as ceftolozane-tazobactam and ceftazidime-avibactam. This study highlights the importance of monitoring the dissemination of such microorganisms across our continent, especially considering the increased usage of ceftazidime-avibactam in the region.

4. Materials and Methods

4.1. Strains, Identification, and Antibiotic Susceptibility Testing

Between August 2021 and July 2022, 43 ceftazidime-resistant *P. aeruginosa* isolates were isolated from 23 patients admitted to the ICU of the University Hospital of Montevideo, Uruguay. The isolates were obtained from various clinical sources, including respiratory secretions, blood culture, catheter tips, bronchoalveolar lavage, urine culture, and surgical wounds.

For this report, we focused on *P. aeruginosa* isolates associated with infections, as determined by the infectious diseases team. We included one isolate per patient for a comprehensive microbiological characterization, and those associated with colonization were not considered.

Bacterial identification was performed using matrix-assisted laser desorption ionization-time-of-flight (MALDI-TOF) mass spectrometry (VITEK MS, bioMérieux, Marcy-l'Étoile, France). Antimicrobial susceptibility testing was assessed using the VITEK 2 system (bioMérieux, Marcy l'Étoile, France). Additionally, susceptibility to cefiderocol (FDC) was studied via disk diffusion and the minimum inhibitory concentration (MIC) to both ceftolozane/tazobactam (C/T) and ceftazidime/avibactam (CZA) were determined via E-tests (bioMérieux, Marcy l'Étoile, France) according to the manufacturer's indications. Results were interpreted based on the Clinical and Laboratory Standards Institute (CLSI) 2022 guidelines. In accordance, when results fell within the intermediate category, they were considered as not susceptible together with the resistant ones [37].

4.2. Resistance Mechanisms Detection

Phenotypic ESBL production was assessed using a double-disk synergy test, using discs of amoxicillin-clavulanic acid and ceftazidime [38]. After analysis of the whole genome sequencing results (see below), and given the epidemiological background of *P. aeruginosa* isolates from the same setting, genes coding for $bla_{\text{PER-1}}$ and *qnrVC* were searched with PCR using specific primers as previously described [39,40] and confirmed using Sanger sequencing (Unidad de Secuenciación, Hospital de Clínicas).

4.3. Pulse-Field Gel Electrophoresis

Pulse-field gel electrophoresis (PFGE) was conducted in accordance with the standard procedures recommended by PulseNet for *Escherichia coli*, *Salmonella*, and *Shigella* (PNL05) [41], with some modifications. Briefly, the isolates were cultured overnight on TSA and colonies were utilized to adjust to an 8.4 McFarland suspension in TE buffer (0.1 M Tris, 0.1 M EDTA, pH8, Sigma-Aldrich, Darmstadt, Germany). Plugs were assembled in molds by mixing 0.15 mL of the cell suspension, 0.15 mL of 1.5% low-melting point agarose (prepared in 1% SDS, TE buffer), and 0.5 g/L of proteinase K. Once solidified, plugs were placed in 2 mL of cell lysis buffer (50 mM Tris, 50 mM EDTA, 1% N-lauryl sarcosine, Sigma-Aldrich, Darmstadt, Germany) with 0.25 g/L proteinase K and incubated for 18 h at 56 °C with 150 rpm shaking in a water bath. Plugs were washed every 15 min, two times with molecular-grade water and three times with TE buffer. A 2 mm slice of each plug was cut, placed in a restriction solution containing 15 U enzyme *Spe*I and 1× buffer (ThermoFisher Scientific, Waltham, MA, USA), and incubated for 18 h at 37 °C. The slices were embedded in a 1% low-melting point agarose gel in 0.5× TBE buffer (45 mM Tris, 45 mM Boric acid, 1 mM EDTA, pH 8). *Salmonella* Braenderup H9812 digested with *Xba*I (ThermoFisher Scientific, Waltham, MA, USA) was used as a standard, as recommended by PulseNet.

PFGE was performed in a CHEF-DR III (Bio-Rad Laboratories, Inc., Life Sciences Group, Hercules, CA, USA) device, at 14 °C, 6 V, initial and final pulse times 4 and 40 s, respectively, for 20 h. Gels were stained with 0.5 g/L ethidium bromide and photographed under UV light. Restriction patterns were analyzed using BioNumerics 6.6 software (Applied Maths, 2011). Comparisons were made by calculating the Dice coefficient (optimization 1%, tolerance 2%) and dendrograms were generated using the UPGMA method (Unweighted Pair Group Method with Arithmetic Mean). Isolates showing 100% similarity were considered identical and those with ≥80% similarity were considered clonally related [42].

4.4. Susceptibility to Combination of Antibiotics

The susceptibility to either meropenem, ciprofloxacin, amikacin, or gentamicin combined with rifampicin was assessed using the checkerboard method in microtiter plates. The fractional inhibitory concentration index (FICI) was calculated as follows: FICI = (MIC_{A-B}/MIC_A) + (MIC_{B-A}/MIC_B), where MIC_A and MIC_B are the minimum inhibitory concentration for antibiotics A and B, respectively, while MIC_{A-B} and MIC_{B-A} represent the MIC to antibiotic A in the presence of antibiotic B and MIC to antibiotic B in the presence of A, respectively. A FICI value ≤0.5 was interpreted as a synergistic effect between antibiotics A and B [43].

4.5. Biofilm Formation and Antibiotic Susceptibility of Mature Biofilm

Biofilm formation capability for two representative strains (HCPa02 and HCPa12) was determined using the crystal violet static model previously described, with few modifications. Briefly, 1/10 aliquots of the overnight cultures in LB broth were placed in 96 flat-bottomed well polystyrene plates at a final volume of 200 µL. After 24 h incubation at 37 °C, the wells were washed with PBS and stained with 1% crystal violet (CV) for 15 min. Excess dye was removed with PBS washes, and CV was solubilized with 95% ethanol. The biofilm biomass was quantified according to the CV optical density (OD) at 590 nm [44]. Biofilm formation categories were defined according to the OD control (ODc) value, corresponding to the OD of wells without bacteria, as follows: OD ≤ Odc = no biofilm producer; Odc < OD ≤ (2 × Odc) = weak biofilm producer; (2 × Odc) < OD ≤ (4 × Odc) = moderate biofilm producer; and (4 × Odc) < OD = strong biofilm producer [45]. *P. aeruginosa* ATCC 27853 was used as a control for strong biofilm production in all assays.

In order to further characterize the isolates, the effects of ciprofloxacin, meropenem, gentamicin, amikacin, and rifampicin on mature biofilms were determined under the same conditions described above. After three washes with PBS to remove planktonic cells, LB alone or with different antibiotic concentrations was added over the 24 h mature biofilms. After 20 h incubation at 37 °C, planktonic cells were quantified according to the determination of DO 600 nm and their viability was confirmed with colony counting; then, they were removed and the CV staining protocol was followed as described above.

All biofilm experiments were performed in triplicate. Differences between treatments were assessed using one-way analysis of variance (ANOVA) and Bonferroni's post-test was used to compare pairs of groups. All analyses and graphics were performed in GraphPad Prism 5.0.

4.6. Short- and Long-Read Genome Sequencing

Five isolates (HCPa01, HCPa02, HCPa10, HCPa12, and HCPa16) belonging to different pulse type variants were subjected to whole genome sequencing (WGS) using short-read genome sequencing, and one of them (HCPa10) was also subjected to long-read genome sequencing, as previously described [17].

Briefly, genomic DNA was extracted using the NZY microbial gDNA Isolation kit (NZYTech Genes & Enzymes, Lisbon, Portugal). Libraries were prepared with the Nextera XT DNA Library Prep kit and Nextera XT Index kit (Illumina Inc., San Diego, CA, USA). Next-generation sequencing was performed using an Illumina MiniSeq system with a MiniSeq High Output reagent kit (Illumina Inc., San Diego, CA, USA) and a 2 × 151 bp paired-end approach. Reads were assembled with SPAdes ver. 3.11.

For the long-read genome sequencing with Oxford Nanopore Technologies, DNA libraries were prepared using a rapid sequencing kit (SQK-RAD004), loaded onto R9.4.1 flow cells (FLOMIN106) and sequenced for 8 h on a MinION device (Oxford Nanopore Technologies, Oxford, UK). Basecalling and data quality determination were assessed as previously described [17]. Genome hybrid assembly, using short and long reads, was performed with Unicycler ver. 0.4.8 [46].

4.7. Sequence Analysis

The prediction of antibiotic resistance genes was performed using AMRFinderPlus v.3.11.18 [47] and ABRicate v.1.01 (https://github.com/tseemann/abricate/, accessed on 1 November 2023) with the ResFinder database. Additionally, ABRicate was used to predict virulence-coding genes using the Virulence Factors Database (VFDB). Sequence type and serotype were determined using MLST 2.0 and PAst, respectively, both available at the Center for Genomics Epidemiology site (https://cge.cbs.dtu.dk/, last accessed on 1 November 2023).

The complete genome of HCPa10 was annotated using the RAST 2.0 suite (Rapid Annotation using Subsystem Technology) [48] and manually curated with Artemis software [49]. Comparisons with publicly available sequences were performed using BLAST (http://blast.ncbi.nlm.nih.gov/, last accessed on 1 November 2023), and physical maps were generated with EasyFig 2.1 using BLAST 2.2.18 (http://mjsull.github.io/Easyfig/, last accessed on 1 November 2023).

Genome data were deposited in the GenBank database under BioProject acc. no. PRJNA1036250, and the assembled HCPa10 chromosome under acc. no. CP139424. The new transposon number was assigned by the Transposon Registry repository [50].

Author Contributions: Methodology, R.P.-E., M.O., N.F.C., L.A., V.G.-F. and P.G.; formal analysis and investigation, R.P.-E., M.O., N.F.C., L.A., V.S., I.B. and R.V.; writing—review and editing, R.P.-E., M.O., N.F.C., L.A., I.B. and R.V.; supervision, V.S., I.B. and R.V. All authors have read and agreed to the published version of the manuscript.

Funding: This research received no external funding.

Institutional Review Board Statement: This work did not require ethical approval.

Informed Consent Statement: Not applicable.

Data Availability Statement: Whole genome sequencing data were deposited in GenBank database under BioProject acc. no. PRJNA1036250.

Conflicts of Interest: The authors declare no conflicts of interest.

References

1. O'Neill, J. *Tackling Drug-Resistant Infections Globally: Final Report and Recommendations*; The Review on Antimicrobial Resistance Chaired by Jim O'Neill; The Government of the United Kingdom: London, UK, 2016.
2. Murray, C.J.; Ikuta, K.S.; Sharara, F.; Swetschinski, L.; Robles Aguilar, G.; Gray, A.; Han, C.; Bisignano, C.; Rao, P.; Wool, E.; et al. Global Burden of Bacterial Antimicrobial Resistance in 2019: A Systematic Analysis. *Lancet* **2022**, *399*, 629–655. [CrossRef] [PubMed]
3. Ikuta, K.S.; Swetschinski, L.R.; Aguilar, G.R.; Sharara, F.; Mestrovic, T.; Gray, A.P.; Weaver, N.D.; Wool, E.E.; Han, C.; Hayoon, A.G.; et al. Global Mortality Associated with 33 Bacterial Pathogens in 2019: A Systematic Analysis for the Global Burden of Disease Study 2019. *Lancet* **2022**, *400*, 2221–2248. [CrossRef] [PubMed]
4. Jian, J.; Yu, P.; Zheng-Li, C.; Huo, L.; Ze Jing, W.; Shao-Shuo, Y.; Yu, S.; Guang-Yi, W.; Shi-Hui, Z.; Bing, M.; et al. Epidemiological Retrospective Analysis in Major Burn Patients: Single Centre Medical Records from 2009 to 2019. *Updates Surg.* **2022**, *74*, 1453–1459. [CrossRef] [PubMed]
5. Qin, S.; Xiao, W.; Zhou, C.; Pu, Q.; Deng, X.; Lan, L.; Liang, H.; Song, X.; Wu, M. *Pseudomonas aeruginosa*: Pathogenesis, Virulence Factors, Antibiotic Resistance, Interaction with Host, Technology Advances and Emerging Therapeutics. *Signal Transduct. Target. Ther.* **2022**, *7*, 199. [CrossRef] [PubMed]
6. Silva, A.; Silva, V.; López, M.; Rojo-Bezares, B.; Carvalho, J.A.; Castro, A.P.; Sáenz, Y.; Igrejas, G.; Poeta, P. Antimicrobial Resistance, Genetic Lineages, and Biofilm Formation in *Pseudomonas aeruginosa* Isolated from Human Infections: An Emerging One Health Concern. *Antibiotics* **2023**, *12*, 1248. [CrossRef] [PubMed]
7. del Barrio-Tofiño, E.; López-Causapé, C.; Oliver, A. Pseudomonas aeruginosa Epidemic High-Risk Clones and Their Association with Horizontally-Acquired β-Lactamases: 2020 Update. *Int. J. Antimicrob. Agents* **2020**, *56*, 106196. [CrossRef]
8. Fonseca, É.L.; Morgado, S.M.; Caldart, R.V.; Freitas, F.; Vicente, A.C.P. Emergence of a VIM-2-Producing Extensively Drug-Resistant (XDR) *Pseudomonas aeruginosa* ST309 in South America: A Comparative Genomic Analysis. *Int. J. Antimicrob. Agents* **2022**, *59*, 106507. [CrossRef] [PubMed]
9. Khan, A.; Tran, T.T.; Rios, R.; Hanson, B.; Shropshire, W.C.; Sun, Z.; Diaz, L.; Dinh, A.Q.; Wanger, A.; Ostrosky-Zeichner, L.; et al. Extensively Drug-Resistant *Pseudomonas aeruginosa* ST309 Harboring Tandem Guiana Extended Spectrum β-Lactamase Enzymes: A Newly Emerging Threat in the United States. *Open Forum Infect. Dis.* **2019**, *6*, ofz273. [CrossRef]
10. Morales-Espinosa, R.; Delgado, G.; Espinosa, L.F.; Isselo, D.; Méndez, J.L.; Rodriguez, C.; Miranda, G.; Cravioto, A. Fingerprint Analysis and Identification of Strains ST309 as a Potential High Risk Clone in a *Pseudomonas aeruginosa* Population Isolated from Children with Bacteremia in Mexico City. *Front. Microbiol.* **2017**, *8*, 313. [CrossRef]

11. Tamma, P.D.; Aitken, S.L.; Bonomo, R.A.; Mathers, A.J.; van Duin, D.; Clancy, C.J. Infectious Diseases Society of America 2022 Guidance on the Treatment of Extended-Spectrum β-Lactamase Producing Enterobacterales (ESBL-E), Carbapenem-Resistant Enterobacterales (CRE), and *Pseudomonas aeruginosa* with Difficult-to-Treat Resistance (DTR-*P. aeruginosa*). *Clin. Infect. Dis.* **2022**, *75*, 187–212. [CrossRef]
12. Botelho, J.; Grosso, F.; Peixe, L. Antibiotic Resistance in *Pseudomonas aeruginosa*—Mechanisms, Epidemiology and Evolution. *Drug Resist. Updat.* **2019**, *44*, 100640. [CrossRef] [PubMed]
13. Richard, E.; Darracq, B.; Loot, C.; Mazel, D. Unbridled Integrons: A Matter of Host Factors. *Cells* **2022**, *11*, 925. [CrossRef] [PubMed]
14. Partridge, S.R.; Kwong, S.M.; Firth, N.; Jensen, S.O. Mobile Genetic Elements Associated with Antimicrobial Resistance. *Clin. Microbiol. Rev.* **2018**, *31*, e00088-17. [CrossRef] [PubMed]
15. Partridge, S.R. Analysis of Antibiotic Resistance Regions in Gram-Negative Bacteria. *FEMS Microbiol. Rev.* **2011**, *35*, 820–855 [CrossRef] [PubMed]
16. Papa-Ezdra, R.; Bado, I.; Cordeiro, N.; Gutierrez, C.; Hitateguy, P.; Seija, V.; Vignoli, R. VIM-2-Producing *Pseudomonas* Spp. in Uruguay: Sequence Types, Pulsotypes, and Class 1 Integrons Including New Variable Regions Featuring blaVIM-2 and blaGES-7. *Antimicrob. Agents Chemother.* **2016**, *60*, 5620–5622. [CrossRef] [PubMed]
17. Papa-Ezdra, R.; Cordeiro, N.F.; Outeda, M.; Garcia-Fulgueiras, V.; Araújo, L.; Seija, V.; Ayala, J.A.; Bado, I.; Vignoli, R. Novel Resistance Regions Carrying Tn*aphA6*, bla_{VIM-2}, and bla_{PER-1}, Embedded in an IS*Pa40*-Derived Transposon from Two Multi-Resistant *Pseudomonas aeruginosa* Clinical Isolates. *Antibiotics* **2023**, *12*, 304. [CrossRef] [PubMed]
18. Sánchez-Diener, I.; Zamorano, L.; Peña, C.; Ocampo-Sosa, A.; Cabot, G.; Gómez-Zorrilla, S.; Almirante, B.; Aguilar, M.; Granados, A.; Calbo, E.; et al. Weighting the Impact of Virulence on the Outcome of *Pseudomonas aeruginosa* Bloodstream Infections. *Clin. Microbiol. Infect.* **2020**, *26*, 351–357. [CrossRef]
19. Fournier, D.; Carrière, R.; Bour, M.; Grisot, E.; Triponney, P.; Muller, C.; Lemoine, J.; Jeannot, K.; Plésiat, P. Mechanisms of Resistance to Ceftolozane/Tazobactam in *Pseudomonas aeruginosa*: Results of the GERPA Multicenter Study. *Antimicrob. Agents Chemother.* **2021**, *65*, e01117-20. [CrossRef]
20. Torrens, G.; Van Der Schalk, T.E.; Cortes-Lara, S.; Timbermont, L.; Del Barrio-Tofiño, E.; Xavier, B.B.; Zamorano, L.; Lammens, C.; Ali, O.; Ruzin, A.; et al. Susceptibility Profiles and Resistance Genomics of *Pseudomonas aeruginosa* Isolates from European ICUs Participating in the ASPIRE-ICU Trial. *J. Antimicrob. Chemother.* **2022**, *77*, 1862–1872. [CrossRef]
21. Ortiz De La Rosa, J.M.; Nordmann, P.; Poirel, L. ESBLs and Resistance to Ceftazidime/Avibactam and Ceftolozane/Tazobactam Combinations in *Escherichia coli* and *Pseudomonas aeruginosa*. *J. Antimicrob. Chemother.* **2019**, *74*, 1934–1939. [CrossRef]
22. Polse, R.F.; Khalid, H.M.; Mero, W.M.S. Distribution of bla_{OXA-10}, bla_{PER-1}, and bla_{SHV} Genes in ESBL-Producing *Pseudomonas aeruginosa* Strains Isolated from Burn Patients. *Sci. Rep.* **2023**, *13*, 18402. [CrossRef] [PubMed]
23. Babouee Flury, B.; Bösch, A.; Gisler, V.; Egli, A.; Seiffert, S.N.; Nolte, O.; Findlay, J. Multifactorial Resistance Mechanisms Associated with Resistance to Ceftazidime-Avibactam in Clinical *Pseudomonas aeruginosa* Isolates from Switzerland. *Front. Cell. Infect. Microbiol.* **2023**, *13*, 1098944. [CrossRef]
24. Ruiz, J. Transferable Mechanisms of Quinolone Resistance from 1998 Onward. *Clin. Microbiol. Rev.* **2019**, *32*, e00007-19. [CrossRef] [PubMed]
25. Hernández-García, M.; García-Castillo, M.; García-Fernández, S.; López-Mendoza, D.; Díaz-Regañón, J.; Romano, J.; Pássaro, L.; Paixão, L.; Cantón, R. Presence of Chromosomal CrpP-like Genes Is Not Always Associated with Ciprofloxacin Resistance in *Pseudomonas aeruginosa* Clinical Isolates Recovered in ICU Patients from Portugal and Spain. *Microorganisms* **2021**, *9*, 388. [CrossRef] [PubMed]
26. Ramirez, S.M.; Tolmasky, E.M. Aminoglycoside Modifing Enzymes. *Drug Resist. Updat.* **2011**, *13*, 151–171. [CrossRef] [PubMed]
27. Mulet, X.; Cabot, G.; Ocampo-Sosa, A.A.; Domínguez, M.A.; Zamorano, L.; Juan, C.; Tubau, F.; Rodríguez, C.; Moyà, B.; Peña, C.; et al. Biological Markers of *Pseudomonas aeruginosa* Epidemic High-Risk Clones. *Antimicrob. Agents Chemother.* **2013**, *57*, 5527–5535. [CrossRef] [PubMed]
28. Zimmerli, W.; Sendi, P. Role of Rifampin against Staphylococcal Biofilm Infections In Vitro, in Animal Models, and in Orthopedic-Device-Related Infections. *Antimicrob. Agents Chemother.* **2019**, *63*, e01746-18. [CrossRef]
29. Hu, Y.-F.; Liu, C.-P.; Wang, N.-Y.; Shih, S.-C. In Vitro Antibacterial Activity of Rifampicin in Combination with Imipenem, Meropenem and Doripenem against Multidrug-Resistant Clinical Isolates of *Pseudomonas aeruginosa*. *BMC Infect. Dis.* **2016**, *16*, 444. [CrossRef]
30. Mikalauskas, A.; Parkins, M.D.; Poole, K. Rifampicin Potentiation of Aminoglycoside Activity against Cystic Fibrosis Isolates of *Pseudomonas aeruginosa*. *J. Antimicrob. Chemother.* **2017**, *72*, 3349–3352. [CrossRef]
31. Nicolas, E.; Lambin, M.; Dandoy, D.; Galloy, C.; Nguyen, N.; Oger, C.A.; Hallet, B. The Tn3-Family of Replicative Transposons. *Microbiol. Spectr.* **2015**, *3*, MDNA3-0060-2014. [CrossRef]
32. Martak, D.; Gbaguidi-Haore, H.; Meunier, A.; Valot, B.; Conzelmann, N.; Eib, M.; Autenrieth, I.B.; Slekovec, C.; Tacconelli, E.; Bertrand, X.; et al. High Prevalence of *Pseudomonas aeruginosa* Carriage in Residents of French and German Long-Term Care Facilities. *Clin. Microbiol. Infect.* **2022**, *28*, 1353–1358. [CrossRef] [PubMed]
33. Abdouchakour, F.; Grau, D.; Aujoulat, F.; Mournetas, P.; Parer, S.; Gibert, P.; Valcarcel, J.; Jumas-bilak, E. Leads to Successive Waves of Contamination of Water in Dental Care. *Appl. Environ. Microbiol.* **2015**, *81*, 7509–7524. [CrossRef]

34. Chilam, J.; Argimon, S.; Limas, M.; Masim, M.; Gayeta, J.; Lagrada, M.; Olorosa, A.; Cohen, V.; Hernandez, L.; Jeffrey, B.; et al. Genomic Surveillance of *Pseudomonas aeruginosa* in the Philippines, 2013–2014. *West. Pacific Surveill. Response J.* **2021**, *12*, 4–18. [CrossRef]
35. Nasrin, S.; Hegerle, N.; Sen, S.; Nkeze, J.; Sen, S.; Permala-Booth, J.; Choi, M.; Sinclair, J.; Tapia, M.D.; Johnson, J.K.; et al. Distribution of Serotypes and Antibiotic Resistance of Invasive *Pseudomonas aeruginosa* in a Multi-Country Collection. *BMC Microbiol.* **2022**, *22*, 13. [CrossRef] [PubMed]
36. Lu, Q.; Eggimann, P.; Luyt, C.E.; Wolff, M.; Tamm, M.; François, B.; Mercier, E.; Garbino, J.; Laterre, P.F.; Koch, H.; et al. *Pseudomonas aeruginosa* Serotypes in Nosocomial Pneumonia: Prevalence and Clinical Outcomes. *Crit. Care* **2014**, *18*, R17. [CrossRef] [PubMed]
37. The Clinical and Laboratory Standards Institute. *Performance Standards for Antimicrobial Susceptibility Testing*, 32nd ed.; CLSI Supplement M100; Clinical and Laboratory Standards Institute: Wayne, PA, USA, 2022.
38. Pagani, L.; Mantengoli, E.; Migliavacca, R.; Nucleo, E.; Pollini, S.; Spalla, M.; Romero, E.; Rossolini, G.M.; Daturi, R. Multifocal Detection of Multidrug-Resistant *Pseudomonas aeruginosa* Producing the PER-1 Extended-Spectrum β -Lactamase in Northern Italy. *J. Clin. Microbiol.* **2004**, *42*, 2523–2529. [CrossRef] [PubMed]
39. Papa-Ezdra, R.; Bado, I.; Caiata, L.; Vignoli, R.; Seija, V. First Report of *Pseudomonas aeruginosa* Co-Harbouring $bla_{\text{VIM-2}}$ and $bla_{\text{PER-1}}$ in Latin America. *J. Glob. Antimicrob. Resist.* **2018**, *15*, 121–122. [CrossRef]
40. Liu, M.; Wong, M.H.Y.; Chen, S. Molecular Characterisation of a Multidrug Resistance Conjugative Plasmid from *Vibrio parahaemolyticus*. *Int. J. Antimicrob. Agents* **2013**, *42*, 575–579. [CrossRef]
41. PulseNet International Standard Operating Procedure for PulseNet PFGE of *Escherichia coli* O157:H7, *Escherichia coli* Non-O157 (STEC), *Salmonella* Serotypes, *Shigella sonnei* and *Shigella flexneri*. Code PNL05. Available online: https://pulsenetinternational.org/assets/PulseNet/uploads/pfge/PNL05_Ec-Sal-ShigPFGEprotocol.pdf (accessed on 1 November 2023).
42. Singh, A.; Goering, R.V.; Simjee, S.; Foley, S.L.; Zervos, M.J. Application of Molecular Techniques to the Study of Hospital Infection. *Clin. Microbiol. Rev.* **2006**, *19*, 512–530. [CrossRef]
43. Synergism Testing: Broth Microdilution Checkerboard and Broth Macrodilution Methods. In *Clinical Microbiology Procedures Handbook*; ASM Press: Washington, DC, USA, 2016; pp. 5.16.1–5.16.23.
44. González, M.J.; Robino, L.; Iribarnegaray, V.; Zunino, P.; Scavone, P. Effect of Different Antibiotics on Biofilm Produced by Uropathogenic *Escherichia coli* Isolated from Children with Urinary Tract Infection. *Pathog. Dis.* **2017**, *75*, ftx053. [CrossRef]
45. Villegas, N.A.; Baronetti, J.; Albesa, I.; Polifroni, R.; Parma, A.; Etcheverría, A.; Becerra, M.; Padola, N.; Paraje, M. Relevance of Biofilms in the Pathogenesis of Shiga-Toxin-Producing *Escherichia coli* Infection. *Sci. World J.* **2013**, *2013*, 607258. [CrossRef]
46. Wick, R.R.; Judd, L.M.; Gorrie, C.L.; Holt, K.E. Unicycler: Resolving Bacterial Genome Assemblies from Short and Long Sequencing Reads. *PLoS Comput. Biol.* **2017**, *13*, e1005595. [CrossRef] [PubMed]
47. Feldgarden, M.; Brover, V.; Gonzalez-Escalona, N.; Frye, J.G.; Haendiges, J.; Haft, D.H.; Hoffmann, M.; Pettengill, J.B.; Prasad, A.B.; Tillman, G.E.; et al. AMRFinderPlus and the Reference Gene Catalog Facilitate Examination of the Genomic Links among Antimicrobial Resistance, Stress Response, and Virulence. *Sci. Rep.* **2021**, *11*, 12728. [CrossRef] [PubMed]
48. Overbeek, R.; Olson, R.; Pusch, G.D.; Olsen, G.J.; Davis, J.J.; Disz, T.; Edwards, R.A.; Gerdes, S.; Parrello, B.; Shukla, M.; et al. The SEED and the Rapid Annotation of Microbial Genomes Using Subsystems Technology (RAST). *Nucleic Acids Res.* **2014**, *42*, 206–214. [CrossRef]
49. Carver, T.; Berriman, M.; Tivey, A.; Patel, C.; Böhme, U.; Barrell, B.G.; Parkhill, J.; Rajandream, M.A. Artemis and ACT: Viewing, Annotating and Comparing Sequences Stored in a Relational Database. *Bioinformatics* **2008**, *24*, 2672–2676. [CrossRef]
50. Tansirichaiya, S.; Rahman, M.A.; Roberts, A.P. The Transposon Registry. *Mob. DNA* **2019**, *10*, 40. [CrossRef]

antibiotics

Article

Characterizing Antibiotic Regimen Modification Behavior, Patient Characteristics, and Outcomes for Patients with Gram-Negative Bacterial Infections, A Retrospective Single-Center Study

Jason Yamaki [1,2,*], Mirna Mikhail [1], Richard Beuttler [1], Philip Robinson [2], Emre Yücel [3] and Alexandre H. Watanabe [3]

[1] Department of Pharmacy Practice, Chapman University School of Pharmacy (CUSP), Irvine, CA 92618, USA; mmikhail@chapman.edu (M.M.); rbeuttle@chapman.edu (R.B.)
[2] Hoag Memorial Hospital, Newport Beach, CA 92663, USA; philip.robinson@hoag.org
[3] Merck & Co., Inc., Rahway, NJ 07065, USA; alexandre.watanabe@merck.com (A.H.W.)
* Correspondence: yamaki@chapman.edu

Abstract: Few studies describe the frequency of antibiotic regimen modification behaviors in the acute care setting. We sought to ascertain patient and treatment characteristics, details of regimen modification, and clinical outcomes with antibiotic modifications. This retrospective study included patients admitted to Hoag Memorial Hospital from 1 January 2019–31 March 2021 with a complicated infection caused by a Gram-negative organism resistant to extended-spectrum cephalosporins or with the potential for resistance (AmpC producers). A total of 400 patients were included. The predominant sources were bloodstream (33%), urine (26%), and respiratory (24%), including patients with multiple sources. The most isolated organisms were *Pseudomonas* spp. and ESBL-producing organisms, 38% and 34%, respectively. A total of 72% of patients had antibiotic regimen modifications to their inpatient antibiotic regimens. In patients where modifications occurred, the number ranged from one to six modifications. The most common reasons for modifications included a lack of patient response (14%), additional history reviewed (9%), and decompensation (7%). No difference in clinical outcomes was observed based on antibiotic modifications. The numerous changes in therapy observed may reflect the limitations in identifying patients with resistant organisms early on in admission. This highlights the need for more novel antibiotics and the importance of identifying patients at risk for resistant organisms.

Keywords: antibiotics; prescribing behaviors; Gram-negative; MDRO; outcomes; stewardship

1. Introduction

Carbapenems are considered one of the last lines of defense in the treatment of infections caused by multidrug-resistant (MDR) Gram-negative bacteria (GNB), including extended-spectrum beta-lactamase (ESBL)- and AmpC beta-lactamase-producing enteric organisms, *Pseudomonas aeruginosa*, *Acinetobacter baumannii*, and other non-lactose fermenting bacteria. The increasing prevalence of these organisms in the patient population coupled with increasing resistance to carbapenems among these organisms has become a public health concern [1]. While progress has recently been made in the development of active antibacterials against the above organisms, there is still a deficit in antimicrobial agents for the varying types and degrees of resistance encountered in Gram-negative organisms.

Due to the severity and/or complexity of GNB infections, patients are often initiated on empiric antibiotic therapy prior to clinicians receiving information on the causative pathogen and its susceptibility to antibiotics. This can lead to a situation wherein the patient is receiving an inappropriate treatment, which has been shown to contribute to worse outcomes [2]. Furthermore, there can be a delay in initiating an appropriate and effective therapy, which increases the risk of poor outcomes in patients with sepsis or septic shock [3,4]. In such cases, when a patient has an ongoing infection and their condition deteriorates or the culture test results indicate the susceptibility of the bacteria to certain antibiotics, clinicians may opt to modify the current treatment or prescribe additional antibiotic agents to specifically target the causative bacteria.

Although studies have been conducted to identify the behavioral and social determinants influencing prescribing practices among physicians and other prescribers [5–8], there is limited documentation in the current literature regarding the frequency and reasons for changing antimicrobial therapy in acute care hospital settings [9]. However, some data do exist in the outpatient setting [5,6,10–12]. Based on clinical practice paradigms, clinicians may change treatment therapies based on patient clinical status changes, for example, if a patient decompensates or does not respond clinically despite antimicrobial treatment. Therapeutic modifications to the patient clinical status can occur in both empiric treatment, when the causative pathogen and its susceptibilities are unknown, and in directed therapy, when the organism and its susceptibilities are known. Therapy changes may also occur once the causative organism's susceptibility is known, which can lead to escalation, de-escalation, or no change.

The primary objectives of this study were to assess the frequency of antibiotic switch and/or add-on, examine the commonly prescribed empiric therapies in our patient population, and identify the documented or potential reasons for modifying treatment regimens in patients infected with aerobic Gram-negative organisms that are resistant to extended-spectrum cephalosporins (e.g., ceftriaxone) or known to carry chromosomally encoded class C beta-lactamases (AmpC). This cohort of patients was chosen because patients infected with these organisms would likely have an increase in inappropriate empiric treatment, or when preliminary culture results identify a genus known to likely be resistant (e.g., *Pseudomonas* or *Serratia*), physicians may modify therapy. Likewise, many of these organisms have the potential to be MDR organisms for which newer agents such as imipenem/relebactam and ceftolozane/tazobactam are available. Hence, understanding prescribing behaviors and the rationale and reasons behind antibiotic regimen modifications, prescribers may be able to make more informed treatment decisions to facilitate positive outcomes.

The secondary objectives of this study were to determine the timing of antimicrobial treatment switch or add-on during the course of treatment, identify the characteristics associated with regimen switch or add-on, examine the commonly used empiric therapies in our patient population, and compare the outcomes of patients who underwent switch or add-on with those who did not undergo such modifications. Patient outcomes based on antimicrobial switching, microbiologic culture and susceptibility results, and the timing of antibiotic administration would also provide information on treatment approaches that may be useful in the future. Moreover, the data collected from this analysis can offer valuable insights for antimicrobial stewardship programs. They can help identify opportunities for optimizing the treatment of patients with antibiotic-resistant organisms, as well as potential opportunities for the de-escalation and escalation of therapy.

Specifically, this analysis focused on assessing the frequency of Gram-negative antibiotic treatment modifications, documenting the timing of therapy modifications during the treatment course, exploring the documented or potential reasons behind these regimen modifications, examining the frequency of prescribers documenting their rationale, identifying the most common Gram-negative organisms resistant to extended-spectrum cephalosporins, and evaluating the outcomes of patients affected by these modifications. The main hypothesis was that modifications made during treatment are likely broadening therapy due to the poor response or possible resistance of infecting organisms, primarily occurring during empiric treatment.

2. Results

A total of 400 patients were included in the study. Only eight patients had repeat admissions accounting for a total of seventeen admissions, one patient accounted for three admissions, and seven patients only had two admissions. The overall median age of patients was 74.5 and ranged from 19 to 101 years, and 243 (61%) of the patients were male. The median Charlson comorbidity index score was 4 (IQR (interquartile range) 3–6). Most patients presented from home (n = 319, 80%), 68 (17%) patients were admitted from a skilled nursing facility, and the remaining patients came from a long-term care facility or were transferred from a different hospital (Table 1). Overall, 194 (48.5%) patients presented with sepsis as defined by the Centers for Medicaid Services (CMS), which can be defined by Systemic Inflammatory Response Syndrome (SIRS). Approximately half of patients had received IV antibiotics or been hospitalized in the past 90 days, and 108 (27%) patients had contracted previous infections with a multidrug-resistant organism. The predominant sources were bloodstream (n = 164, 33%), urine (n = 128, 26%), and respiratory (n = 117, 24%), including patients with multiple sources. Only 18 (4.5%) patients had a concurrent SARS-CoV2 infection.

Table 1. Patient demographics and clinical characteristics.

Variables—Median (IQR [#])	Overall n = 400 (%) (IQR)	Modifications n = 287 (%) (IQR)	No Modifications n = 113 (%) (IQR)	p-Value
Age (y)	74.5 (64–82)	75 (62–83)	74 (66.5–81.5)	0.99
Male (%)	243 (61)	181 (63)	62 (54.9)	0.1306
Previous hospitalization in past the 90-days	216 (54)	161 (56)	55 (48.7)	0.1798
Previous antibiotics in the past 90-days	192 (48)	145 (50.5)	47 (41.6)	0.1075
Previous GNR MDRO	108 (27)	74 (25.8)	34 (30)	0.3826
COVID-19 infection	18 (4.5)	14 (4.9)	4 (3.5)	0.7894
Residency Prior to Admission				
Home	319 (80)	224 (78)	95 (84)	0.3964
Skilled Nursing Facility	68 (17)	51 (18)	17 (15)	0.5135
Long-Term Acute Care Facility	2 (0.5)	2 (0.01)	0	NS
Other Hospital	10 (2.5)	9 (3)	1 (0.01)	0.2941
Homeless	1 (0.25)	1 (0.003)	0	NS
Sepsis (SIRS, Severe, Shock)	194 (48.5)	147 (51.2)	47 (41.6)	0.0828
ICU admission	74 (18.5)	65 (22.6)	9 (8)	0.0005
Time (h) to active therapy	3.375 (1.5–20)	4 (1.5–24.75)	2.5 (1.5–4.875)	0.002
Time (h) to empiric ABX ED	2 (1–4)	2 (1–3.5)	2.5 (1.5–4)	0.0019
Empiric #1 DOT	2 (2–4)	2 (2–4)	--	--
Time (h) to Empiric #2	23.38 (16.13–42)	23.38 (16.13–42)	--	--
Empiric #2 DOT	2 (2–3)	2 (2–3)	--	--
Total Empiric DOT	3 (3–5)	3 (3–5)	--	--
Time (h) to Directed ABX	64 (43.5–86.5)	64 (43.5–86.5)	--	--
Directed ABX #1 DOT	3 (2–5)	3 (2–5)	--	--
Time (h) from Directed #1 to #2	74 (46–98)	74 (46–98)	--	--
Directed ABX total DOT	3 (2–6)	3 (2–6)	--	--

Table 1. *Cont.*

Variables—Median (IQR [#])	Overall n = 400 (%) (IQR)	Modifications n = 287 (%) (IQR)	No Modifications n = 113 (%) (IQR)	p-Value
Total ABX DOT	6 (4–9))	6 (5–10)	5 (4–7)	<0.0001
ABX changes	1 (0–2)	1 (0–2)	--	--
MD consults	2 (1–3)	2 (2–3)	2 (1–3)	0.0098
Charlson Co-morbidity Index	4 (3–6)	4 (3–6)	4 (3–6)	0.6402
ABX at Discharge DOT	7 (5–10)	7 (5–10)	7 (4–7)	0.0060
Time (h) to culture send out	2 (0.75–19.5)	1.75 (0.75–10.5)	3.75 (0.25–26)	0.0128
Time (h) to prelim result	26.25 (17–51)	23 (16.5–47)	40 (22–66)	0.0004
Time (h) to culture and susceptibility	61.25 (47.63–83)	60 (46.3–80.7)	63.88 (49.25–92)	0.1381
Number of ABX classes culture was resistant to	5 (2–9)	2 (0–6)	1 (0–3)	0.0082
MDRO organism isolated	181 (45.3)	143 (49.8)	38 (33.6)	0.0034

[#] IQR—interquartile range, comparison between modification and no modification groups; ABX—antibiotics; DOT—days of therapy; MDRO—multidrug-resistant organism; ED—Emergency Department; ICU—intensive care unit; NS = not significant; p-value $\leq$ 0.05 = statistically significant.

The median time from suspected infection to the first antibiotic empiric therapy was 2 h (IQR 1–4), with most initial antibiotic administrations occurring in the Emergency Department. In half of the cases, inpatient empiric regimens were prescribed via order sets. If an order set was not used, 37% of the time, the ED regimen was started inpatient. The median time to an antibiotic with activity against the eventual isolated organism was 3.375 h (IQR 1.5–20) and ranged from 0.25 to 360 h, with 14 (3.5%) patients never receiving an antibiotic with activity against the cultured organism(s) prior to discharge.

A total of 287 (72%) patients had antibiotic regimen modifications to their inpatient antibiotic regimens. As mentioned above, modifications of ED regimens and changes involving agents specifically for Gram-positive organisms were not included in this definition of "modification". The total number of antibiotic modifications occurring in these patients was 415, with 80% (n = 330) occurring during empiric therapy, and the remaining were modifications in directed therapy. For patients where modifications occurred, the number of changes ranged from one to six modifications. The time to the first modification ranged from 2 to 196 h, with a median time of 23.4 h. Of the initial empiric therapy modifications, 84% (n = 122) were considered broadening in the spectrum of activity, and most were based on preliminary culture results mainly due to rapid diagnostic nucleic acid amplification test results (n = 71, 58%). Additional reasons for modifications were a lack of patient response (n = 17, 14%); additional history reviewed (n = 11, 9%); decompensation (n = 9, 7%); and a different suspected infection, adverse drug reactions, or undocumented reasons (n = 15, 12%).

Specifically looking at directed therapy, the median time to starting directed therapy was 64 h IQR (43.5–86.5) and the time to first modification in directed therapy was 74 h IQR (46–98). In 59% (n = 168) of the 287 patients that had therapy modifications, the modification occurred based on culture and susceptibility results, where, in 51% (n = 146) of these cases, the change resulted in a narrower-spectrum agent being used. Changes that occurred while the patient was already receiving directed therapy (a shift from one directed therapy to another) were predominantly de-escalations (n = 49, 58%). This primarily happened in patients who remained on the same broad-spectrum empiric therapy despite culture results and subsequently underwent a switch in treatment. For the remaining patients wherein escalation occurred, 27% (n = 11) had new culture results become available, leading to modifications; and 15% (n = 6) experienced a poor clinical response; and the remaining were modified due to adverse drug reactions (ADRs), convenience of dosing at discharge, or undocumented reasons. Directed treatment was only modified twice at most, which indicates that more regimen changes occur during empiric treatment, which could see up to four different empiric treatments.

Overall, for empiric and directed therapy modifications, infectious diseases physicians were the most common prescribers to make modifications during treatment (n = 201, 70%), and only in 4% (n = 16) of medication instances was the rationale not documented by a physician. Patients for which modification occurred were more likely to have a higher number of physician consultants, longer time to active therapy, higher likelihood of being admitted to the ICU, longer duration of antibiotic therapy, shorter times to preliminary culture results, and more likely to have a multidrug-resistant organism (MDRO) isolated. (Table 1).

The most common inpatient initial empiric therapies were piperacillin/tazobactam (n = 127, 32%), cefepime (n = 94, 24%), ceftriaxone (n = 67, 17%), and meropenem (n = 59, 15%). The most used antibiotic for directed treatment was ertapenem, followed by meropenem, together accounting for 57% (n = 244) of directed treatments used. Directed treatment was continued in 71% (n = 195) of patients at discharge, where 29% (n = 79) of patients were transitioned to an oral treatment (this change was not counted in regimen changes). The median duration of outpatient-directed treatment was 7 days. Patients that experienced regiment modifications during their inpatient treatment were more likely to have slightly longer median durations of treatment as an outpatient (median (IQR): 7 (5–10) vs. 7 (4–7), p = 0.006).

The median time to microbiology culture collection was 2 h (IQR 0.75–19.5) from initial arrival to the hospital. Comparing the time to culture collection between patients that had modifications to those that did not, those without regimen changes had slightly longer median times: 3.75 vs. 1.75 h (p = 0.0128). The median time to preliminary results overall was 26.25 h, and the time to preliminary results in patients with regimen modifications was shorter, 23 vs. 40 h (p = 0.0004). The time to susceptibility results did not differ between groups and ranged from 22.5 to 554 h with a median of 61.25 h (IQR 47.63–83). The most common organisms isolated were *Pseudomonas* spp. and ESBL-producing organisms, with 38% (n = 159) and 34% (n = 146) of isolates, respectively. *Escherichia coli* made up the largest portion of ESBL-producing organisms (n = 118, 81%), followed by *Klebsiella* spp. (n = 19, 13%) and *Proteus* spp. (n = 9, 6%). The remaining 28% (n = 119) of organisms were comprised primarily of *Enterobacter* spp., *Serratia* spp., and *Citrobacter* spp. Overall, 109 (27%) patients had more than one organism grow out or multiple sites of isolation on cultures. Categorically, 188 (47%) organisms were multidrug resistant (MDR), including to ESBL organisms; 42 (11%) were carbapenem resistant; and 2 organisms were extensively drug resistant. Specifically looking at the 158 *Pseudomonas* spp. isolates, 93 (59%) were pan-susceptible, 22 (14%) were multidrug resistant, and 28 (18%) *Pseudomonas* isolates were carbapenem resistant. Interestingly, of the carbapenem-resistant *Pseudomonas* spp. isolates, 11 (39%) were not considered MDR, with carbapenems being the only beta-lactams they were resistant to. Susceptibility testing for ceftazidime/avibactam was only carried out for 10 organisms overall, and 150 organisms were tested for ceftolozane/tazobactam. Out of the 150 organisms tested for susceptibility to ceftolozane/tazobactam, only 5 were resistant, 4 of which were ESBL *E. coli* and 1 of which was carbapenem-resistant *Klebsiella pneumoniae*. Patients infected with an MDRO were more likely to have their antibiotic regimen modified compared to those that did not, 50% vs. 34% (p = 0.0034).

In comparing outcomes between patients that had changes to antimicrobials versus those where no modifications occurred, no statistically significant differences in clinically meaningful outcome measures were observed. These outcomes included 30- and 90-day readmission and 30- and 90-day mortality. Other outcomes included time to clinical stability, which was 1 day, and the median time between the patients where modification occurred vs. not was also 1 day in each group, but with a slightly longer time in IQR (p = 0.0128) for the modification group. The longer time to reach clinical stability also led to longer lengths of stay (LOSs) in patients in the modification group with a median LOS of 7 days vs. 5 days in the no modification group (p < 0.0001). The 30-day readmission rate for all patients was approximately 21% of patients being readmitted; the majority of the time, this was infection related, with over half of patients being readmitted for the same type of infection, which was usually caused by the same organism (72%). However, as mentioned above, there was no difference between patient groups in terms of 30- or 90-day readmission rates. The 30- and 90-day mortality rates also did not differ between groups (Table 2).

Table 2. Patient outcomes based on regimen modification.

Patient Outcomes	Overall $n = 400$ (%)	Modifications $n = 287$ (%)	No Modifications $n = 113$ (%)	p-Value
Days to clinical stability (IQR)	1 (0–2)	1 (0–3)	1 (0–2)	0.0137
Hospital LOS (IQR)	6 (4.25–10)	7 (5–11)	5 (4–8)	<0.0001
ICU admission	74 (18.5)	65 (22.6)	9 (8)	0.0005
30-day mortality	36 (9)	28 (9.8)	8 (7.1)	0.4450
90-day mortality	7 (1.8)	4 (1.4)	3 (2.7)	0.4086
30-day readmission	85 (21.3)	55 (19.1)	30 (26.5)	0.1041
90-day readmission	130 (32.5)	91 (31.7)	39 (34.5)	0.5896

LOS—length of stay; p-value ≤ 0.05 = statistically significant.

Clinical outcomes were assessed for receipt of active antibiotic within 48 h compared to those who received an active antibiotic after 48 h. There was no difference in 30- or 90-day mortality in patients that received an active ABX within 48 h vs. not. A difference was observed in the length of stay for the <48 h group as the median (IQR) was 6 (4–10) vs. 8 (5.75–14.75), $p = 0.0015$ (Table 3). When exploring for variables associated with therapy modifications utilizing multiple logistic regression, ICU admission and hospital length of stay were the variables associated with therapy modification using a mixed-effects model and with a generalized estimating equations (GEE) model. When utilizing the regression model with only index events (removing subsequent visits from patients admitted multiple times, $n = 382$), IV antibiotic use in the previous 90 days became an additional significant variable (Table 4). The GEE model was selected for representation in a forest plot in Figure 1.

Table 3. Outcomes by receipt of active antibiotic within 48 h of suspected infection.

Patient Outcomes	<48 h $n = 342$ (%)	≥48 $n = 44$ (%)	p-Value
Days to clinical stability (IQR)	1 (0–2)	0 (0–1)	0.0056
Hospital LOS (IQR)	6 (4–10)	8 (5.75–14.75)	0.0015
30-day mortality	29 (8.48)	4 (9.09)	0.7704
90-day mortality	6 (1.75)	1 (2.27)	0.5555
30-day readmission	76 (22.22)	5 (11.36)	0.1160
90-day readmission	41 (11.99)	3 (6.81)	0.4498

p-value ≤ 0.05 = statistically significant.

Table 4. Multiple logistic regression analysis.

	Regression (Mixed-Effects Model)		Regression (GEE Model)		Regression (Index Only Model)	
Variable	Odds Ratio	95% CI	Odds Ratio	95% CI	Odds Ratio	95% CI
Charlson Co-morbidity Index	0.690	0.397 to 1.2	0.699	0.412 to 1.18	0.648	0.377 to 1.11

Table 4. *Cont.*

	Regression (Mixed-Effects Model)		Regression (GEE Model)		Regression (Index Only Model)	
Previous GNR MDRO infection	0.807	0.457 to 1.42	0.809	0.462 to 1.41	0.829	0.467 to 1.43
IV antibiotics in past 90-days	1.63	0.991 to 2.69	1.612	0.998 to 2.60	1.64	1.02 to 2.68
Community acquired	0.973	0.407 to 2.32	0.969	0.399 to 2.35	0.949	0.410 to 2.31
ICU admission	2.49	1.12 to 5.56	2.451	1.111 to 5.40	2.83	1.30 to 6.86
Sepsis present	1.35	0.814 to 2.23	1.337	0.815 to 2.19	1.26	0.772 to 2.07
MDRO infection	0.951	0.861 to 1.05	0.952	0.866 to 1.04	0.955	0.867 to 1.05
Hospital length of stay	1.08	1.02 to 1.14	1.083	1.02 to 1.14	1.08	1.02 to 1.14
Gender	1.24	0.768 to2.00	1.24	0.778 to 1.97	1.21	0.753 to 1.94

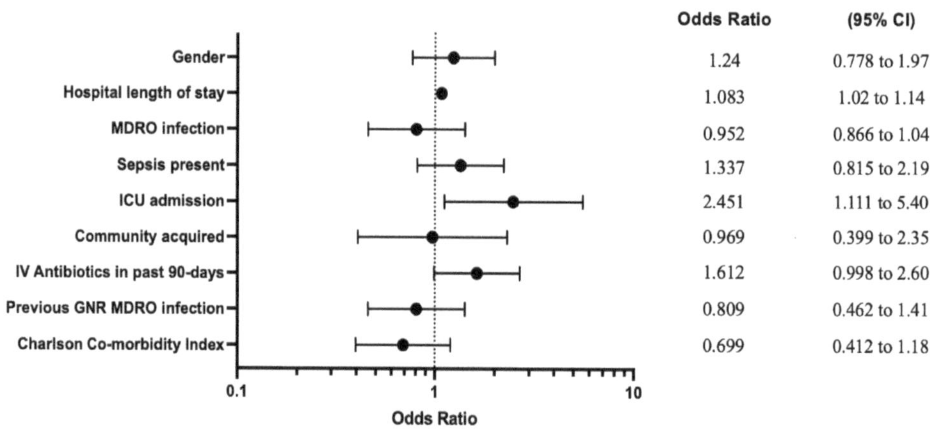

Figure 1. Regimen modification associations. Forest plot of the multiple logistic regression analysis utilizing a generalized estimating equations model, illustrating independent factors associated with antibiotic therapy modification. Calculated odds ratios with 95% confidence intervals are depicted.

3. Discussion

This study sought to describe the number of antibiotic regimen modifications patients with Gram-negative infections experienced during their treatment course and reasons for modifications, specifically in patients infected with resistant or potentially resistant organisms. While practicing clinicians are aware that antibiotic regimens undergo frequent changes in the inpatient setting due to various reasons, few data have been published on how often modifications occur, the rationale for modifications, how often the rationale is documented, when in therapy changes often occur, and the associated outcomes with antimicrobial regimen modifications. Previous studies have described antibiotic prescribing characteristics in the outpatient setting; our study adds to the literature by specifically focusing on patients with Gram-negative infections. To our knowledge, this is the first study to examine the frequency, rationale, timing, and associated outcomes with antimicrobial therapy modifications in the inpatient setting in patients with organisms that were resistant or had the potential to be resistant to extended-spectrum cephalosporins.

In our study, the patient population was predominantly elderly, consistent with the overall demographic composition of our institution. Despite the advanced age of the patients, the overall mortality rate was found to be 9%, a figure lower than reported in previous studies focusing on patients with ESBL-producing organisms [13,14]. The median time to any antibiotic being administered was 2 h, indicating that upon presentation to the Emergency Department, patients were promptly evaluated for infection and received antibiotics. This rapid response aligns with the recommended timeline for initiating antibiotic therapy, particularly for patients with sepsis [15]. The median time to an active antibiotic was considerably longer at ~24 h. This delay can be attributed to the fact that our patient population was infected with organisms that were or had the potential to be resistant to extended-spectrum cephalosporins, including *Pseudomonas* spp. and ESBL organisms. As a result, prescribers or order sets may not have initially included carbapenems or broader-spectrum anti-pseudomonal agents as empiric options. Notably, the time to active antibiotic therapy, though relatively delayed, occurred much earlier than the availability of culture and susceptibility results, which took an average of approximately 61 h from when infection was first suspected. This was primarily due to the availability of rapid diagnostics for blood cultures utilizing nucleic acid amplification testing as part of patient standard of care. When the DNA of resistant pathogens such as *Pseudomonas aeruginosa* or a resistance gene such as CTX-M was detected, the therapy was promptly adjusted to cover for these organisms. This likely explains the relatively short median time to the first antibiotic regimen modification, which occurred at ~23 h, consistent with the time to an active antibiotic and the median time to preliminary microbiology culture results of approximately 26 h. This underscores the crucial role and clinical utility of rapid diagnostics in starting active antimicrobial therapy, particularly for invasive infections, which has previously been described in the literature [16–18].

Regimen modification was common with 72% of patients experiencing modification specifically to Gram-negative targeting agents, with most of the modifications occurring during empiric therapy. The changes occurring during empiric treatment are not unexpected as, in practice, microbiology laboratories provide daily, if not multiple times daily, updates to the patient chart with new or preliminary culture results, including rapid diagnostic results, thus leading to empiric therapy adjustments. Furthermore, in line with antimicrobial stewardship best practices, once a patient was on directed therapy, regimen modifications were more likely to narrow treatment regimens. Significantly, the study only accounted for modifications made to antibiotics targeting Gram-negative organisms. It did not encompass regimen changes involving antimicrobials targeting Gram-positive organisms or other microorganisms. Additionally, modifications made to the initial regimen administered in the Emergency Department upon patient admission were also not included in the analysis. Thus, patients would have an even higher number of antimicrobial modifications if these aspects were also included.

Our study examined the implications of antibiotic regimen modifications in the context of MDRO-infected patients, focusing on patient outcomes. We found an association between regimen modifications and MDRO infections, but major patient outcomes such as 30- and 90-day mortality and 30- and 90-day readmission did not show significant differences. However, patients that experienced regimen modifications did have prolonged hospital stays (6 vs. 8 days ($p < 0.0001$), possibly influenced by factors such as inadequate initial therapy, delayed culture results, or the perceived need for extending antibiotic therapy when MDROs were isolated. Further exploration of outcomes based on the initiation of active antibiotics within 48 h of suspected infection revealed that only length of hospital stay and time to clinical stability were significantly different (Table 3). Consistent with what would be expected, patients that received an active antibiotic > 48 h from the onset of a suspected infection had longer lengths of hospital stay, likely due to either not responding to initial empiric therapy that was not active or possibly a delay in preliminary culture results. Bonine et al. used a similar definition of where a delay in therapy was defined as no receipt of an active agent within 2 days of the index date. They found that a delay in active antibiotic therapy was also associated with an increased length of stay and increased hospital cost, as would be expected. Furthermore, in their patient population, an increase in mortality was observed with delay in active antibiotic therapy [19]. Thus, our findings align with the existing literature, reinforcing the notion that delays in initiating active antibiotic therapy can be associated with increased hospital stays and costs, highlighting the critical importance of timely active antibiotic administration in the management of infections.

To determine any factors that might be predictive of patients requiring regimen modifications, multiple logistic regression with various models was used to identify risk factors that could potentially influence therapy modifications, including the MDRO status, which was univariately associated with modifications. In this analysis, the only variables associated with therapy modification were infection-related admission to the ICU and hospital length of stay, per the GEE and mixed-effects models. The representative forest plot (Figure 1) from the GEE model is likely the best fit model for this study population, as we are able to include all patient encounters and the GEE model makes fewer assumptions about the correlation structure between repeated measurements compared to traditional and mixed models [20]. Patient-specific risk factors that would be available to review and act upon at admission to the hospital, including previous antibiotic use and hospitalization in the past 90 days and previous MDRO infection, were included in the multiple logistic regression, but there were no significant associations.

As our study focused on patients with extended-spectrum cephalosporin-resistant or potentially resistant Gram-negative organisms, 47% of the isolates were identified as multidrug-resistant organisms (MDROs). Among these, a significant majority (82%) comprised organisms producing extended-spectrum beta-lactamases (ESBL). MDR *Pseudomonas* spp. constituted a substantial portion of the remaining MDROs; however, notably, 59% of *Pseudomonas* spp. were pan-susceptible. As mentioned above, patients infected with an MDRO were more likely to undergo antibiotic regimen modification, as indicated by univariate comparison. Despite the high prevalence of MDROs, only two patients were infected with carbapenem-resistant Enterobacterales (CRE), and two patients were infected with extensively drug-resistant (XDR) *Acinetobacter* spp. All four of the patients survived at 30 and 90 days, which may not have been expected considering the high mortality rates observed with CRE and XDR organisms, which have been reported to be 30–80%. In the CRE cases, the time to active therapy was within 10 h as the patients were also started on an aminoglycoside because of their previous history of CRE and previous known susceptibility to these agents. In both cases of XDR *Acinetobacter* infection, the time to active treatment was delayed to over 72 h. Overall, nine patients received ceftazidime/avibactam or ceftolozane/tazobactam, which were our formulary antibiotics reserved for the treatment of MDROs and were used in this study for patients with carbapenem-resistant *Pseudomonas* spp. or CRE. While the overall number of patients with CRE or XDR organisms was low during this time period, the observed association with regimen modification, potential increased length of hospital stay, and potential delay in active therapy within this patient population warrant close consideration of the likelihood a patient is infected with an MDRO in order to start prompt active therapy with a newer broad-spectrum antibiotic, such as ceftazidime/avibactam, ceftolozane/tazobactam, or imipenem/relebactam. As multidrug-resistant organisms continue to increase in incidence, including *Clostridioides difficile* infection (CDI), exposure to various antibiotics and antibiotic classes may increase the incidence of both CDI and resistant organism generation. In theory, frequent changing to therapy may lead to exposure to numerous different antibiotics, resulting in potential resistance or increased patient microbiome disruption. However, in practice, changes to antimicrobial therapy frequently occur despite this risk. Importantly, in many instances, modification to antimicrobial regimens is warranted, particularly in patients not responding to therapy or when additional culture information becomes available, which were the primary reasons within our patient population for antimicrobial regimen modifications. However, in our study, the impact on the patient microbiome, CDI risk, and resistance generation were not explored. Future research on the unintended consequences of regimen changes on resistance pressure may be warranted.

As a single-center retrospective observational study, there are inherent limitations associated with these data. This includes the potential generalizability of these data and our findings relevant to other centers both in the US and outside the US. Our patient population was specific to patients infected with extended-spectrum cephalosporin-resistant or potentially resistant organisms and did not include patients solely infected with non-resistant organisms, which can impact generalizability. Limitations associated with free text from medical records may include variability in the quality of documentation by providers in notes and missing information or explicit justification of treatment approaches. Potential inherent selection bias was mitigated by setting a priori enrollment criteria, definitions for exposures and outcomes, and randomization prior to patient selection and review. The outcomes of interest, however, were minimally susceptible to misclassification. Invasive confirmatory culture samples may not always be obtained, and this limits our ability to definitively identify the causative pathogen and susceptibility profile. This limitation mirrors the limitation in clinical practice. However, the inclusion requirement is an infection caused by Gram-negative organisms resistant or with the potential to be resistant to extended-spectrum cephalosporins; thus, this would minimize the potential of missing an organism due to the inability to acquire an invasive culture. Confounding is a potential issue in observational studies; we attempted to minimize this through regression analyses when analyzing factors associated with regimen modifications. Finally, when exploring the secondary objectives of the study in relation to clinical outcomes, these estimates were generated without adjusted models; as such, they are not adjusted estimates. We acknowledge this as a limitation, as adjusting for potential confounding variables would provide a more accurate estimation of the association between regimen modifications and clinical outcomes.

4. Materials and Methods

This was a retrospective descriptive study conducted at Hoag Hospital, Orange County, California, USA, a 584-bed community hospital. Hoag hospital has had an antimicrobial stewardship program for the past 16 years. The study included all hospitalized adult patients (aged $\geq$ 18 years) with complicated Gram-negative infections as defined above (complicated urinary tract infection, bacteremia, intra-abdominal infection, and/or pneumonia), as confirmed by culture, which grew an aerobic Gram-negative extended-spectrum cephalosporin-resistant organism. Organisms included all aerobic Enterobacterales and non-fermenting Gram-negative organisms. Patients were excluded if they did not meet the inclusion criteria or if patients were infected with Gram-positive bacteria only, were co-infected with Gram-positive organisms other than in intra-abdominal infections, did not receive Gram-positive targeting antibiotics within 24 h of intra-abdominal infection onset, received less than 24 h of total treatment, or had less than 48 h of hospitalization. Charts of patients with Gram-negative infections from a microbiology report based on organism and source from 1 January 2019 to 31 March 2021 were reviewed. Patients were randomly chosen by selecting every tenth patient, whose electronic medical record was reviewed for inclusion and exclusion criteria, and all pertinent clinical data were collected if the selected patient met the criteria. Inclusion of patients was not limited to index episodes as patients were still included in the study if they had more than one admission to the hospital during the time period.

MDR strains were defined as those organisms non-susceptible to an antimicrobial agent in three or more classes of antibiotics: carbapenems, penicillins (piperacillin/tazobactam), cephalosporins (ceftazidime, ceftriaxone, and/or cefepime); monobactams; aminoglycosides; and fluoroquinolones. In cases of Gram-negative polymicrobial infection caused by an extended-spectrum cephalosporin-resistant (ESCR) organism and a non-ESCR isolate, this was defined as having an ESCR organism and the patient included in the study. Infections were defined as nosocomially acquired if the onset of infection was >48 h after admission or community acquired if <48 h after admission. At our hospital, nucleic acid-based rapid diagnostic tests are routinely used in standard of care to provide molecular information within ~3 h of testing a micrology sample. Currently, order sets are available to guide prescribers in the treatment of infections. These order sets were developed based on the type of infection, formulary of antimicrobials, and annual cumulative antibiogram data. Restricted antimicrobials can only be ordered by an Infectious Diseases physician.

In addressing the primary objective, patient data were collected and were analyzed to determine the frequency of switch/add of Gram-negative antimicrobial therapy, the overall time in hours to the start of any empiric antimicrobial therapy, directed antimicrobial therapy, and the timing of when switch/add occurred. When classifying modifications to antimicrobial regimens, the initial ED regimen was not considered an inpatient empiric and, thus, deviations from ED regimens were not classified as modifications. The most used empiric and directed agents were calculated. In patients where switch/add on occurred, the frequency of these changes was calculated based on the reasons for changing therapy: (1) decompensation (worsening of symptoms/vitals leading to a higher level of care), (2) lack of response, (3) culture results, (4) adverse reaction, (5) preliminary culture results, and (6) "other".

For the purposes of the secondary objectives, patients were divided into two groups: those that had switch/add occur and those that did not; and patient covariates were compared to identify any associations with therapy switch/add. Patient covariates included age and gender, source of infection, infection type, previous hospital admission and antibiotic use 90 days prior to admission; nosocomial vs. community-acquired infection, comorbid conditions, Charlson Comorbidity Index, ICU admission, and active initial therapy versus delayed active therapy. Outcomes of interest for comparison included duration of antibiotic therapy post culture, length of stay in hospital post culture, discharge disposition, readmission within 90 days, and the composite outcome of in-hospital death or discharge to hospice.

Continuous variables were compared using the Mann–Whitney U test and the *t*-test. Qualitative categorical variables were compared using the Chi-square test, odds ratios, and 95% confidence intervals (CIs). For univariate analysis of outcomes, adjusted models were not used. Multiple logistic regression analysis was conducted on patient factors potentially associated with switch/add-on and included all statistically significant variables from univariate analysis, gender, age, ICU admission, MDR status, type of infection, and any other clinically relevant variables, whether they were statistically significant or not. Measures of independence were obtained to assess the performance of the models. Backward elimination was used to remove each successive least significant variable. Each variable was then checked by itself using linear regression models. Multicollinearity was assessed amongst covariates. Both a generalized estimating equations model and a mixed-effects model with a random effect of patients to control for multiple admissions by the same patients were utilized [21]. An additional multiple logistic regression was utilized with only index encounters. All three models were compared as part of a sensitivity analysis to look at the effects of the small number of multiple events from the same patients. The analysis was performed with the stepwise logistic regression model of R version 4.3.2, RStudio statistical package.

5. Conclusions

A large portion of patients with complicated Gram-negative infections caused by organisms that are resistant or potentially resistant to extended-spectrum cephalosporins had modifications to their antibiotic treatment regimens. Univariate analysis revealed that modifications to antimicrobial regimens were associated with MDR organism isolation, which was also associated with longer lengths of stay. Furthermore, not receiving an active antibiotic within 48 h was also associated with longer lengths of stay. From an antimicrobial stewardship perspective, it is critical to have a balanced approach in ensuring the appropriate patient populations receive optimal therapy, as too narrow a spectrum of agents may be ineffective, leading to detrimental effects on the patient outcomes, while overly broad-spectrum agents may lead to adverse drug reactions [22–25] and/or the development of resistance. As the prevalence of MDR organisms continues to increase, access to broader and more active agents such as ceftolozane/tazobactam, ceftazidime/avibactam, and imipenem/relebactam and newer agents will become more necessary. Thus, further studies on identifying patient risk factors associated with MDR organisms are necessary to identify patients with MDROs and thus start empiric active agents sooner, as our results demonstrated that this has an impact on hospital length of stay, and in other studies, poor clinical outcomes were associated with delays in active treatment.

Author Contributions: Conceptualization, J.Y. and P.R.; methodology, J.Y., R.B., A.H.W. and E.Y.; formal analysis, J.Y. and R.B.; investigation, J.Y. and M.M.; writing—original draft preparation, J.Y., P.R., A.H.W. and E.Y.; writing—review and editing, J.Y., M.M., P.R., A.H.W. and E.Y.; funding acquisition, J.Y. All authors have read and agreed to the published version of the manuscript.

Funding: This research was funded by MERCK SHARP & DOHME CORP., grant# VEAP 8908.

Institutional Review Board Statement: The study was conducted in accordance with the Declaration of Helsinki and given exemption status by the Institutional Review Board of Providence Health System (STUDY#2021000728, approved 18 November 2021.

Informed Consent Statement: Patient consent was waived due to the IRB study exemption status.

Data Availability Statement: Data are unavailable due to privacy and/or ethical restrictions.

Conflicts of Interest: Emre Yücel and Alexandre H. Watanabe are employees of Merck Sharp & Dohme LLC, a subsidiary of Merck & Co., Inc., Rahway, NJ, USA, who may own stock and/or hold stock options in Merck & Co., Inc., Rahway, NJ, USA. Funding for this research was provided by Merck Sharp & Dohme LLC, a subsidiary of Merck & Co., Inc., Rahway, NJ, USA. Although the sponsors formally reviewed a penultimate draft, the opinions expressed are those of the authors and may not necessarily reflect those of the sponsors. All co-authors approved the final version of the manuscript.

References

1. Prestinaci, F.; Pezzotti, P.; Pantosti, A. Antimicrobial resistance: A global multifaceted phenomenon. *Pathog. Glob. Health* **2015**, *109*, 309–318. [CrossRef] [PubMed]
2. Bassetti, M.; Rello, J.; Blasi, F.; Goossens, H.; Sotgiu, G.; Tavoschi, L.; Zasowski, E.J.; Arber, M.R.; McCool, R.; Patterson, J.V.; et al. Systematic review of the impact of appropriate versus inappropriate initial antibiotic therapy on outcomes of patients with severe bacterial infections. *Int. J. Antimicrob. Agents* **2020**, *56*, 106184. [CrossRef]
3. Paterson, D.L. Impact of antibiotic resistance in gram-negative bacilli on empirical and definitive antibiotic therapy. *Clin. Infect. Dis.* **2008**, *47*, S14–S20. [CrossRef] [PubMed]
4. Liu, V.X.; Fielding-Singh, V.; Iwashyna, T.J.; Bhattacharya, J.; Escobar, G.J. The Timing of Early Antibiotics and Hospital Mortality in Sepsis: Playing Devil's Advocate Reply. *Am. J. Resp. Crit. Care* **2017**, *196*, 935–936. [CrossRef]
5. Papoutsi, C.; Mattick, K.; Pearson, M.; Brennan, N.; Briscoe, S.; Wong, G. Social and professional influences on antimicrobial prescribing for doctors-in-training: A realist review. *J. Antimicrob. Chemother.* **2017**, *72*, 2418–2430. [CrossRef] [PubMed]
6. Sydenham, R.V.; Jarbol, D.E.; Hansen, M.P.; Justesen, U.S.; Watson, V.; Pedersen, L.B. Prescribing antibiotics: Factors driving decision-making in general practice. A discrete choice experiment. *Soc. Sci. Med.* **2022**, *305*, 115033. [CrossRef] [PubMed]
7. Warreman, E.B.; Lambregts, M.M.C.; Wouters, R.H.P.; Visser, L.G.; Staats, H.; van Dijk, E.; de Boer, M.G.J. Determinants of in-hospital antibiotic prescription behaviour: A systematic review and formation of a comprehensive framework. *Clin. Microbiol. Infect.* **2019**, *25*, 538–545. [CrossRef]
8. Wright, S.K.; Neill, K.M. Factors influencing the antibiotic-prescribing decisions of nurse practitioners. *Clin. Excell. Nurse Pract.* **2001**, *5*, 159–167. [CrossRef]
9. Elhanan, G.; Sarhat, M.; Raz, R. Empiric antibiotic treatment and the misuse of culture results and antibiotic sensitivities in patients with community-acquired bacteraemia due to urinary tract infection. *J. Infect.* **1997**, *35*, 283–288. [CrossRef]
10. King, L.M.; Fleming-Dutra, K.E.; Hicks, L.A. Advances in optimizing the prescription of antibiotics in outpatient settings. *BMJ—Brit. Med. J.* **2018**, *363*, k3047. [CrossRef]
11. Krishnakumar, J.; Tsopra, R. What rationale do GPs use to choose a particular antibiotic for a specific clinical situation? *BMC Fam. Pract.* **2019**, *20*, 178. [CrossRef] [PubMed]
12. Coutinho, D.; Vaz, D.; Brito, M.; Shiang, T. Modification of empirical antibiotic therapy in community acquired pneumonia. *Eur. Respir. J.* **2014**, *44*, P2574.
13. Hsieh, C.J.; Shen, Y.H.; Hwang, K.P. Clinical Implications, Risk Factors and Mortality Following Community-onset Bacteremia Caused by Extended-spectrum beta-lactamase (ESBL) and non-ESBL Producing Escherichia coli. *J. Microbiol. Immunol.* **2010**, *43*, 240–248. [CrossRef]
14. Lim, C.L.; Spelman, D. Mortality impact of empirical antimicrobial therapy in ESBL- and AmpC-producing Enterobacteriaceae bacteremia in an Australian tertiary hospital. *Infect. Dis. Health* **2019**, *24*, 124–133. [CrossRef]

15. Evans, L.; Rhodes, A.; Alhazzani, W.; Antonelli, M.; Coopersmith, C.M.; French, C.; Machado, F.R.; Mcintyre, L.; Ostermann, M.; Prescott, H.C.; et al. Surviving Sepsis Campaign: International Guidelines for Management of Sepsis and Septic Shock 2021. *Crit. Care Med.* **2021**, *49*, E1063–E1143. [CrossRef] [PubMed]
16. Claeys, K.C.; Schlaffer, K.E.; Heil, E.L.; Leekha, S.; Johnson, J.K. Validation of an Antimicrobial Stewardship-Driven Verigene Blood-Culture Gram-Negative Treatment Algorithm to Improve Appropriateness of Antibiotics. *Open Forum Infect. Dis.* **2018**, *5*, ofy233. [CrossRef]
17. Pogue, J.M.; Heil, E.L.; Lephart, P.; Johnson, J.K.; Mynatt, R.P.; Salimnia, H.; Claeys, K.C. An Antibiotic Stewardship Program Blueprint for Optimizing Verigene BC-GN within an Institution: A Tale of Two Cities. *Antimicrob. Agents Chemother.* **2018**, *62*, 10–1128. [CrossRef]
18. Lockwood, A.M.; Perez, K.K.; Musick, W.L.; Ikwuagwu, J.O.; Attia, E.; Fasoranti, O.O.; Cernoch, P.L.; Olsen, R.J.; Musser, J.M. Integrating Rapid Diagnostics and Antimicrobial Stewardship in Two Community Hospitals Improved Process Measures and Antibiotic Adjustment Time. *Infect. Control Hosp. Epidemiol.* **2016**, *37*, 425–432. [CrossRef]
19. Bonine, N.G.; Berger, A.; Altincatal, A.; Wang, R.; Bhagnani, T.; Gillard, P.; Lodise, T. Impact of Delayed Appropriate Antibiotic Therapy on Patient Outcomes by Antibiotic Resistance Status From Serious Gram-negative Bacterial Infections. *Am. J. Med. Sci.* **2019**, *357*, 103–110. [CrossRef]
20. Hubbard, A.E.; Ahern, J.; Fleischer, N.L.; Van der Laan, M.; Lippman, S.A.; Jewell, N.; Bruckner, T.; Satariano, W.A. To GEE or not to GEE: Comparing population average and mixed models for estimating the associations between neighborhood risk factors and health. *Epidemiology* **2010**, *21*, 467–474. [CrossRef]
21. Bates, D.; Mächler, M.; Bolker, B.M.; Walker, S.C. Fitting Linear Mixed-Effects Models Using lme4. *J. Stat. Softw.* **2015**, *67*, 1–48. [CrossRef]
22. Hagiya, H.; Kokado, R.; Ueda, A.; Okuno, H.; Morii, D.; Hamaguchi, S.; Yamamoto, N.; Yoshida, H.; Tomono, K. Association of Adverse Drug Events with Broad-spectrum Antibiotic Use in Hospitalized Patients: A Single-center Study. *Intern. Med.* **2019**, *58*, 2621–2625. [CrossRef]
23. Gerber, J.S.; Ross, R.K.; Bryan, M.; Localio, R.; Szymczak, J.E.; Wasserman, R.; Barkman, D.; Odeniyi, F.; Conaboy, K.; Bell, L.; et al. Association of Broad-vs Narrow-Spectrum Antibiotics with Treatment Failure, Adverse Events, and Quality of Life in Children With Acute Respiratory Tract Infections. *JAMA* **2017**, *318*, 2325–2336. [CrossRef] [PubMed]
24. Talpaert, M.J.; Rao, G.G.; Cooper, B.S.; Wade, P. Impact of guidelines and enhanced antibiotic stewardship on reducing broad-spectrum antibiotic usage and its effect on incidence of Clostridium difficile infection. *J. Antimicrob. Chemother.* **2011**, *66*, 2168–2174. [CrossRef] [PubMed]
25. Slawson, D. Broad-Spectrum Antibiotics Increase Adverse Events in Children with Acute Respiratory Infections with Minimal Benefit. *Am. Fam. Phys.* **2018**, *97*, 474–475.

antibiotics

MDPI

Article

From Epidemiology of Community-Onset Bloodstream Infections to the Development of Empirical Antimicrobial Treatment-Decision Algorithm in a Region with High Burden of Antimicrobial Resistance

Darunee Chotiprasitsakul [1,*], Akeatit Trirattanapikul [1], Warunyu Namsiripongpun [1], Narong Chaihongsa [2] and Pitak Santanirand [2]

1 Division of Infectious Diseases, Department of Medicine, Faculty of Medicine Ramathibodi Hospital, Mahidol University, Bangkok 10400, Thailand; akeatit.tri@mahidol.ac.th (A.T.); warunyu.nam@mahidol.ac.th (W.N.)

2 Microbiology Laboratory, Department of Pathology, Faculty of Medicine Ramathibodi Hospital, Mahidol University, Bangkok 10400, Thailand; narong.chh@mahidol.ac.th (N.C.); pitak.san@mahidol.ac.th (P.S.)

* Correspondence: darunee.cho@mahidol.ac.th

Abstract: Antimicrobial-resistant (AMR) infections have increased in community settings. Our objectives were to study the epidemiology of community-onset bloodstream infections (BSIs), identify risk factors for AMR-BSI and mortality-related factors, and develop the empirical antimicrobial treatment-decision algorithm. All adult, positive blood cultures at the emergency room and outpatient clinics were evaluated from 08/2021 to 04/2022. AMR was defined as the resistance of organisms to an antimicrobial to which they were previously sensitive. A total of 1151 positive blood cultures were identified. There were 450 initial episodes of bacterial BSI, and 114 BSIs (25%) were AMR-BSI. Non-susceptibility to ceftriaxone was detected in 40.9% of 195 *E. coli* isolates and 16.4% among 67 *K. pneumoniae* isolates. A treatment-decision algorithm was developed using the independent risk factors for AMR-BSI: presence of multidrug-resistant organisms (MDROs) within 90 days (aOR 3.63), prior antimicrobial exposure within 90 days (aOR 1.94), and urinary source (aOR 1.79). The positive and negative predictive values were 53.3% and 83.2%, respectively. The C-statistic was 0.73. Factors significantly associated with 30-day all-cause mortality were Pitt bacteremia score (aHR 1.39), solid malignancy (aHR 2.61), and urinary source (aHR 0.30). In conclusion, one-fourth of community-onset BSI were antimicrobial-resistant, and one-third of *Enterobacteriaceae* were non-susceptible to ceftriaxone. Treatment-decision algorithms may reduce overly broad antimicrobial treatment.

Keywords: bacteremia; antimicrobial resistance; epidemiology; community; Asia

1. Introduction

Bloodstream infections (BSIs) are a leading cause of morbidity and mortality globally [1,2]. The epidemiology of BSI has evolved and differs considerably between developed and developing countries. Asia is considered a high burden region of antimicrobial resistance [3]. Multidrug resistance was presented in 30% of cases of Gram-negative bacteremia in a community hospital in Thailand [4]. The mortality rate of BSI varies from 12% for community-onset BSI [5], 22% in a population-based cohort study [6], 30% for patients who have severe comorbidities, and 40–60% for intensive care unit patients [2]. Increasing antimicrobial resistance (AMR) is a significant cause of death worldwide. The highest burden is in low-resource settings [7]. Mortality in *Escherichia coli* and *Klebsiella* BSI strongly depends on resistance to fluoroquinolones or third-generation cephalosporins and on adequate therapy [8]. The median turnaround times were 0.80, 1.81, and 2.71 days for Gram stain, identification of organism, and antimicrobial susceptibility test (AST) results, respectively [9]. A prompt selection of empirical antimicrobial treatment without knowing whether an infection is antimicrobial-resistant, while balancing the risk of ineffective treatment versus excessively broad antimicrobial therapy is difficult. Understanding the extent of community-onset BSIs would help address the magnitude and impact of AMR and develop a solution.

We aimed to describe the contemporary epidemiology and outcomes of bacteremia in a community setting at emergency room (ER) and outpatient clinics. Our secondary objective was to identify independent factors associated with AMR in patients with BSI and develop the empirical antimicrobial treatment-decision algorithm for patients with suspected community-onset bloodstream infections.

2. Results

2.1. Distribution of Pathogenic Organisms Associated with Community-Onset Bloodstream and Antimicrobial Susceptibility

A total of 1151 positive blood cultures were identified. Of these, 410 specimens were contaminants. There were 460 initial episodes of BSI; 336 monomicrobial Gram-negative, 13 polymicrobial Gram-negative, 97 monomicrobial Gram-positive, 4 polymicrobial Gram-positive, 8 mixed Gram-negative and Gram-positive, 5 *cryptococcus*, and 3 *candida* (Figure 1). Of 487 unduplicated bacterial isolates with AST, the most common organisms were *Escherichia coli* (40.0%), followed by *Klebsiella pneumoniae* (13.8%), *Staphylococcus aureus* (7.6%), beta-hemolytic streptococcus (5.3%), *Salmonella* spp. (3.7%), and *Pseudomonas aeruginosa* (3.7%). Non-susceptibility to ceftriaxone was identified in 40.9% of 195 *E. coli* isolates and 16.4% among 67 *K. pneumoniae* isolates. All 18 *P. aeruginosa* isolates were susceptible to ceftazidime. Among the Gram-positive bacteria, 5.4% of 37 *S. aureus* isolates were methicillin-resistant, and 17.6% of 17 viridans streptococci isolates were non-susceptible to penicillin and ampicillin. Pathogens with ≥5 first isolate counts and antibiograms are summarized in Figure S1 in the Supplementary Data.

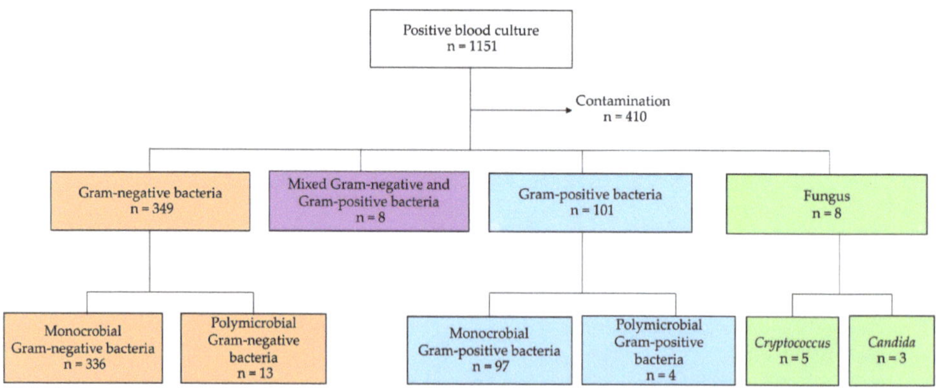

Figure 1. Distribution of pathogenic organisms associated with community-onset bloodstream infection.

2.2. Clinical Characteristics of Patients with Antimicrobial-Resistant BSIs (AMR-BSIs) and Non-Antimicrobial-Resistant BSIs (NAMR-BSIs)

A cohort of 450 unique adult patients with bacterial BSIs with AST were identified during the study period and included in the comparative analysis for risk factors associated with AMR. Of these, 114 BSIs (25%) were antimicrobial-resistant. The baseline characteristics of AMR-BSI compared to NAMR-BSI patients are shown in Table 1. The majority of bacterial BSIs were detected at ER (70%). The median age in the AMR-BSI group and NAMR-BSI group was 74 years (interquartile range [IQR] 57–83 years) and 71 years (IQR 59–80 years), respectively. The severity of the acute illness index was not different between the two groups. Both groups had a Pitt bacteremia score of 1 (IQR 0-2). Some differences between the two groups were observed. Patients with AMR-BSI were more likely to have a neurological disease, connective tissue disease prior admission, colonization or infection with multidrug-resistant organisms (MDROs), previous antimicrobial exposure within 90 days, and a higher proportion of urinary sources.

Table 1. Baseline characteristics of 450 unique adult patients with AMR-BSI and NAMR-BSI.

Variables	AMR-BSI n = 114 (%)	NAMR-BSI n = 336 (%)	*p* Value
Emergency room	72 (63.2%)	241 (71.7%)	
Outpatient clinic	42 (36.8%)	95 (28.3%)	
Age (years), median (IQR)	74 (57–83)	71 (59–80)	0.56
Male	47 (41.2%)	147 (43.8%)	0.64
Preexisting medical conditions			
Chronic pulmonary disease	7 (6.1%)	17 (5.1%)	0.66
Cardiovascular disease	35 (30.7%)	75 (22.3%)	0.07
Chronic liver disease	9 (7.9%)	25 (7.4%)	0.87
Chronic kidney disease	16 (14.0%)	34 (10.1%)	0.25
Neurologic disease	30 (26.3%)	58 (17.3%)	0.04
Diabetes mellitus	41 (36.0%)	127 (37.8%)	0.73
Hypertension	60 (52.6%)	168 (50%)	0.63
Active solid tumor	29 (25.4%)	64 (19.1%)	0.15
Active hematologic malignancies	3 (2.6%)	19 (5.7%)	0.20
HIV	0 (0%)	5 (1.5%)	0.19
Kidney transplantation	8 (7.0%)	11 (3.3%)	0.09
Stem cell transplantation	0 (0%)	2 (0.6%)	0.41
Connective tissue diseases	11 (9.7%)	15 (4.5%)	0.04
Chemotherapy in 6 months	7 (6.1%)	27 (8.0%)	0.51
Corticosteroids at ≥20 mg of prednisone daily or equivalent for ≥14 days	4 (3.5%)	12 (3.6%)	0.98
Post-COVID-19 within 60 days	9 (7.9%)	13 (3.9%)	0.09
Presence of hemodialysis or central venous catheters	9 (7.9%)	35 (10.4%)	0.43
Severity of acute illness index			
qSOFA score, median (IQR)	1 (0–2)	1 (0–2)	0.40
Pitt bacteremia score	1 (0–2)	1 (0–2)	0.97
ICU admission following BSIs	21 (18.4%)	80 (23.8%)	0.23
On mechanical ventilator	14 (12.3%)	51 (15.2%)	0.45
On vasopressor	14 (12.3%)	60 (17.9%)	0.17
Epidemiological risks			
Prior admission within 90 days	57 (50.0%)	94 (28.0%)	<0.001
Colonization or infection with MDROs during preceding 90 days	40 (35.1%)	27 (8.0%)	<0.001
Ceftriaxone-resistant *Enterobactericeae*	47 (41.2%)	16 (4.8%)	
Carbapenem-resistant *Enterobactericeae*	11 (9.6%)	6 (1.8%)	
Extremely drug-resistant *P. aeruginosa*	6 (5.3%)	3 (0.9%)	
Extremely drug-resistant *A. baumannii*	4 (3.5%)	3 (0.9%)	
Previous antibiotic exposure within 90 days	67 (58.8%)	95 (28.3%)	<0.001
Carbapenems	21 (18.4%)	14 (4.2%)	
Ceftriaxone	22 (19.3%)	22 (6.5%)	
Piperacillin-tazobactam	16 (14.0%)	26 (7.7%)	
Fluoroquinolones	15 (13.2%)	16 (4.8%)	
Amoxicillin-clavulanate	10 (8.8%)	17 (5.1%)	
Vancomycin	4 (3.5%)	4 (1.2%)	
Oral third generation cephalosporins	4 (3.5%)	3 (0.9%)	
Ceftazidime	3 (2.6%)	0	
Cefepime	1 (0.9%)	5 (1.5%)	
Type of identification			
Unknown primary source	14 (12.3%)	71 (21.1%)	0.04
Pneumonia	4 (3.5%)	16 (4.8%)	0.58
Skin and soft tissue infection	3 (7.5%)	8 (6.0%)	0.72

Table 1. *Cont.*

Variables	AMR-BSI n = 114 (%)	NAMR-BSI n = 336 (%)	p Value
Bone and joint infection	1 (0.9%)	12 (3.6%)	0.14
Urinary tract infection	58 (50.9%)	100 (29.8%)	<0.001
Hepatobiliary infection	12 (10.5%)	42 (12.5%)	0.58
Intra-abdominal infection	17 (14.9%)	38 (11.3%)	0.31
Infective endocarditis	1 (0.9%)	13 (3.9%)	0.11
Central nervous system infection	0 (0%)	8 (2.4%)	0.10
Catheter-associated bloodstream infection	2 (1.8%)	20 (6.0%)	0.07

Abbreviations: IQR, interquartile range; AMR-BSI, antimicrobial-resistant bloodstream infection; NAMR-BSI, non-antimicrobial-resistant bloodstream infection; MDROs, multidrug-resistant organisms.

2.3. Analysis of Risk Factors Associated with AMR and 30-Day All-Cause Mortality in Patients with Community-Onset Bloodstream Infections

In the multivariate analysis, presence of MDROs during the preceding 90 days (adjusted odds ratio [aOR] 3.62; 95% CI 1.95–6.75; $p < 0.001$), prior antimicrobial exposure within 90 days (aOR 1.94; 95% CI 1.08–3.50; $p = 0.03$), and urinary source of bacteremia (aOR 1.78; 95% CI 1.06–3.01; $p = 0.03$) were independent factors for antimicrobial-resistant infection (Table 2).

Table 2. Univariable analysis and multivariable analysis of risk factors for antimicrobial resistance.

Variables	Univariable Analysis OR (95% CI)	p	Multivariable Analysis aOR (95% CI)	p
Cardiovascular disease	1.54 (0.96–2.48)	0.07	1.51 (0.88–2.59)	0.13
Neurologic disease	1.71 (1.03–2.83)	0.04	1.40 (0.79–2.46)	0.25
Kidney transplantation	2.23 (0.87–5.69)	0.09	1.65 (0.56–4.80)	0.36
Connective tissue diseases	2.29 (1.02–5.13)	0.05	2.26 (0.91–5.61)	0.08
Prior admission within 90 days	2.57 (1.66–3.99)	<0.001	1.30 (0.73–2.31)	0.37
Presence of MDROs during preceding 90 days	6.19 (3.57–10.72)	<0.001	3.63 (1.95–6.75)	<0.001
Previous antibiotic exposure within 90 days	3.61 (2.32–5.63)	<0.001	1.94 (1.08–3.50)	0.03
Unknown primary source	0.52 (0.28–0.96)	0.04	0.75 (0.37–1.53)	0.43
Urinary source	2.44 (1.58–3.78)	<0.001	1.79 (1.06–3.01)	0.03
CLABSI	0.08 (0.06–1.23)	0.09	0.33 (0.07–1.62)	0.17

Abbreviations: aOR, adjusted odds ratio; CI, confidence interval; MDROs, multidrug-resistant organisms; CLABSI, central line-associated bloodstream infection.

The 30-day all-cause mortality in the AMR-BSI and NAMR-BSI groups were 6.5% and 9.1%, respectively (HR 0.71; 95% CI 0.31–1.62; $p = 0.41$). Inactive empirical treatment within 24 h was not associated with 30-day all-cause mortality (HR 0.67; 95% CI 0.20–2.19; $p = 0.51$). Factors significantly associated with 30-day all-cause mortality in the multivariate model were Pitt bacteremia score (HR 0.71; 95% CI 1.20–1.62; $p < 0.001$), solid malignancies (HR 2.62; 95% CI 1.30–5.24; $p = 0.007$), and urinary source (HR 0.30; 95% CI 0.11–0.79; $p = 0.015$) (Table 3).

Table 3. Hazard ratio for 30-day all-cause mortality for adult patients with community-onset bacterial bloodstream infection.

Variables	Univariable Analysis HR (95% CI)	p	Multivariable Analysis aHR (95% CI)	p
Antimicrobial resistance	0.71 (0.31–1.62)	0.41		
Inactive empirical treatment within 24 h	0.67 (0.20–2.19)	0.51		
Pitt bacteremia score	1.42 (1.23–1.63)	<0.001	1.39 (1.20–1.62)	<0.001
Solid malignancy	3.03 (1.53–5.99)	0.001	2.61 (1.30–5.24)	0.01
Hypertension	0.47 (0.23–0.95)	0.04	0.49 (0.24–1.01)	0.054
Urinary source	0.27 (0.10–0.69)	0.006	0.30 (0.11–0.79)	0.02
Pneumonia source	4.86 (1.88–12.57)	0.001	2.05 (0.75–5.56)	0.16

Abbreviations: HR, hazard ratio, CI, confidence interval.

2.4. Appropriateness of Antimicrobial Use

There were 441 unique adult patients with bacterial BSI receiving at least one dose of an antimicrobial. Of 328 patients with NAMR-BSI, 190 (57.9%) received empirical treatment with broad spectrum active coverage. In total, 34 (30.1%) patients in 113 AMR-BSI patients were empirically treated with inactive spectrum coverage. Appropriate definitive treatment was not significantly different between AMR-BSI and NAMR-BSI patients. Optimal drug and duration in the AMR-BSI and NAMR-BSI groups were 33.0% and 36.3%, respectively. The median duration of treatment was 13 days in both groups. The possibility of shortening treatment duration in the AMR-BSI group and NAMR-BSI group were 57.8% and 50.2%, respectively (Table 4).

Table 4. Appropriateness of antimicrobial treatment of 441 unique adult patients with AMR-BSI vs. NAMR-BSI receiving at least one dose of an antimicrobial agent.

Variables	AMR-BSI	NAMR-BSI	*p* Value
Empirical antimicrobial treatment	n = 113 (%)	n = 328 (%)	
Optimally active coverage (appropriate)	79 (69.9)	121 (36.9)	<0.001
Broad spectrum active coverage	0	190 (57.9)	
Inactive spectrum coverage	34 (30.1)	17 (5.2)	
Multiple antimicrobial change in 48 h (range 2–6 times)	27 (23.9)	92 (28.1)	0.39
Unnecessary double coverage	2 (1.8)	21 (6.4)	0.06
Definitive antimicrobial treatment	n = 109 (%)	n = 311 (%)	
Optimal drug and duration (appropriate)	36 (33.0)	113 (36.3)	0.71
Narrower/simpler antimicrobial is available	14 (12.8)	57 (18.3)	0.33
Inadequate spectrum coverage	4 (3.7)	2 (0.6)	0.06
Step down to oral treatment is possible	8 (7.3)	42 (12.8)	0.18
Unnecessary double coverage	2 (1.8)	10 (3.2)	0.59
Duration of antimicrobial treatment [a], median (IQR)	13 (8–14)	13 (9–14)	0.74
Shorter duration is possible [a]	52/90 (57.8)	115/229 (50.2)	0.22

Abbreviation: IQR, interquartile range. [a] In patients with uncomplicated BSI, and not in palliative care within 5 days.

2.5. Proposed Empirical Antimicrobial Treatment Algorithm for Patients with Suspected Community-Onset Bloodstream Infections

Important risk factors of AMR from the analysis were integrated into treatment-decision algorithms. The following triage steps were created: identifying patients with clinical symptoms and signs suspecting bacterial BSI and stratifying patients by risk of AMR. Patients with the highest risk were defined as those who had MDROs during the preceding 90 days. These patients would be treated with broad spectrum antimicrobials. Patients not meeting this definition would be reviewed for prior antimicrobial exposure within 90 days; those with no exposure would be treated with narrower spectrum antimicrobials. Patients previously exposed to antimicrobials within 90 days would be evaluated for the suspected source of infection; those with suspected urinary source would be treated with broad- spectrum antimicrobials, and those with suspected non-urinary source would be treated with narrower spectrum antimicrobials. The development of a treatment-decision algorithm is depicted in Figure 2. The sensitivity and specificity of the algorithm for predicting AMR-BSI were 49.1% (95% CI 36.9–58.7%) and 85.4% (95% CI 81.2–89.0%), respectively. The PPV and NPV were 53.3% (95% CI 45.4–61.1%) and 83.2% (95% CI 80.4–85.6%), respectively. A receiver operating characteristic curve (ROC) curve derived from a logistic regression comprising the three most important variables yielded a C-statistic of 0.73.

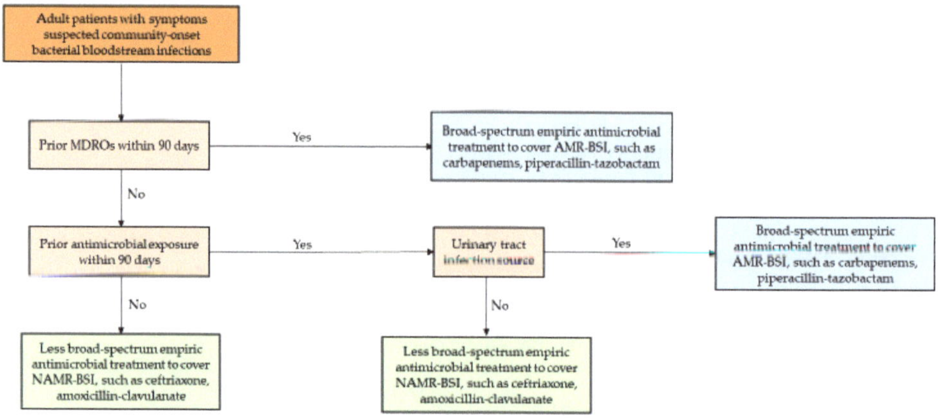

Figure 2. Proposed empirical antimicrobial treatment algorithm for patients with suspected community-onset bloodstream infections. Abbreviations: AMR-BSI, antimicrobial-resistant bloodstream infection; NAMR-BSI, non-antimicrobial-resistant bloodstream infection, MDROs, multidrug-resistant organisms.

Empirical treatment following the algorithm resulted in 14.9% of patients with NAMR-BSI receiving broad spectrum active coverage, and 17.7% of patients with AMR-BSI receiving inactive spectrum coverage.

Sensitivity analyses were performed on the subset of 298 patients who had *Enterobacteriaceae* BSIs. The three significant risk factors for AMR remained similar to the entire dataset. The sensitivity and specificity of the algorithm for predicting ceftriaxone-resistant BSI were 57.5% (95% CI 46.8–67.6%) and 81.9% (95% CI 75.9–86.9%), respectively. The PPV and NPV were 59.3% (95% CI 50.7–67.2%) and 80.7% (95% CI 76.6–84.2%), respectively. The C-statistic was 0.76. The internal validation of 227 bacterial BSIs revealed a C-statistic of 0.77 (Table 5).

Table 5. Sensitivity, specificity, positive predictive value, negative predictive value, and C-statistics of the algorithm in predicting antimicrobial-resistant infection.

	All BSIs (n = 450)	*Enterobacteriaceae* BSIs (n = 298)	Validation Cohort (n = 227)
Sensitivity (95% CI)	49.1% (36.9–58.7%)	57.5% (46.8–67.6%)	55.2% (41.5–68.3%)
Specificity (95% CI)	85.4% (81.2–89.0%)	81.9% (75.9–86.9%)	93.5% (88.7–96.7%)
PPV (95% CI)	53.3% (45.4–61.1%)	59.3% (50.7–67.2%)	74.4% (58.8–86.5%)
NPV (95% CI)	83.2% (80.4–85.6%)	80.7% (76.6–84.2%)	85.9% (80.0–90.6%)
C-statistic	0.73	0.76	0.77

Abbreviations: CI, confidence interval; positive predictive value, PPV; negative predictive value, NPV; BSIs, bloodstream infections.

3. Discussion

The most common pathogens of community-onset BSI at our institution included *E. coli*, *K. pneumoniae*, and *S. aureus*. This is consistent with previous reports of community-acquired bacteremia in Thailand and other countries [10–12]. These top three pathogens accounted for 61.4% of all bacterial isolates. Other leading pathogens were beta-hemolytic streptococcus and *Salmonella* spp., which were different from previous studies [10–12]. A surveillance of community-acquired bacteremia in Thailand during 2016–2017 found that *E. coli* and *K. pneumoniae* had susceptibility rates to ceftriaxone of 73% and 98%, respectively [10]. In contrast, our study revealed lower susceptibility rates for these bacteria with susceptibility rates to ceftriaxone of 60.1% and 83.6%, respectively. The incidence of antimicrobial resistance continues to rise with a change driven by an increase in community-onset cases. The incidence of ceftriaxone-resistant *Enterobacteriaceae* infection increased by 53% from 2012 to 2017 according to the US Centers for Disease Control and Prevention [13].

In the present study, the independent risk factors for antimicrobial-resistant BSI included the presence of MDROs within 90 days, prior antimicrobial exposure within 90 days, and urinary sources in our study. These factors were similar to those identified in previous studies [3,14]. Antimicrobial selective pressure was linked to bacterial resistance [15]. High rates of community-onset antimicrobial-resistant infection have been occurring worldwide, predominantly in urinary tract infection with *E. coli* [16]. Ceftriaxone-resistant uropathogens were isolated in 21.3% of patients with acute cystitis in Thai general practice clinics from 2014 to 2016 [17]. Similarly, a study at a tertiary care hospital in Thailand reported that 22.3% of *E. coli* causing community-acquired UTI were ceftriaxone-resistant in 2017 [18].

The Pitt bacteremia score and solid malignancies were associated with an increase in the overall 30-day mortality in our study. The severity of illness has been well established in predicting mortality [19]. All patients with solid tumors who died during the 30-day follow-up period in our study had advanced-stage malignancy. Despite no survival advantage, antibiotics were administered in 82% of patients with terminal cancer within three days of death at an academic hospital in a retrospective study conducted in Korea [20]. Appropriately directed palliative care can reduce aggressive antimicrobial use near the end of life. It would benefit individual patients' quality of life and decrease selection pressure that can lead to MDROs. The majority of inactive empirical antimicrobials in our study were against ceftriaxone-resistant Gram-negative BSI. Studies have shown mixed results regarding the association between the effectiveness of empirical antimicrobial treatment and ceftriaxone-resistant Gram-negative bacteremia [14,21,22]. The finding that inadequate empirical antibiotic treatment does not significantly impact the mortality in our study is consistent with previous studies [14,22]. The mortality in our study was below 10%, which is comparable to previous studies [14,22]. The patients in our study were not severely ill (median Pitt bacteremia score of 1). Relatively low mortality in community-onset bacteremia could be primarily driven by underlying conditions and disease severity [22]. The impact of empirical antimicrobial choice on mortality may be limited in this scenario. Urinary source was significantly associated with lower mortality in our study. A multi-center study in English acute hospitals found that patients with urinary tract-related bacteremia were less acutely unwell [22]. Piperacillin-tazobactam (TZP) is commonly used as an empirical treatment in our setting. The multinational, randomized, controlled trial of patients with ESBL-producing bacteremia (MERINO study) showed that definitive treatment with TZP increased 30-day mortality compared to meropenem, and no difference in mortality between urinary versus non-urinary source [23]. However, Sharara et al. reported no differences between TZP versus carbapenems in the clinical resolution or mortality for the treatment of ESBL-producing pyelonephritis [24]. A urinary pharmacokinetics study found that high TZP concentrations in urine and could result in treatment efficacy [25]. A small randomized trial showed that indwelling catheter replacement before initiating antimicrobial therapy significantly decreased bacteriuria and time to clinical improvement [26]. High TZP concentration in urine and biofilm removal in catheter-associated UTI may contribute to better outcomes compared to non-urinary-source infections.

The most frequent inappropriate prescribing was empirical broad spectrum antimicrobial treatment was 57.9% among NAMR-BSI in our study. A study evaluating practice at ER reported that inappropriate antimicrobial prescription in adult patients was 36.9% [27]. Short courses of antimicrobial therapy (6–10 days) have been shown to have comparable clinical outcomes as prolonged courses of therapy (11–16 days) for Gram-negative bacteremia [28]. The median duration of antimicrobial treatment for the uncomplicated bacterial BSIs was 13 days in our study; shorter courses were possible in more than half of cases in our study. Although we have institutional empirical treatment guidelines and weekly handshake stewardship in the ER, there is a more pressing need to develop initiatives to improve ER-based antimicrobial prescribing and emphasis on optimal treatment duration [29].

Potential risk factors driving the emergence of antimicrobial-resistant bacterial infections have been identified in various studies. However, integration of multiple risk factors into actual practice is scarce. Clinicians continue to face a significant challenge when treating serious Gram-negative infections due to the difficult balance between the risk of ineffective agents versus overly broad empiric antimicrobial treatment. A prior study developed an easy-to-use clinical decision algorithm to determine the probability of an extended-spectrum beta-lactamase (ESBL)-producing bacterial BSI in a bacteremic patient that could aid in selecting appropriate empiric treatment [3]. However, it could not be applied in regions with high ESBL prevalence.

From the analysis of risk factors for antimicrobial resistance, we developed a decision tree algorithm with three predictors; the presence of MDROs within 90 days, antibiotic exposure in the previous 90 days, and urinary tract infection source. There is always a trade-off between sensitivity and specificity. The ability to correctly predict NAMR-BSI cases (specificity) is essential to ensure the lowest risk of ineffective therapy. Patients classified as AMR-BSI cases by the algorithm were 53.3% more likely to be true AMR-BSI cases (PPV), and patients classified as NAMR-BSI cases were 83.2% more likely to be true NAMR-BSI (NPV). The subset of patients with *Enterobacteriaceae* BSIs yielded 6% higher PPV and 2.5% lower NPV. Empirical treatment following the algorithm would reduce broad spectrum antimicrobial therapy in NAMR-BSI cases by 43.0% and inactive spectrum antimicrobial in AMR-BSI cases by 12.4% in this dataset. This easy-to-use algorithm could improve the prediction of AMR-BSIs and reduce inappropriate empirical antimicrobial use. However, this algorithm cannot replace clinical judgment. Relevant components, such as clinical appearance, underlying conditions, and concern level of clinicians should be incorporated into decision-making.

There are several limitations in our study. First, this study was conducted at a single center. Our results may not be generalizable to patients in other settings with different prevalences of antimicrobial-resistant bacteria. Our findings should be validated in other cohorts. Second, the presence of MDRO was reviewed from our clinical microbiology laboratory reports, and previous antimicrobial exposure was retrospectively reviewed from the medical records. However, cases visiting outpatient clinics and ER were usually established patients who had received healthcare services at our hospital, and missing data on these two independent factors were likely small. Finally, the broad clinical approach included both Gram-positive and Gram-negative bacteria in our algorithm. Ceftriaxone is recommended and the most commonly used empirical treatment for community-acquired sepsis at our institution. Many Gram-positive bacteria non-susceptible to penicillin are susceptible to ceftriaxone. A subset of AMR-BSI predicted by this proposed algorithm would include ceftriaxone-susceptible Gram-positives, in which ceftriaxone is reasonable to use as an empirical treatment. We performed the sensitivity analyses on *Enterobacteriaceae* BSIs, which yielded 6% increased PPV and 2.5% decreased NPV. This finding suggests that the algorithm is robust to Gram-positive and *Enterobacteriaceae*. This broad-range approach would better represent real-world practice when the initial presentation cannot distinguish between Gram-positive versus Gram-negative organisms.

4. Material and Methods

4.1. Study Population

We conducted a retrospective cohort study at Ramathibodi Hospital, a 1300-bed tertiary-care hospital in Bangkok, Thailand, between 1 August 2021 and 15 April 2022. All positive blood cultures from patients aged > 18 years at ER and outpatient clinics were identified. Only the initial episode of bacterial BSI with AST results was included in the comparative analysis of risk factors for antimicrobial-resistant infection.

4.2. Data Collection

Information regarding demographics, pre-existing medical conditions, and the severity of acute illness on day 1 of BSI, including Quick Sequential Organ Failure Assessment (qSOFA) score, Pitt bacteremia score, intensive care unit (ICU) admission, mechanical ventilation, vasopressor administration, receipt of antimicrobial treatment during preceding 90 days, antimicrobial-resistant bacterial colonization or infection during the prior 90 days, and microbiological and mortality data were obtained from medical records. Mortality and cause of death were assessed at 30 days. Duplicate isolates of the same species with the identical AST profile recovered from consecutive blood cultures on the same patient after the index BSI were excluded from cumulative AST.

4.3. Definitions

Community-onset BSI refers to the location of the onset of BSI episodes which includes community and long-term healthcare facilities. BSI is defined as the positive growth of the organism (s) from blood specimen (s) in $\geq$1 blood culture bottle taken from a patient with compatible clinical features of infection. The isolated bacteria are classified as contaminants if they are common commensal organisms on the skin or environment e.g., coagulase-negative staphylococci, *Bacillus* spp., *Corynebacterium* spp., *Propionibacterium* spp., *Aerococcus* spp., *Micrococcus* spp., and the patient has no compatible clinical syndrome that could be caused by such organisms. Polymicrobial BSI is defined as isolation of $\geq$2 different pathogens from the same blood sample. The source of BSI is determined based on clear clinical evidence that the BSI was linked to focal infection at another body site.

Antimicrobial resistance is the resistance of organisms to an antimicrobial to which they were previously sensitive. For the purpose of this study, antimicrobial-resistant bacteria are defined as follows; Gram-negative bacteria other than *P. aeruginosa* that exhibit in vitro non-susceptibility to ceftriaxone, *P. aeruginosa* that exhibits in vitro non-susceptibility to ceftazidime, and Gram-positive bacteria that exhibit in vitro non-susceptibility to penicillin-class drugs (penicillin, ampicillin, or oxacillin).

4.4. Microbiological Testing

All isolates were tested for their antimicrobial susceptibility by an automated microbroth dilution testing system (Sensititre™; Thermo Fisher Scientific, Cleveland, OH, USA). Clinical and Laboratory Standards Institute (CLSI) clinical breakpoints were used to interpret the minimum inhibitory concentration (MIC) values [30].

4.5. Assessment of the Appropriateness of Antimicrobial Use

Independent adjudication of the appropriateness of antimicrobial treatment was performed based on institutional antimicrobial guidelines by three infectious disease specialists. The local antimicrobial guidelines for common bacterial infections consisted of the recommended antimicrobials for empirical therapy and the recommended dosage of each antimicrobial [29]. Antimicrobial treatment was considered active when isolated pathogens were susceptible in vitro to the prescribed antimicrobial. Antimicrobial de-escalation or escalation encompassed the antimicrobial change within 48–72 h of available culture and susceptibility test results.

All BSI episodes wherein at least one antimicrobial was prescribed were randomized, and each was assigned to two specialists. Each expert independently assessed the antimicrobial prescription into specific categories for appropriate antimicrobial use modified from a previous study [31] (Table S1 in the Supplementary Data).

4.6. Statistical Analysis

Continuous variables were summarized as the mean (standard deviation) for normally distributed data or median (interquartile range) for non-normally distributed data. Categorical variables were displayed using absolute counts and percentages. Comparative analysis of variables associated with antimicrobial resistance was conducted using the

Student's t-test or Wilcoxon rank-sum tests for continuous variables. The Chi-square test or Fisher exact test was used to compare categorical variables. Logistic regression was used to calculate the odds ratios (ORs) and 95% confidence intervals (CIs) for factors associated with AMR. The Cox regression model was utilized to estimate the unadjusted hazard ratios (HRs) and 95% CIs of risk factors for mortality within 30 days from bacteremic onset in patients who received antimicrobial treatment at least three days following the start of infection.

Variables with univariate p-values < 0.10 or clinical plausibility were included in the multivariable models to identify independent factors associated with AMR and 30-day mortality. Differences were considered statistically significant at a p-value of less than 0.05. All statistical analyses were performed using Stata, version 18.0 (Stata Corp., College Station, TX, USA).

4.7. *Empirical Antimicrobial Treatment-Decision Algorithm Development*

To generate clinically practical algorithms, we placed the significant risk factors for AMR from the multivariate model into several triage steps. The AMR is a serious concern of delayed active empirical treatment. The most optimal algorithm should have high specificity, yet it should lessen the inappropriate empirical broad spectrum antimicrobial treatment. The sensitivity, specificity, positive predictive value (PPV), negative predictive value (NPV), and C-statistic of the algorithm in predicting AMR-BSI were calculated among patients who had bacterial BSI. The algorithm was evaluated in the internal validation cohort from 16 April 2022 to 30 June 2022.

5. Conclusions

One-fourth of community-onset BSIs were antimicrobial-resistant. Almost one-third of *Enterobacteriaceae* were non-susceptible to ceftriaxone. Development of a treatment-decision algorithm based on the independent risk factors for AMR-BSI consisting of the presence of MDROs within 90 days, prior antimicrobial use within 90 days, and urinary source could aid in empiric treatment decisions. Reducing excessive broad antimicrobial treatment in NAMR-BSI and increasing useful broad treatment in AMR-BSI would optimize clinical outcomes while reducing the risk of further resistance emergence.

Supplementary Materials: The following supporting information can be downloaded at: https://www.mdpi.com/article/10.3390/antibiotics12121699/s1, Figure S1: Antimicrobial susceptible percentage of clinical isolated bacteria with ≥5 first isolate counts; Table S1: Categories of appropriate and inappropriate antimicrobial use.

Author Contributions: Conceptualization, D.C.; methodology, D.C. and P.S.; software, D.C. and P.S.; validation, D.C., A.T., W.N. and N.C.; formal analysis, D.C.; data curation, D.C., A.T., W.N. and N.C.; writing—original draft preparation, D.C.; writing—review and editing, D.C. and P.S. All authors have read and agreed to the published version of the manuscript.

Funding: This research received no external funding.

Institutional Review Board Statement: This study was approved by the Institutional Review Board (IRB) of the Faculty of Medicine Ramathibodi Hospital (approval number: MURA2022/492).

Informed Consent Statement: A waiver of informed consent was granted due to the retrospective nature of the study.

Data Availability Statement: The study dataset is available from the corresponding author upon reasonable request.

Acknowledgments: The authors gratefully acknowledge Alexander D Pyden for editing a proof of this manuscript. Partial results of this study were presented as a poster presentation at IDWeek 2023, 11–15 October 2023 (Boston, MA, USA).

Conflicts of Interest: The authors declare no conflict of interest.

References

1. Dat, V.Q.; Vu, H.N.; The, H.N.; Nguyen, H.T.; Hoang, L.B.; Viet, D.V.T.; Bui, C.L.; Van Nguyen, K.; Nguyen, T.V.; Trinh, D.T.; et al. Bacterial bloodstream infections in a tertiary infectious diseases hospital in Northern Vietnam: Aetiology; drug resistance; and treatment outcome. *BMC Infect. Dis.* **2017**, *17*, 493. [CrossRef] [PubMed]
2. Santoro, A.; Franceschini, E.; Meschiari, M.; Menozzi, M.; Zona, S.; Venturelli, C.; Digaetano, M.; Rogati, C.; Guaraldi, G.; Paul, M.; et al. Epidemiology and risk factors associated with mortality in consecutive patients with bacterial bloodstream infection: Impact of MDR and XDR bacteria. *Open Forum Infect. Dis.* **2020**, *7*, ofaa461. [CrossRef] [PubMed]
3. Goodman, K.E.; Lessler, J.; Cosgrove, S.E.; Harris, A.D.; Lautenbach, E.; Han, J.H.; Milstone, A.M.; Massey, C.J.; Tamma, P.D.; Antibacterial Resistance Leadership Group. A clinical decision tree to predict whether a bacteremic patient is infected with an extended-spectrum beta-lactamase-producing organism. *Clin. Infect. Dis.* **2016**, *63*, 896–903. [CrossRef] [PubMed]
4. Ponyon, J.; Kerdsin, A.; Preeprem, T.; Ungcharoen, R. Risk factors of infections due to multidrug-resistant gram-negative bacteria in a community hospital in rural Thailand. *Trop. Med. Infect. Dis.* **2022**, *7*, 328. [CrossRef] [PubMed]
5. Laupland, K.B.; Svenson, L.W.; Gregson, D.B.; Church, D.L. Long-term mortality associated with community-onset bloodstream infection. *Infection* **2011**, *39*, 405–410. [CrossRef] [PubMed]
6. Nielsen, S.L.; Lassen, A.T.; Gradel, K.O.; Jensen, T.G.; Kolmos, H.J.; Hallas, J.; Pedersen, C. Bacteremia is associated with excess long-term mortality: A 12-year population-based cohort study. *J. Infect.* **2015**, *70*, 111–126. [CrossRef] [PubMed]
7. Antimicrobial Resistance Collaborators. Global burden of bacterial antimicrobial resistance in 2019: A systematic analysis. *Lancet* **2022**, *399*, 629–655. [CrossRef]
8. Kern, W.V.; Rieg, S. Burden of bacterial bloodstream infection-a brief update on epidemiology and significance of multidrug-resistant pathogens. *Clin. Microbiol. Infect.* **2020**, *26*, 151–157. [CrossRef]
9. Tabak, Y.P.; Vankeepuram, L.; Ye, G.; Jeffers, K.; Gupta, V.; Murray, P.R. Blood culture turnaround time in U.S. acute care hospitals and implications for laboratory process optimization. *J. Clin. Microbiol.* **2018**, *56*, e00500-18. [CrossRef]
10. Sirijatuphat, R.; Sripanidkulchai, K.; Boonyasiri, A.; Rattanaumpawan, P.; Supapueng, O.; Kiratisin, P.; Thamlikitkul, V. Implementation of global antimicrobial resistance surveillance system (GLASS) in patients with bacteremia. *PLoS ONE* **2018**, *13*, e0190132. [CrossRef]
11. Lee, C.C.; Wang, J.L.; Lee, C.H.; Hung, Y.P.; Hong, M.Y.; Chang, C.M.; Ko, W.C. Age-related trends in adults with community-onset bacteremia. *Antimicrob. Agents Chemother.* **2017**, *61*, e01050-17. [CrossRef] [PubMed]
12. Rothe, K.; Wantia, N.; Spinner, C.D.; Schneider, J.; Lahmer, T.; Waschulzik, B.; Schmid, R.M.; Busch, D.H.; Katchanov, J. Antimicrobial resistance of bacteraemia in the emergency department of a German university hospital (2013–2018): Potential carbapenem-sparing empiric treatment options in light of the new EUCAST recommendations. *BMC Infect. Dis.* **2019**, *19*, 1091. [CrossRef] [PubMed]
13. Jernigan, J.A.; Hatfield, K.M.; Wolford, H.; Nelson, R.E.; Olubajo, B.; Reddy, S.C.; McCarthy, N.; Paul, P.; McDonald, L.C.; Kallen, A.; et al. Multidrug-resistant bacterial infections in U.S. hospitalized patients; 2012–2017. *N. Engl. J. Med.* **2020**, *382*, 1309–1319. [CrossRef] [PubMed]
14. Lin, W.P.; Huang, Y.S.; Wang, J.T.; Chen, Y.C.; Chang, S.C. Prevalence of and risk factor for community-onset third-generation cephalosporin-resistant *Escherichia coli* bacteremia at a medical center in Taiwan. *BMC Infect. Dis.* **2019**, *19*, 245. [CrossRef] [PubMed]
15. Kolar, M.; Urbanek, K.; Latal, T. Antibiotic selective pressure and development of bacterial resistance. *Int. J. Antimicrob. Agents* **2001**, *17*, 357–363. [CrossRef] [PubMed]
16. Rogers, B.A.; Sidjabat, H.E.; Paterson, D.L. *Escherichia coli* O25b-ST131: A pandemic; multiresistant; community-associated strain. *J. Antimicrob. Chemother.* **2011**, *66*, 1–14. [CrossRef] [PubMed]
17. Pruetpongpun, N.; Khawcharoenporn, T.; Damronglerd, P.; Suwantarat, N.; Apisarnthanarak, A.; Rutjanawech, S. Inappropriate empirical treatment of uncomplicated cystitis in Thai women: Lessons learned. *Clin. Infect. Dis.* **2017**, *64*, S115–S118. [CrossRef]
18. Sirijatuphat, R.; Pongsuttiyakorn, S.; Supapueng, O.; Kiratisin, P.; Thamlikitkul, V. Implementation of global antimicrobial resistance surveillance system (GLASS) in patients with bacteriuria. *J. Glob. Antimicrob. Resist.* **2020**, *20*, 60–67. [CrossRef]
19. Tumbarello, M.; Viale, P.; Viscoli, C.; Trecarichi, E.M.; Tumietto, F.; Marchese, A.; Spanu, T.; Ambretti, S.; Ginocchio, F.; Cristini, F.; et al. Predictors of mortality in bloodstream infections caused by *Klebsiella pneumoniae* carbapenemase-producing *K. pneumoniae*: Importance of combination therapy. *Clin. Infect. Dis.* **2012**, *55*, 943–950. [CrossRef]
20. Kim, J.H.; Yoo, S.H.; Keam, B.; Heo, D.S. The impact of palliative care consultation on reducing antibiotic overuse in hospitalized patients with terminal cancer at the end of life: A propensity score-weighting study. *J. Antimicrob. Chemother.* **2022**, *78*, 302–308. [CrossRef]
21. Scarsi, K.K.; Feinglass, J.M.; Scheetz, M.H.; Postelnick, M.J.; Bolon, M.K.; Noskin, G.A. Impact of inactive empiric antimicrobial therapy on inpatient mortality and length of stay. *Antimicrob. Agents Chemother.* **2006**, *50*, 3355–3360. [CrossRef] [PubMed]
22. Fitzpatrick, J.M.; Biswas, J.S.; Edgeworth, J.D.; Islam, J.; Jenkins, N.; Judge, R.; Lavery, A.J.; Melzer, M.; Morris-Jones, S.; Nsutebu, E.F.; et al. Gram-negative bacteraemia; a multi-centre prospective evaluation of empiric antibiotic therapy and outcome in English acute hospitals. *Clin. Microbiol. Infect.* **2016**, *22*, 244–251. [CrossRef] [PubMed]
23. Harris, P.N.A.; Tambyah, P.A.; Lye, D.C.; Mo, Y.; Lee, T.H.; Yilmaz, M.; Alenazi, T.H.; Arabi, Y.; Falcone, M.; Bassetti, M.; et al. Effect of piperacillin-tazobactam vs meropenem on 30-day mortality for patients with *E coli* or *Klebsiella pneumoniae* bloodstream infection and ceftriaxone resistance: A randomized clinical trial. *JAMA* **2018**, *320*, 984–994. [CrossRef] [PubMed]

24. Sharara, S.L.; Amoah, J.; Pana, Z.D.; Simner, P.J.; Cosgrove, S.E.; Tamma, P.D. Is piperacillin-tazobactam effective for the treatment of pyelonephritis caused by extended-spectrum beta-lactamase-producing organisms? *Clin. Infect. Dis.* **2020**, *71*, e331–e337. [CrossRef] [PubMed]
25. Gould, M.; Ginn, A.N.; Marriott, D.; Norris, R.; Sandaradura, I. Urinary piperacillin/tazobactam pharmacokinetics in vitro to determine the pharmacodynamic breakpoint for resistant *Enterobacteriaceae*. *Int. J. Antimicrob. Agents.* **2019**, *54*, 240–244. [CrossRef] [PubMed]
26. Raz, R.; Schiller, D.; Nicolle, L.E. Chronic indwelling catheter replacement before antimicrobial therapy for symptomatic urinary tract infection. *J. Urol.* **2000**, *164*, 1254–1258. [CrossRef] [PubMed]
27. Denny, K.J.; Gartside, J.G.; Alcorn, K.; Cross, J.W.; Maloney, S.; Keijzers, G. Appropriateness of antibiotic prescribing in the emergency department. *J. Antimicrob. Chemother.* **2019**, *74*, 515–520. [CrossRef]
28. Chotiprasitsakul, D.; Han, J.H.; Cosgrove, S.E.; Harris, A.D.; Lautenbach, E.; Conley, A.T.; Tolomeo, P.; Wise, J.; Tamma, P.D.; Antibacterial Resistance Leadership Group. Comparing the outcomes of adults with *Enterobacteriaceae* bacteremia receiving short-course versus prolonged-course antibiotic therapy in a multicenter; propensity score-matched cohort. *Clin. Infect. Dis.* **2018**, *66*, 172–177. [CrossRef]
29. Chotiprasitsakul, D.; Bruminhent, J.; Watcharananan, S.P. Current state of antimicrobial stewardship and organ transplantation in Thailand. *Transpl. Infect. Dis.* **2022**, *24*, e13877. [CrossRef]
30. Clinical and Laboratory Standards Institute. *Performance Standards for Antimicrobial Susceptibility Testing, M100*, 32nd ed.; Clinical and Laboratory Standards Institute: Wayne, PA, USA, 2022.
31. Apisarnthanarak, A.; Lapcharoen, P.; Vanichkul, P.; Srisaeng-Ngoen, T.; Mundy, L.M. Design and analysis of a pharmacist-enhanced antimicrobial stewardship program in Thailand. *Am. J. Infect. Control.* **2015**, *43*, 956–959. [CrossRef]

antibiotics

MDPI

Article

Multi-Model Strategies for Prevention of Infection Caused by Certain Multi-Drug Resistant Organisms in A Rehabilitation Unit: A Semi-Experimental Study

Shiyu Li [1,2], Ji Lin [2], Siyuan Tao [2], Linwen Guo [2], Wenzhi Huang [2], Jingwen Li [2], Chunping Du [1], Zhiting Wang [3], Liwen Liu [4], Yi Chen [2,*] and Fu Qiao [2,*]

[1] Department of Rehabilitation Medicine, West China Hospital, Sichuan University, Chengdu 610041, China; lishiyu@wchscu.cn (S.L.); duchunping118@wchscu.cn (C.D.)

[2] Department of Infection Control, West China Hospital, Sichuan University, Chengdu 610041, China; linji@scu.edu.cn (J.L.); taosiyuan@wchscu.cn (S.T.); guolw@scu.edu.cn (L.G.); huangwenzhi@wchscu.cn (W.H.); lijingwen@wchscu.cn (J.L.)

[3] Department of Operation Management, West China Hospital, Sichuan University, Chengdu 610041, China; wangzhiting@wchscu.cn

[4] Department of Equipment and Materials, West China Hospital, Sichuan University, Chengdu 610041, China; liuliwen@scu.edu.cn

[*] Correspondence: toddy@scu.edu.cn (Y.C.); qiaofu2005@scu.edu.cn (F.Q.)

Abstract: Objective: To assess the effectiveness of multi-model strategies on healthcare-associated infections (HAIs) caused by multi-drug resistant organisms (MDROs) in rehabilitation units. **Methods:** A semi-experimental study was conducted in a rehabilitation unit with 181 beds from January 2021 to December 2022 in a teaching hospital with 4300 beds in China. In 2021, many basic prevention and control measures were conducted routinely. Based on the basic measures, strengthening multi-model strategies for the prevention and control of MDROs was pursued year-round since 1 January 2022. **Results:** A total of 6206 patients were enrolled during the study period. The incidence density of HAIs caused by MDROs decreased from 1.22 (95% CI, 0.96~1.54) cases/1000 patient-days in the pre-intervention period to 0.70 (95% CI, 0.50~0.95) cases/1000 patient-days ($p = 0.004$). Similarly, the incidence of HAIs in the intervention period was 50.85% lower than that in the pre-intervention period (2.02 (95% CI, 1.50~2.72) vs. 4.11 (95% CI, 3.45–4.85) cases/100 patients, $p < 0.001$). The rate of MDROs isolated from the environment decreased by 30.00%, although the difference was not statistically significant ($p = 0.259$). **Conclusion:** Multi-model strategies can reduce the incidence of HAIs and HAIs caused by certain MDROs in the rehabilitation unit.

Keywords: multi-model strategies; multi-drug resistant organisms; healthcare-associated infections; rehabilitation unit

1. Introduction

In recent years, healthcare-associated infections (HAIs) caused by multi-drug resistant organisms (MDROs) have become severe challenges to clinical treatment, often prolonging the length of stay (LOS), increasing mortality, and increasing medical costs [1–3]. The prevalence of MDROs poses a threat to public health worldwide. A report published by the World Health Organization (WHO) estimates that there are more than 25,000 and 23,000 MDRO infection-related deaths every year in Europe and the United States, respectively. The annual economic burden caused by the infection of MDROs is EUR 1.5 billion and USD 3.4 billion in Europe and the United States, respectively [4]. Several recent studies also confirmed that the medical cost of patients with MDRO infections was 1.4~1.7 times that of patients without [5–7]. In China, one study conducted in 68 hospitals showed that the average increase in LOS due to infections caused by studied MDROs was 14 days [8].

In recent decades, with the development of early rehabilitation, many surgical patients are transferred to the rehabilitation unit within one or two days after surgery, and some critical patients are transferred to the rehabilitation unit directly from the intensive care unit (ICU). Moreover, many rehabilitation instruments are shared between different patients, and it is almost impossible to disinfect the instruments in time after each use [9]. All of these factors increase the risk of HAIs for patients hospitalized in rehabilitation units, especially the infection of MDROs. A study revealed that the MDRO test-positive rate was 16.70% (96/575) in neurorehabilitation ward patients [10]. Patients colonized or infected with MDROs may severely or moderately limit their rehabilitation outcome [5], and the infection during rehabilitation will increase the length of hospitalization, reduce the rehabilitative outcomes and increase the mortality rate significantly [7]. A bundle of interventions, including education and training, hand hygiene promotion, contact precaution, and environmental disinfection, have been confirmed to reduce the incidence of HAIs caused by MDROs [11–14]. However, there is less evidence on the strategies for MDRO prevention and control in rehabilitation units. To assess the effectiveness of multi-model strategies on infections caused by MDROs in the rehabilitation unit, especially in carbapenem-resistant *Acinetobacter baumannii* (CRAB) and carbapenem-resistant *Enterobacteriaceae* (CRE) endemic areas, we, therefore, conducted this study.

2. Results

2.1. Characteristics of Patients in The Pre-Intervention and Intervention Periods

A total of 7200 patients were admitted to the rehabilitation unit from January 2021 to December 2022, and 6206 patients of them were eligible (Figure 1). A total of 3262 patients and 2944 patients were included in the pre-intervention and intervention periods, respectively (S). The mean age was 53.87 ± 16.13 years in the intervention period, which was significantly higher than that in the pre-intervention period ($p < 0.001$). Most of the underlying diseases were no significant difference between the two groups, although the proportion of patients with respiratory failure and malignant tumors was statistically different between the two groups, the intervention period was higher than that in the pre-intervention period, 2.65% vs. 1.69% ($p = 0.009$) and 6.69% vs. 4.20% ($p < 0.001$), respectively. A total of 19.13% (1187/6206) of the patients had a history of surgery during their hospitalization, but there was no significant difference between the two groups. Forty-five patients had at least one MDRO infection at admission in the intervention period, which was slightly higher than the pre-intervention group, but there was no significant difference ($p = 0.552$). The details of the two groups are shown in Table 1.

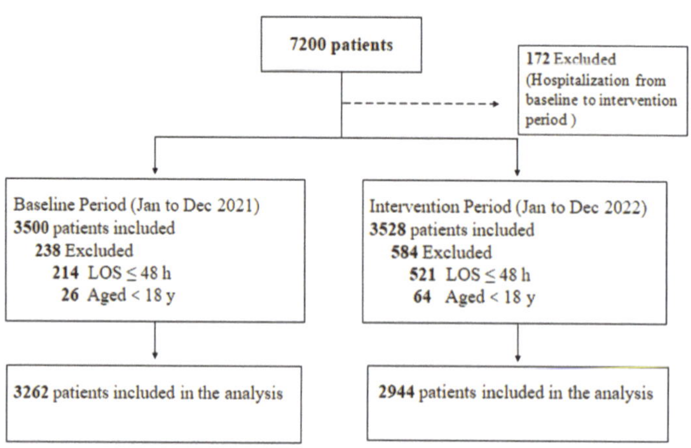

Figure 1. Flow chart and cohorts for analyses.

Table 1. Characteristics of the study population in each period.

	All Patients (n = 6206)	Pre-Intervention Period (n = 3262)	Intervention Period (n = 2944)	p Value
Age, mean (SD), y	52.96 (16.75)	52.14 (17.25)	53.87 (16.13)	<0.001
Sex (n, %)				
Female	2646 (42.63)	1313 (40.25)	1333 (45.28)	<0.001
Male	3560 (57.36)	1949 (59.75)	1611 (54.72)	
Underlying diseases (n, %)				
Hypertension	1797 (28.96)	959 (29.40)	838 (28.46)	0.418
Diabetes	770 (12.41)	404 (12.39)	366 (12.43)	0.955
Respiratory failure	133 (2.14)	55 (1.69)	78 (2.65)	0.009
Tuberculosis	40 (0.64)	24 (0.74)	16 (0.54)	0.345
Heart failure	103 (1.66)	46 (1.41)	57 (1.94)	0.105
Renal failure	93 (1.5)	50 (1.53)	43 (1.46)	0.815
Hepatic insufficiency	82 (1.32)	40 (1.23)	42 (1.43)	0.490
Malignant tumors	334 (5.38)	137 (4.20)	197 (6.69)	<0.001
Hematological diseases	5 (0.08)	2 (0.06)	3 (0.10)	0.909
COPD	59 (0.95)	27 (0.83)	32 (1.09)	0.293
HIV	5 (0.08)	2 (0.06)	3 (0.1)	0.909
Hemiplegia/Paraplegia	1306 (21.04)	687 (21.06)	619 (21.03)	0.973
Mortality (n, %)	15 (0.24)	9 (0.28)	6 (0.20)	0.564
Surgery (n, %)	1187 (19.13)	600 (18.39)	587 (19.94)	0.065
LOS, median (IQR), d	19 (12–22)	20 (13–22)	18 (11–23)	0.467
MDROs at admission (n, %)	89 (1.43)	44 (1.35)	45 (1.53)	0.552

Abbreviations: COPD, chronic obstructive pulmonary disease; HIV, human immunodeficiency virus; LOS, length of stay.

2.2. The Compliance of Infection Control Measures

The rate of family caregivers for patients decreased to 38.48% in the intervention period. Hand hygiene compliance during the intervention period was 89.08% (1771/1988) while it was 90.54% in the pre-intervention period (p = 0.071). The differences in the consumption of liquid soap and alcohol-based hand rub between the intervention and pre-intervention periods were also not statistically significant (S). The consumption of gloves in the intervention period was 2.15 ± 0.26 pairs per patient day which was statistics significantly higher than that in the pre-intervention period (p = 0.010), while the consumption of gowns was 64.67 ± 10.01 pieces per 1000 patient days, which was higher than that in the pre-intervention period but not statistical significance (p = 0.892).

At the end of December 2021, 84 environmental samples, including bed rails and tables, were collected randomly from the wards, and 18 (21.34%) of them were positive for MDROs, including 14 CRKP and 4 CRAB strains. In March 2023, environmental sampling was conducted again, and 100 samples were taken randomly from the wards. Of the 100 samples, 15 (15.00%) were positive for MDROs, including 10 CRKP strains and 5 CRAB strains. No other target MDRO was isolated from the environmental sampling. The environmental pollution rate of MDROs decreased by 30.00%, although the difference was not statistical significance (p = 0.259). The details are shown in Table 2.

Table 2. The process index in each period.

Process Indicators	Pre-Intervention Period	Intervention Period	p Value
Compliance with hand hygiene (n, %)	3937 (90.54%)	1771 (89.08%)	0.071
HH equipment consumption (mL per patient day)	27.66	27.38	0.923
Liquid soap (mL per patient day)	7.84 ± 3.18	10.92 ± 6.53	0.156
ABHR (mL per patient day)	19.83 ± 5.36	16.46 ± 4.67	0.115
PPE usage			
Gloves (pairs per patient day)	1.42 ± 0.11	2.15 ± 0.26	0.010
Gowns (pieces per 1000 patient day)	60.75 ± 14.72	64.67 ± 10.01	0.892
Environment surveillance culture (n, %)	18 (21.43%)	15 (15.00%)	0.259
CRKP	14 (16.67%)	10 (10.00%)	0.182
CRAB	4 (4.76%)	5 (5.00%)	0.940

Abbreviations: HH, hand hygiene; ABHR, alcohol-based hand rub; PPE, personal protective equipment; CRKP, carbapenem-resistant *Klebsiella pneumoniae*; CRAB, carbapenem-resistant *Acinetobacter baumannii*.

2.3. Healthcare-Associated Infections Caused by MDROs

In total, 134 cases and 60 cases of HAIs were confirmed in the pre-intervention period and intervention period, respectively, and the incidence of HAIs was significantly lower in the intervention period than in the pre-intervention period (2.04% (95% CI, 1.56~2.62) vs. 4.11% (95% CI, 3.45~4.85), $p < 0.001$). In those infection cases, 113 cases of MDROs were isolated from 109 patients more than 48 h after admission during the study period, and the total incidence density of HAIs caused by MDROs was 0.97 (95% CI, 0.96–1.54) cases per 1000 patient days. CRKP was the most common pathogen, accounting for 35.40% of HAIs caused by MDROs. Lower respiratory tract infections were the most common sites of HAIs, followed by urinary tract infections, which accounted for 62.83% (71/113) and 30.09% (34/113) of the whole infection site, respectively.

During the intervention period, only 39 cases of MDROs were isolated from eligible patients, and the incidence density of HAIs caused by MDROs was significantly decreased compared with that during the pre-intervention period (0.70 (95% CI, 0.50–0.95) cases per 1000 patient-days vs. 1.22 (95% CI, 0.96–1.54) cases per 1000 patient days, $p = 0.004$). HAIs caused by CRKP decreased from 0.48 (95% CI, 0.32–0.69) cases per 1000 patient days to 0.20 (95% CI, 0.10–0.35) cases per 1000 patient days, $p = 0.008$. However, HAIs caused by CRAB decreased by 47.37%, but the difference was not statistically significant (0.20 (95% CI, 0.10–0.35) cases per 1000 patient days vs. 0.38 (95% CI, 0.24–0.57) cases per 1000 patient days). The incidence density of HAIs caused by other MDROs was also not significantly decreased between the intervention period and pre-intervention period because of low incidence density. The monthly variation of the incidence density is shown in Figure 2.

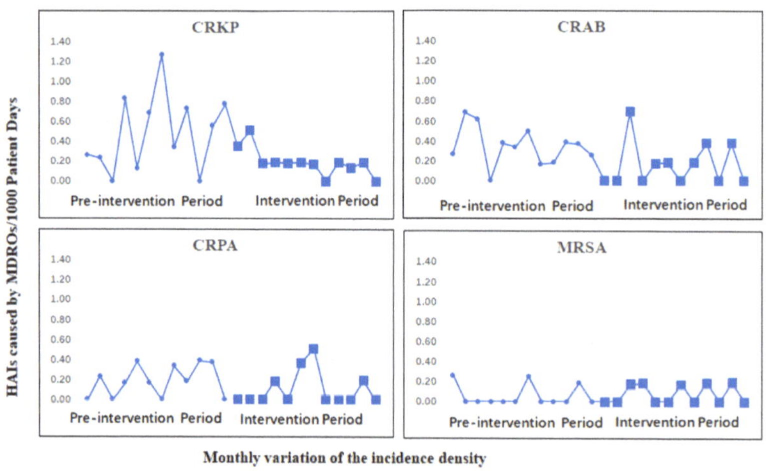

Figure 2. Monthly incidence density of HAIs caused by CRKP, CRAB, CRPA, and MRSA. Abbreviations: HAIs, healthcare-associated infections; CRKP, carbapenem-resistant *Klebsiella pneumoniae*; CRAB, carbapenem-resistant *Acinetobacter baumannii*; CRPA, carbapenem-resistant *Pseudomonas aeruginosa*; MRSA, methicillin-resistant *Staphylococcus aureus*.

In terms of infection site, both the incidence density of lower respiratory tract infection and urinary tract infection caused by MDROs were decreased statistically significantly between the intervention period and pre-intervention period, that is, 0.43 (95% CI, 0.27–0.64) cases per 1000 patient-days vs. 0.78 (95% CI, 0.57–1.03) cases per 1000 patient days and 0.18 (95% CI, 0.09–0.33) cases per 1000 patient days vs. 0.40 (95% CI, 0.25–0.59) cases per 1000 patient days, separately. The details of the incidence density between the intervention period and the pre-intervention period are shown in Table 3.

Table 3. Effect of multi-strategies on HAIs caused by MDROs.

	Pre-Intervention Period		Intervention Period		*RR* (95% CI)	*p* Value
	Case (*n*)	Incidence Density [a] (95% CI)	Case (*n*)	Incidence Density [a] (95% CI)		
MDROs	74	1.22 (0.96, 1.54)	39	0.70 (0.50, 0.95)	0.57 (0.39, 0.84)	0.004
CRE	34	0.56 (0.39, 0.79)	16	0.29 (0.16, 0.46)	0.51 (0.28,0.92)	0.022
CRKP	29	0.48 (0.32, 0.69)	11	0.20 (0.10, 0.35)	0.41 (0.20, 0.82)	0.008
CR-*E. cloacae*	4	0.07 (0.02, 0.17)	1	0.02 (0.01,0.10)	0.27 (0.30, 2.42)	0.199
CR-*K. oxytoca*	1	0.02 (0, 0.09)	2	0.04 (0.01, 0.13)	2.16 (0.19, 23.86)	0.262
CR-*E.coli*	0	0 (0, 0.06)	2	0.04 (0.01, 0.13)	--	--
CRAB	23	0.38 (0.24, 0.57)	11	0.20 (0.10, 0.35)	0.52 (0.25, 1.06)	0.064
CRPA	13	0.21 (0.11, 0.37)	7	0.13 (0.05, 0.26)	0.58 (0.23, 1.46)	0.249
MRSA	3	0.05 (0.01, 0.14)	5	0.09 (0.03, 0.21)	1.80 (0.43, 7.54)	0.209
VRE	1	0.02 (0.01, 0.09)	0	0 (0, 0.07)	--	--
MDRO Infection site	71	1.17 (0.92, 1.48)	34	0.61 (042, 0.85)	0.52 (0.34, 0.78)	
LRTI	47	0.78 (0.57, 1.03)	24	0.43 (0.27, 0.64)	0.55 (0.34, 0.90)	0.015
UTI	24	0.40 (0.25, 0.59)	10	0.18 (0.09, 0.33)	0.45 (0.22, 0.94)	0.027

Notes: [a] cases per 1000 patient days. Abbreviations: CI, confidence interval; *RR*, relative risk; MDROs, multi-drug resistant organisms; CRE, carbapenem-resistant *Enterobaceteriacae;* CRKP, carbapenem-resistant *Klebsiella pneumoniae;* CRAB, carbapenem-resistant *Acinetobacter baumannii;* CRPA, carbapenem-resistant *Pseudomonas aeruginosa;* VRE, vancomycin-resistant *Enterococci;* MRSA, methicillin-resistant *Staphylococcus aureus;* LRTI, lower respiratory tract infection; UTI, urinary tract infection.

3. Materials and Methods

3.1. Study Design

A semi-experimental study was conducted in a rehabilitation unit with 181 beds from 1 January 2021 to 31 December 2022 in a teaching hospital with 4300 beds in mainland China. This unit was major for spinal cord injury disease, musculoskeletal disease, and nervous system disease. In 2021, a bundle of basic prevention and control measures was conducted routinely. Based on the basic measures, strengthening multi-model strategies for the prevention and control of infections caused by MDROs have been implemented all year round since 1 January 2022, which was the intervention period.

3.2. Inclusion and Exclusion Criteria

All the patients (Age ≥18 years) older than 18 years and above admitted to the rehabilitation unit in the study period were included. Patients were excluded if they were (1) Age < 18 years; (2) stayed in the unit for less than 48 h; or (3) had a length of stay between 2021 and 2022.

3.3. Prevention and Control Measures in The Pre-Intervention Period

(1) Notification. A notification phone call was given to the doctors in charge by microbiology laboratory workers immediately when the MDRO was isolated, and the results were uploaded to the laboratory information system (LIS) immediately. A contact precaution marker was labeled automatically on the hospital information system (HIS). (2) Contact precautions. Patients with MDROs were isolated when the nurse received the notification, and a contact precaution label was posted on the patient's bed. All healthcare workers must wear gloves and gowns during care work. The infection control link nurses in the ward checked the compliance of contact precautions, and the infection control practitioners did an audit within 24 h to ensure that the clinicians correctly understood the communication and, therefore, consequently acted. (3) Reduce the number of people. Only three doctors were allowed to take the turnaround at the same time. One fixed family member or one professional nurse assistant was permitted to accompany, and only one visitor was enabled each time. (4) Routine disinfection. All of the rehabilitation instruments were cleaned and disinfected with 500 mg/L effective chlorine disinfectant twice a day.

3.4. Multi-Model Strategies for Prevention of HAIs Caused by MDROs in The Intervention Period

3.4.1. System Change

An MDRO prevention and control work group was established in the unit, which was composed of a vice director, chief rehabilitation technician, head nurse, medical team leader, nursing team leader, infection control link nurse, and infection control professional. The responsibilities of the group included (1) coordinating and allocating beds in the whole unit to cohort patients with MDROs; (2) analyzing and discussing the reason for infection case by case; and (3) drafting prevention and control guidelines.

3.4.2. Personnel and Behavior Management

(1) Family visits were canceled during hospitalization. (2) Whether the patients needed to be cared for by family members was assessed by the doctor in charge based on the patient's condition. None of the family members were permitted to care for patients in the quasi-intensive care unit. (3) One doctor and one nurse were fixed in the quasi-intensive care unit to conduct the ward rounds and care work per shift. The treatment and ward rounds were carried out at different intervals. Only one technician was allowed to work at the same time except in emergencies.

3.4.3. Education and Training

(1) All of the new healthcare workers must finish the course of prevention and control measures for MDROs before they engage in clinical work. (2) Regular education was carried out for different types of professions. (3) A training and test of infection prevention and control measures must be completed for doctors and nurses before they work in a quasi-intensive care unit.

3.4.4. Communication and Data Feedback

(1) Weekly communication meetings were built between infection control practitioners and link nurses to feedback on the audit results and discuss work plans for next step since January 2022. (2) All of MDROs data in the pre-intervention period were feedback to the management group in January 2022. Then, a quarterly data feedback meeting was conducted with the directors, head nurse, and other management group members. (3) Statistical data of HAIs of MDROs was feedback to all of the healthcare workers on morning shift irregularly.

3.4.5. Environmental Control

(1) All instruments and equipment were disinfected by the rehabilitation therapist immediately after use by the MDRO's patient. (2) The concentration of disinfectant was checked by a designated nurse every day. (3) A fluorescent marker was used to check compliance with cleaning twice a week. (4) Environmental surveillance culture was conducted to assess the environmental burden of MDROs. The environment swabbing was taken by sterile rayon swabs (Copan, Brescia, Italy) moistened with tryptic soy broth (TSB, Hopebio, Qingdao, China). The swabs were immediately placed into 15 mL sterile tubes containing 6 mL of TSB. The tubes were incubated at 37 °C overnight and centrifuged. The supernatant was discarded, and precipitates were resuspended in 1 mL of TSB. A 50 μL suspension was streaked onto Acinetobacter selected-agar plates (CHROMagarTM, Paris, France) containing 4 μg/mL meropenem to screen CRAB, Pseudomonas selected-agar plates (CHROMagarTM, Paris, France) containing 4 μg/mL meropenem to screen CRPA, Simmon citrate agar plates containing 2 μg/mL meropenem to screen CRK, Orientation selected-agar plates (CHROMagarTM, Paris, France) containing 2 μg/mL meropenem to screen CRE, MRSA, and VRE selected-agar plates (CHROMagarTM, Paris, France) to screen MRSA and VRE separately. The plates were incubated at 37 °C for 18~24 h, and suspected colonies were subjected to preliminary species identification based on matrix-assisted laser desorption/ionization–time of flight mass spectrometry (Bruker, Billerica, MA, USA).

3.5. Definitions

The diagnostic criteria for HAIs were the diagnostic standards for HAIs published by the Ministry of Health in 2001 [15]. The certain MDROs in this study included Carbapenem-resistant *Acinetobacter baumannii* (CRAB), Carbapenem-resistant *Enterobaceteriacae* (CRE), Carbapenem-resistant *Pseudomonas aeruginosa* (CRPA), Vancomycin-resistant *Enterococci* (VRE), and Methicillin-resistant *Staphylococcus aureus* (MRSA).

3.6. Main Outcomes

The primary study outcome was the incidence density of HAIs caused by MDROs in inpatients. The secondary outcome was the incidence density of HAIs. We also report the compliance of hand hygiene and usage of protective equipment during two periods.

3.7. Clinical Data Collection and Statistical Analysis

Demographic and clinical data of eligible patients were collected, including gender, age, diagnosis, underlying disease, infection of MDROs at admission, and the length of stay.

For descriptive analysis, qualitative data were expressed in terms of frequency, and the incidence of MDROs was expressed as incidence density (cases per 1000 patient days). Quantitative variables with a normal distribution were expressed as the mean ± standard deviation (SD), which were expressed as median (25% percentile, 75% percentile) if the normal distribution was not met. Qualitative data were analyzed by the Chi-square test or Fisher's exact test, Kolmogorov–Smirnov normality test was used for normality test. *T*-test was used if quantitative variables were normal distribution; otherwise, Mann–Whitney U-test was conducted. The difference of MDROs incidence density among interventions was analyzed by Poisson analysis, and RR was used to indicate the relative risk. Analyses were performed using R v. 4.3.0.

4. Discussion

Patients colonized or infected with MDROs may affect their rehabilitation programs. One survey conducted in Germany showed that 27% of rehabilitation facilities refused to accept patients with MDROs, and only 27% of the rehabilitation centers allowed patients with MDROs to participate in full rehabilitation programs [16]. Another survey conducted in Europe showed that patients with MDROs wait longer for admission in 36% of facilities and have been refused admission in 11% of facilities [17]. However, the prevalence of MDROs in rehabilitation units is still very high and sometimes causes outbreaks [9]. A multicenter observation study for neurorehabilitation showed that 55% of HAIs needed functional isolation due to multidrug-resistant germs [18]. Another study in Germany showed that 2.2% of general rehabilitation patients were colonized with VRE, and 6.7% of them had multi-drug-resistant gram-negative pathogens [19]. In this study, we found out that the incidence of HAIs caused by MDROs in the pre-intervention period was 2.27% (1.22 cases per 1000 patient days), and the most common MDROs were CRKP, CRAB, and CRPA. The incidence is lower than that in a previous study conducted in Southwest China and some other studies [10], which may be because of the difference in case definition. However, the concept of MDROs was similar.

How patients with MDRO infections should be managed in a rehabilitation setting is still lacking due to a survey in Europe [17]. Effective strategies to prevent HAIs caused by MDROs are still limited. In our study, multi-model strategies, including system change, personnel and behavior management, education and training, communication and data feedback, and environmental control, were carried out in the intervention period in the rehabilitation unit, and the incidence of HAIs caused by MDROs decreased significantly through the strategies during the intervention period, even though the infection rate at admission was slightly higher. We indeed explored a way to decrease HAIs caused by MDROs in the rehabilitation department. System change may play a key role in our multi-strategies, and one study conducted in the China Rehabilitation Research Centre decreased

HAIs caused by MDROs through the PDCA cycle. In their study, the detection rate of MRSA and CRPA decreased significantly, but that of MDRAB did not [20].

In this study, the average age in the pre-intervention period was lower than that in the intervention period, but the difference was only 1.73 years. Several studies reported that age was an independent risk factor for MDROs infection [21,22]. The older the age, the higher the risk of infection. Although the average age in the intervention group was higher than that in the control group in this study, there was still a decreased incidence of MDROs, indicating a relatively robust result. Respiratory failure, which reflected the severity of the patients' illness, has been reported as an independent risk factor for MDROs infection [23]. In this study, the proportion of respiratory failure in the intervention period was higher than that in the pre-intervention period; there was still a decrease in the infection rate of MDROs, which indicated a relatively robust result.

Personal and behavior management are always hard to execute. In China, the rate of family caregivers is as high as 90% for cultural reasons [24], and compliance with hand hygiene and other infection control measures is very low, even when the patients they care for have communicable diseases [25,26]. In this study, to decrease the rate of family caregivers and reduce the number of persons in each ward, family visits were canceled, and each family caregiver was evaluated by the doctor in charge. The family caregiver rate decreased to 38.48% in 2022. However, we did not collect the details of family caregivers all year round in 2021, and the exact rate of family caregivers was not acquired, but we estimate that the rate was as high as 65% based on limited data collected by the head nurse.

Compliance with environmental cleaning for the immediate surrounding area is also a key recommendation to decrease infection caused by MDROs [27]. One study showed that the MDRO-positive rate was 7.7% in the common area and rehabilitation gym environment [28]. In this study, the contamination rate of common areas in the rehabilitation unit was 21.43%, which is much higher than that in a previous study. However, the contamination rate decreased by 30% through the intervention. This may contribute to reducing the risk of MDRO transmission. Although the consumption of gloves increased during the intervention period, there might have been misused gloves, so the consumption of gloves only partially reflects compliance.

There are several limitations to this study. First, the study is a single-center study in an area where MDROs are notably prevalent, especially CRAB and CRKP, which limits the extrapolation of the results to primary hospitals; however, it has significance for areas with high MDRO prevalence. Second, this study was conducted in the rehabilitation unit that received many severe patients transferred from ICUs or surgical units in a large teaching hospital, which also limits the reference to other units and other rehabilitation units receiving traditional patients, but it has significance for the rehabilitation units that are developing rapid rehabilitation medicine. Finally, we did not conduct active surveillance screening of MDROs for the patients on admission and during their hospitalization, and we could not evaluate the effect of multi-strategies on colonized organisms.

5. Conclusions

This semi-experimental study found that the comprehensive multi-model strategies reduced the incidence of HAIs and the HAIs caused by MDROs. It also reduced the contamination rate of MDROs in patients' room environments. These findings demonstrate that these interventions can effectively decrease the burden of MDROs in rehabilitation units.

Author Contributions: Conceptualization, S.L., Y.C. and F.Q.; methodology, S.L., F.Q. and J.L. (Ji Lin); formal analysis, J.L. (Ji Lin), S.T., L.G. and W.H.; investigation, S.L., S.T., J.L. (Jingwen Li) and C.D.; data curation, S.L. and J.L. (Ji Lin); resources, Z.W. and L.L.; writing—original draft, S.L. and F.Q.; writing—review and editing, S.L., F.Q., J.L. (Ji Lin) and W.H.; supervision, F.Q. and S.L.; project administration, Y.C. and F.Q.; visualization, S.L. All authors have read and agreed to the published version of the manuscript.

Funding: This work was supported by grants from the China National Health Development Research Center (No. 2020-53).

Institutional Review Board Statement: This study was approved by the ethics committee of West China Hospital of Sichuan University (No. 2023 (651)).

Informed Consent Statement: Patient consent was waived due to the ethical review.

Data Availability Statement: The data presented in this study are available on request from the corresponding author. The data are not publicly available due to privacy.

Acknowledgments: We would like to acknowledge the physicians and allied health staff in the Department of Rehabilitation Medicine of West China Hospital and the Department of Electronic Medical Record system of West China Hospital.

Conflicts of Interest: The authors declare no confict of interest.

References

1. Zong, Z.; Wu, A.; Hu, B. Infection Control in the Era of Antimicrobial Resistance in China: Progress, Challenges, and Opportunities. *Clin. Infect. Dis.* **2020**, *71*, S372–S378. [CrossRef] [PubMed]
2. European Antimicrobial Resistance Collaborators. The burden of bacterial antimicrobial resistance in the WHO European region in 2019: A cross-country systematic analysis. *Lancet Public Health* **2022**, *7*, e897–e913. [CrossRef] [PubMed]
3. Global Antimicrobial Resistance and Use Surveillance System (GLASS) Report: 2021. Available online: https://www.who.int/publications-detail-redirect/9789240027336 (accessed on 5 June 2023).
4. Johnston, K.J.; Thorpe, K.E.; Jacob, J.T.; Murphy, D.J. The incremental cost of infections associated with multidrug-resistant organisms in the inpatient hospital setting-A national estimate. *Health Serv. Res.* **2019**, *54*, 782–792. [CrossRef] [PubMed]
5. Stewardson, A.J.; Allignol, A.; Beyersmann, J.; Graves, N.; Schumacher, M.; Meyer, R.; Tacconelli, E.; De Angelis, G.; Farina, C.; Pezzoli, F.; et al. The health and economic burden of bloodstream infections caused by antimicrobial-susceptible and non-susceptible Enterobacteriaceae and *Staphylococcus aureus* in European hospitals, 2010 and 2011: A multicentre retrospective cohort study. *Eurosurveillance* **2016**, *21*, 30319. [CrossRef] [PubMed]
6. Vasudevan, A.; Memon, B.I.; Mukhopadhyay, A.; Li, J.; Tambyah, P.A. The costs of nosocomial resistant gram negative intensive care unit infections among patients with the systemic inflammatory response syndrome—A propensity matched case control study. *Antimicrob. Resist. Infect. Control* **2015**, *4*, 3. [CrossRef]
7. Morales, E.; Cots, F.; Sala, M.; Comas, M.; Belvis, F.; Riu, M.; Salvadó, M.; Grau, S.; Horcajada, J.P.; Montero, M.M.; et al. Hospital costs of nosocomial multi-drug resistant *Pseudomonas aeruginosa* acquisition. *BMC Health Serv. Res.* **2012**, *12*, 122. [CrossRef]
8. Jia, H.; Li, L.; Li, W.; Hou, T.; Ma, H.; Yang, Y.; Wu, A.; Liu, Y.; Wen, J.; Yang, H.; et al. Impact of Healthcare-Associated Infections on Length of Stay: A Study in 68 Hospitals in China. *BioMed Res. Int.* **2019**, *2019*, 2590563. [CrossRef]
9. Boonstra, M.B.; Spijkerman, D.C.M.; Voor In 't Holt, A.F.; van der Laan, R.J.; Bode, L.G.M.; van Vianen, W.; Klaassen, C.H.W.; Vos, M.C.; Severin, J.A. An outbreak of ST307 extended-spectrum beta-lactamase (ESBL)-producing *Klebsiella pneumoniae* in a rehabilitation center: An unusual source and route of transmission. *Infect. Control Hosp. Epidemiol.* **2020**, *41*, 31–36. [CrossRef]
10. Jiang, W.; Li, L.; Wen, S.; Song, Y.; Yu, L.; Tan, B. Gram-negative multidrug-resistant organisms were dominant in neurorehabilitation ward patients in a general hospital in southwest China. *Sci. Rep.* **2022**, *12*, 11087. [CrossRef]
11. Allegranzi, B.; Pittet, D. Role of hand hygiene in healthcare-associated infection prevention. *J. Hosp. Infect.* **2009**, *73*, 305–315. [CrossRef]
12. Chen, W.; Li, S.; Li, L.; Wu, X.; Zhang, W. Effects of daily bathing with chlorhexidine and acquired infection of methicillin-resistant *Staphylococcus aureus* and vancomycin-resistant Enterococcus: A meta-analysis. *J. Thorac. Dis.* **2013**, *5*, 518–524. [CrossRef]
13. Septimus, E.J.; Schweizer, M.L. Decolonization in Prevention of Health Care-Associated Infections. *Clin. Microbiol. Rev.* **2016**, *29*, 201–222. [CrossRef]
14. Hayden, M.K.; Lin, M.Y.; Lolans, K.; Weiner, S.; Blom, D.; Moore, N.M.; Fogg, L.; Henry, D.; Lyles, R.; Thurlow, C.; et al. Prevention of colonization and infection by *Klebsiella pneumoniae* carbapenemase-producing enterobacteriaceae in long-term acute-care hospitals. *Clin. Infect. Dis.* **2015**, *60*, 1153–1161. [CrossRef]
15. Ministry of Health of the People's Republic of China Diagnostic criteria for nosocomial infections(proposed). *Natl. Med. J. China* **2001**, *81*, 314–320. [CrossRef]
16. Heudorf, U.; Berres, M.; Hofmann, S.; Steul, K. Management of patients with multidrug-resistant organisms in rehabilitation facilities. Results of a survey in the Rhine-Main region, Germany, 2019. *GMS Hyg. Infect. Control* **2020**, *15*, Doc15. [CrossRef]
17. Doherty, A.; McNicholas, S.; Burger, H.; Boldrini, P.; Delargy, M. European survey of management of patients with multidrug-resistant organisms in rehabilitation facilities. *Eur. J. Phys. Rehabil. Med.* **2019**, *55*, 418–423. [CrossRef]
18. Bartolo, M.; Zucchella, C.; Aabid, H.; Valoriani, B.; Copetti, M.; Fontana, A.; Intiso, D.; Mancuso, M. Impact of healthcare-associated infections on functional outcome of severe acquired brain injury during inpatient rehabilitation. *Sci. Rep.* **2022**, *12*, 5245. [CrossRef]

19. Steul, K.; Schmehl, C.; Berres, M.; Hofmann, S.; Klaus-Altschuck, A.; Hogardt, M.; Kempf, V.A.; Pohl, M.; Heudorf, U. Multidrug Resistant Organisms (MDRO) in Rehabilitation: Prevalence and Risk Factors for MRGN and VRE. *Die Rehabil.* **2020**, *59*, 366–375. [CrossRef]
20. Li, C.; Liu, Z.; Xie, J.; Zhang, H.; Bai, Z. Effects of PDCA Cycle Quality Management Model on Multi-drug Resistant Organism Infection Control in Reha- bilitation Hospital. *Chin. J. Rehabil. Theory Pract.* **2016**, *22*, 1476–1479.
21. Forde, C.; Stierman, B.; Ramon-Pardo, P.; Dos Santos, T.; Singh, N. Carbapenem-resistant *Klebsiella pneumoniae* in Barbados: Driving change in practice at the national level. *PLoS ONE* **2017**, *12*, e0176779. [CrossRef]
22. Guo, N.; Xue, W.; Tang, D.; Ding, J.; Zhao, B. Risk factors and outcomes of hospitalized patients with blood infections caused by multidrug-resistant *Acinetobacter baumannii* complex in a hospital of Northern China. *Am. J. Infect. Control* **2016**, *44*, e37–e39. [CrossRef] [PubMed]
23. Kim, S.Y.; Jung, J.Y.; Kang, Y.A.; Lim, J.E.; Kim, E.Y.; Lee, S.K.; Park, S.C.; Chung, K.S.; Park, B.H.; Kim, Y.S.; et al. Risk factors for occurrence and 30-day mortality for carbapenem-resistant *Acinetobacter baumannii* bacteremia in an intensive care unit. *J. Korean Med. Sci.* **2012**, *27*, 939–947. [CrossRef] [PubMed]
24. Zhu, H.; Zhou, Y.; Yang, X.; Zheng, S.; Wu, Y.; Chen, G. Construction and application of the closed caregivers closed-loop management scheme of the ward under the pandemic prevention and control. *Chin. J. Emerg. Crit. Care Nurs.* **2022**, *3*, 458–463.
25. Chang, W.; Chen, X.; Li, Y.; He, W.; Lin, T. Hand hygiene compliance of caregivers for patients with multidrug-resistant bacteria: A study based on contact path. *J. Med. Theory Pract.* **2021**, *34*, 3831–3833.
26. Chang, W.; Chen, X.; Li, Y.; He, W.; Lin, T. Hand contact behaviors among caregivers of patients with multidrug- resistant organism infection. *Chin. J. Infect. Control* **2021**, *20*, 1088–1093.
27. WHO. Guidelines for the Prevention and Control of Carbapenem-Resistant Enterobacteriaceae, *Acinetobacter baumannii* and *Pseudomonas aeruginosa* in Health Care Facilities. 2017. Available online: http://www.who.int/infection-prevention/publications/guidelines-cre/en/ (accessed on 7 August 2019).
28. Gontjes, K.J.; Gibson, K.E.; Lansing, B.; Cassone, M.; Mody, L. Contamination of Common Area and Rehabilitation Gym Environment with Multidrug-Resistant Organisms. *J. Am. Geriatr. Soc.* **2020**, *68*, 478–485. [CrossRef]

antibiotics

MDPI

Review

Multidrug-Resistant Sepsis: A Critical Healthcare Challenge

Nishitha R. Kumar [1], Tejashree A. Balraj [2], Swetha N. Kempegowda [1] and Akila Prashant [1,3,*]

1 Department of Biochemistry, JSS Medical College and Hospital, JSS Academy of Higher Education & Research, Mysuru 570004, India; nishitharkumar@jssuni.edu.in (N.R.K.); swethank@jssuni.edu.in (S.N.K.)
2 Department of Microbiology, JSS Medical College and Hospital, JSS Academy of Higher Education & Research, Mysuru 570004, India; tejashreea@jssuni.edu.in
3 Department of Medical Genetics, JSS Medical College and Hospital, JSS Academy of Higher Education & Research, Mysuru 570004, India
* Correspondence: akilaprashant@jssuni.edu.in; Tel.: +91-9008097970

Abstract: Sepsis globally accounts for an alarming annual toll of 48.9 million cases, resulting in 11 million deaths, and inflicts an economic burden of approximately USD 38 billion on the United States healthcare system. The rise of multidrug-resistant organisms (MDROs) has elevated the urgency surrounding the management of multidrug-resistant (MDR) sepsis, evolving into a critical global health concern. This review aims to provide a comprehensive overview of the current epidemiology of (MDR) sepsis and its associated healthcare challenges, particularly in critically ill hospitalized patients. Highlighted findings demonstrated the complex nature of (MDR) sepsis pathophysiology and the resulting immune responses, which significantly hinder sepsis treatment. Studies also revealed that aging, antibiotic overuse or abuse, inadequate empiric antibiotic therapy, and underlying comorbidities contribute significantly to recurrent sepsis, thereby leading to septic shock, multi-organ failure, and ultimately immune paralysis, which all contribute to high mortality rates among sepsis patients. Moreover, studies confirmed a correlation between elevated readmission rates and an increased risk of cognitive and organ dysfunction among sepsis patients, amplifying hospital-associated costs. To mitigate the impact of sepsis burden, researchers have directed their efforts towards innovative diagnostic methods like point-of-care testing (POCT) devices for rapid, accurate, and particularly bedside detection of sepsis; however, these methods are currently limited to detecting only a few resistance biomarkers, thus warranting further exploration. Numerous interventions have also been introduced to treat MDR sepsis, including combination therapy with antibiotics from two different classes and precision therapy, which involves personalized treatment strategies tailored to individual needs. Finally, addressing MDR-associated healthcare challenges at regional levels based on local pathogen resistance patterns emerges as a critical strategy for effective sepsis treatment and minimizing adverse effects.

Keywords: sepsis; drug resistance; microbial; critical illness; mortality; healthcare costs

1. Background

Sepsis is a critical medical condition associated with significant biological and chemical abnormalities that pose a high death rate. Unlike superficial and confined infections, sepsis is a complex disturbance of the delicate immunologic equilibrium between inflammatory and anti-inflammatory responses. This interaction demonstrates the fragile connection between the immune system and the clinical signs of sepsis. Over the past few decades, a comprehensive definition of "sepsis" has continuously evolved and improved [1]. Significantly, the current definition of sepsis (Sepsis-3) was proposed by the Third International Consensus, which defined it as "organ dysfunction caused by a dysregulated host response to infection". This description is the first to stress the vital function played by the natural and acquired immune system response at the onset of a medical illness [2].

During the initial stages of sepsis, the immune system mediates the activation of pro- and anti-inflammatory cytokines, pathogen-related molecules, and mediators, leading to the initiation of the complement cascade and coagulation [3]. For instance, numerous endogenous host-derived signals like damage-associated molecular patterns (DAMPs) or exogenous stimulations like pathogen-derived molecular patterns (PAMPs), such as DNA fragments, lipids, exotoxins, and endotoxins, are the starting signals for sepsis. These molecules interact with toll-like receptors (TLRs) present on the surface of antigen-presenting cells (APCs) and monocytes, leading to the expression of genes associated with pro-inflammatory interleukins (IL, IL-1, IL-6, IL-8, IL-12, and IL-18), tumor necrosis factor-alpha (TNF-α), and interferons (IFNs like IFN-y) and anti-inflammatory (IL-10) pathways and acquired immunity [4,5]; these processes are usually observed during the initial stages of sepsis [6–8]. This upregulated inflammation progresses to concomitant immunosuppression, leading to progressive tissue damage, multi-organ failure, increased immune cell apoptosis, and T cell exhaustion, which all together result in "immunoparalysis", thereby making sepsis patients prone to opportunistic and nosocomial infection [6,9]. A signal transduction caused by PAMPs- and DAMPs-mediated activation of monocytes and APCs causes the translocation of nuclear factor-kappa-light-chain-enhancer of activated B cells (NF-κB) into the cell nuclei. However, in short, the overall impact of the dysregulated immune response, whether hyper- or hypo-responsiveness, on the individual's immunological response is highly personalized, leading to significant challenges in diagnosis [1].

Sepsis is a worldwide public health concern characterized by high rates of morbidity and mortality and a significant financial burden [10,11]. For instance, Rudd and coworkers recently revealed the alarming worldwide estimations of sepsis, as 48.9 million cases of sepsis were reported in 2017, with 11 million deaths attributable to sepsis [11]. In 2011, sepsis substantially burdened healthcare facilities in the United States with USD 20 billion in annual costs [12]. Additionally, numerous indirect expenses might dramatically impact the quality of life of patients with sepsis. For instance, older patients with sepsis may experience long-term severe health issues, such as cognitive impairment and functional disability [13]. Furthermore, a study on the sepsis burden in the Indian intensive care unit (ICU) revealed that the elderly population is more prone to sepsis due to multiple comorbidities caused by compromised immunity. The study found that 132 out of 387 patients with sepsis had septic shock, with the lungs (45.5%) being the most common site of infection. The mortality rate was 60.7% and 78.9% in old and very old patients, compared to a 45.6% mortality rate observed in younger adults [14]. Similarly, another study identifying sepsis burden in Malaysian ICUs revealed that aging was significantly associated with a 30-day mortality rate among elderly sepsis patients (particularly patients aged ≥65 years), with a high 30-day mortality rate (28.9%) among elderly sepsis patients [15].

Like acute myocardial infarction and cerebrovascular stroke, sepsis is a critical and persistent chronological condition. In the case of sepsis, early and correct usage of antimicrobial drugs is of utmost significance within the first hour of detection, concurrently with organ support. If the microbial pathogen emerges as an MDR, including the methicillin-resistant *Staphylococcus aureus* (MRSA), carbapenem-resistant *Enterobacteriaceae* (CRE), and MDR *Pseudomonas aeruginosa*, the therapeutic efficacy of currently available antimicrobial drugs is compromised, which hinders treatment success. Additionally, multidrug resistance poses a substantial risk of developing numerous sepsis-related adverse effects [16], necessitating prompt administration of the most appropriate antimicrobial therapy. However, while antibiotic resistance in bacteria is continuously growing globally, it poses a critical challenge to treating clinical infectious diseases, particularly those leading to life-threatening sepsis, septic shock, and multi-organ failure [17]. Additionally, as bacteria evolve, new mechanisms of resistance are emerging regularly and spreading worldwide, which are restricting current treatment options and making it challenging to treat prevalent infectious diseases [18]. Despite the persistent need for new antimicrobial drugs, major pharmaceutical industries have withdrawn from this field due to the rising costs of clinical trials, demanding approval criteria, and a general lack of economic viability [19,20]. This has widened the gap between the urgent public health need for effective antibiotics and the diminishing prospects of developing new antibacterial medications, resulting in a concerning situation [19].

Most patients with sepsis are given empirical antibiotic treatment without a prior confirmed diagnosis. This may raise the likelihood of developing multidrug resistance, accompanied by significant ecological adverse effects. Moreover, sepsis patients receive higher initial doses of empirical antimicrobial therapy regardless of organ failure, which may increase the synthesis of circulating pro-inflammatory and anti-inflammatory mediators, negatively impacting their overall health and well-being [21]. Additionally, the widespread misuse of antibiotics contributes significantly to increased mortality rates [22] and the surge in antimicrobial resistance (AMR). This misuse jeopardizes individual health and overburdens national healthcare systems financially [23]. One major contributor to antibiotic overuse is the unethical sale of antibiotics without proper prescriptions or diagnostic tests [24]. Similarly, self-medication practices often driven by economic constraints result in an incomplete antibiotic course, which promotes antibiotic resistance development due to suboptimal dosing [25]. Additionally, economic incentives for vendors to promote antibiotic sales make changing such practices challenging [26]. Furthermore, we are only now starting to understand the implications of antibiotic restrictions on outcomes and costs. We are hindered by the absence of universal ethical guidelines and comprehensive data on outcomes. Additionally, the concept of "best" and "effective" therapy varies significantly among groups, which makes the decision to select antibiotics difficult. Moreover, suboptimal antibiotic therapy cannot eradicate the infectious agent from the body, exposing affected individuals to the risk of adverse outcomes and wider antimicrobial resistance. Therefore, rational antibiotic usage primarily relies on identifying patients who, in fact, require treatment or optimizing treatment for a faster recovery [27]. In this respect, this review aims to comprehensively analyze the current burden of sepsis, the factors responsible for its development and increasing severity, and sepsis-associated healthcare challenges to reduce sepsis risk and improve MDR sepsis therapy, particularly in critically ill hospitalized patients.

2. Epidemiology and Burden of MDR Sepsis

Sepsis is a worldwide severe health issue. Septic shock is a subclass of sepsis distinguished by metabolic, cellular, and circulatory defects that increase mortality risk among sepsis patients. Due to increased prevalence and pathobiological, molecular, genomic, and medical complications, sepsis and septic shock pose a growing worldwide burden and a formidable challenge for emergency doctors [28]. Since the first consensus definition (Sepsis-1) of sepsis in 1991, the occurrence and prevalence of sepsis and septic shock have steadily increased, reaching about 49 million confirmed cases with 11 million confirmed sepsis-related mortalities worldwide in 2017 [29,30]. According to a 2016 systematic review conducted in well-developed countries, over 30 million hospital-treated sepsis cases were reported annually worldwide, and 5.3 million individuals died from sepsis [31]. Sepsis is also vital in the ICU, affecting around 30% of ICU patients, with significant regional differences [32]. A Chinese study reporting national incidence and mortality of hospitalized sepsis revealed an annual increase in hospitalized sepsis from 328.25 to 421.85 cases per 100,000 during 2017–2019. In light of these findings, the World Health Organization (WHO) confirmed sepsis as a worldwide health priority [30].

Incidence and fatality rates of sepsis vary significantly, with the most significant burden in Oceania, sub-Saharan Africa, and the South, Southeast, and East Asian regions. An Indian study (2007) identifying the epidemiology of sepsis identified 176 out of 230 cases of systemic inflammatory response syndrome (SIRS) caused by sepsis in patients in intensive therapy units (ITUs). The mean age of patients was 54.9 years, and 67% were male. Patients with severe sepsis had significantly high ITU mortality, hospital mortality, and 28-day mortality, which were 54.1%, 59.3%, and 57.6%, respectively. Additionally, the percentage of cases with infection being the primary cause of hospital admission was 89.8% [32]. Another

Indian study conducted in 2016 identified the clinical microbiological profile of elderly sepsis patients. It revealed that 28.75% of cases were blood culture positive, of which 51.7% had Gram-negative infection and 48.30% had Gram-positive infection. Similarly, *Staphylococcus aureus* (49 patients) and *Escherichia coli* (36 patients) were the most prevalent pathogens isolated from sepsis patients. Subsequently, a 2017 study conducted in India identifying the sepsis burden in the adult population demonstrated that 282 of the total patients (4711) admitted to the hospital had severe sepsis, with 63.6%, followed by 62.8% and 56% hospital mortality, 28-day mortality, and ICU mortality, respectively [33]. The respiratory tract was the predominant site of infection among sepsis patients. Similarly, Gram-negative bacteria were the dominant cause of sepsis, with *Acinetobacter baumanni* being the most isolated pathogen. Additionally, researchers also found a significantly high mortality rate for sepsis patients, which was 85% [34]. A most recent study (2023) determining the clinical and demographic profile of elderly patients admitted to medical ICUs at a Tertiary Care Center demonstrated that sepsis was the most common cause of death among elderly patients. Moreover, bloodstream infections with Gram-negative pathogens were more prevalent than those caused by Gram-positive pathogens [35,36].

Besides India, a retrospective study by the National Mortality Surveillance System (NMSS) reported approximately one million sepsis-related fatalities in China [37]. Another Chinese study reported an estimated incidence of 328.25 cases per 100,000 populace in 2017 [38], slightly less than the previously reported incidence rate of 352.10 in the Western Pacific region. Still, it is significantly less than the incidence rate of 415.13 cases per 100,000 population in the Pan-American region [39]. Similarly, around 85% of cases and deaths occurred globally due to sepsis in low- and middle-income countries [11]. Moreover, sepsis can afflict people of any age or gender, and considerable differences exist in the burden of the illness. A three-year study from 2017 to 2019 found that sepsis afflicted the elderly population over 65 with a 57.5% incidence rate, followed by children under ten with a 20% incidence rate [38]. Similarly, in 2017, the global age-standardized incidence of sepsis was higher among females (716.5 cases per 100,000 population) than males (642.8 cases per 100,000 population) [11].

Numerous studies have found a correlation between the frequency and incidence of MDR sepsis and hospital stay within the last 90 days, a history of stroke, aging, and infection with MDR organisms (MDRO). These observations may be explained by the growing incidence of MDROs in hospital wards caused by the widespread antibiotic usage and transmission between healthcare staff and patients [40–43]. ESBL-producing *Enterobacteriaceae* appear to be the most common (9.7%) among all MDROs, with ESBL-producing *E. coli* and *Klebsiella pneumoniae* accounting for 35% of all *E. coli* and *K. pneumoniae* isolates [44]. The percentage of ESBL production among *Enterobacteriaceae* varies from nation to nation; however, it is on the rise throughout Europe, with Italy having one of the highest prevalences of ESBL-producing *Enterobacteriaceae* [45]. Another study found that being hospitalized within the past 90 days is a particular risk factor for ESBL *Enterobacteriaceae*. This finding demonstrates the significance of contact with the healthcare setting, necessitating the empiric administration of carbapenems to sepsis patients who have this risk factor [46]. However, with the emergence of CRE, treating sepsis patients has become a formidable challenge for physicians [16]. Similarly, stroke is another risk factor linked to the emergence of ESBL-positive bacteria, which may be attributed to extended hospital stays, nursing home stays, and indwelling invasive devices like gastrostomies, bladder catheters, and nasogastric tubes. Given the frequency of ESBL *Enterobacteriaceae* and if an infection with MDROs is suspected (for example, previous hospitalization), an adequate administration of selective antibiotic therapy can be considered for ESBL+ pathogens while awaiting culture results [46].

Sepsis also poses a significant economic burden on healthcare systems. The annual healthcare costs of sepsis in the United States were USD 20 billion in 2011 [12] and USD 24 billion in 2013–2014 [47], which were increased to USD 27 billion in 2019. Overall, sepsis costs the US healthcare system over USD 38 billion annually, making it the most expensive illness linked to hospitalization [48]. In India, an estimated sepsis cost per patient was USD 55 in 2005 [49], while a 2008 study proposed a projected estimate of USD 53 million for the Indian healthcare system in 2012 [50]. Similarly, before the COVID-19 pandemic, the annual costs of sepsis were about USD 1.3 billion per year in Ontario and Canada [51]. According to a nationwide study conducted in Japan, the adjusted annual gross medical cost of sepsis rose from USD 3.04 billion to USD 4.38 billion during the study period, which was linked to an increasing number of patients with sepsis (indicating 67,318 cases in 2010 to 233,825 in 2017). Another study discovered that shorter hospital stays were related to lower medical expenses [52]. These escalating healthcare-associated expenses have been attributed to prolonged hospital stays, expensive medications, and, regrettably, restricted access to treatment for sepsis patients, contributing to an alarming number of misdiagnosed sepsis-related fatalities. For instance, recent research estimated that sepsis affected 48.9 million people, and around 11 million people died globally in 2017, making sepsis responsible for approximately 20% of all global deaths [11]. Moreover, an eight-year Japanese study found that sepsis caused 18.9% of in-hospital mortality [52].

3. Pathogenesis and Mechanisms of Drug Resistance

In an ecological environment, bacteria are believed to strive for resources for existence, equipping various microbes with the complex chemical compounds yielded through metabolic activity that can inhibit or kill other microbes [53]. For instance, Penicillins and Cephalosporins are metabolic products of *Penicillium* and *Cephalosporium* species. A study identifying antimicrobial resistance (AMR) in archaea demonstrated that 30,000-year-old archaea were resistant to aminoglycoside (streptomycin) and β-lactam antibiotics (penicillin) [54]. Similarly, bacteria have evolved with time and developed resistance to antimicrobial drugs, a self-defense mechanism achieved through natural selection. Many antibiotic resistance mechanisms within bacterial metabolic pathways have additional functions to perform. For instance, the efflux pump that transfers particular antibiotics outside the bacterial cell membrane may also export toxic compounds like heavy metal ions to protect the bacterial cell [55]. Other antibiotic resistance mechanisms in bacteria involve adapting a latent state, structural and morphological changes, reduced permeability of bacterial cell walls and cell membranes, decreasing drug uptake, inactivating drugs, regulation of metabolism, target site modification, secreting target-protecting proteins, initiation of self-repair systems, and biofilm production, which all collectively constitute the defense system of bacteria against antibiotics [56,57]. Hence, rapidly spreading AMR across microbial populations cannot be caused by a single factor; instead, it involves multiple complex mechanisms.

Additionally, there are a few difficult-to-treat AMR pathogens categorized under the well-known abbreviation "ESKAPE", which include *Enterococcus faecium*, *S. aureus*, *A. baumannii*, *K. pneumonia*, *Enterobacter* species, and *P. aeruginosa* [58]. As mentioned previously, all these pathogens have different mechanisms of resistance and thus cause varied degrees of infection. For example, *A. baumannii*, which causes hospital-acquired AMR infections, confers resistance to antibiotics by producing β-lactamases (all four classes: A–D) to degrade beta-lactam antibiotics, activating drug efflux pumps, producing modified porins to reduce drug permeability through bacterial outer membranes, and altering drug targeting sites [59]. Similarly, *P. aeruginosa* causes both acute and chronic hospital-acquired and severe respiratory infections. Like *A. baumannii*, *P. aeruginosa* can produce all four classes (A–D) of β-lactamases. Moreover, this pathogen can confer resistance through gene mutation, resulting in overexpression of *AmpC* β-lactamases. It can produce transferable aminoglycoside modifying enzymes (AMEs), which reduce the binding affinity of aminoglycosides to their target site in the bacterial cell [60]. Additionally, *S. aureus*, which causes mild and severe life-threatening skin and soft tissue infections, pleuropulmonary, bacterial endocarditis, and device-related infections, has decades of AMR history [61] attributed to the presence of penicillin-binding proteins (PBP and PBP2a) and genes, including *mecA*, *mecC*, *VanA*, *gyrA*, *gyrB*, and *erm* (*ermA*, *ermB*, *ermC*, and *ermF*) [62,63].

Besides AMR mechanisms, complex immune reactions involving the production and utilization of pro- and anti-inflammatory molecules, although aimed at protecting organisms from internal and external threats, lead to the excessive production of these inflammatory molecules. This, in turn, results in the rapid and simultaneous display of immune activation and immunosuppression signs in sepsis patients [64], as illustrated in Figure 1.

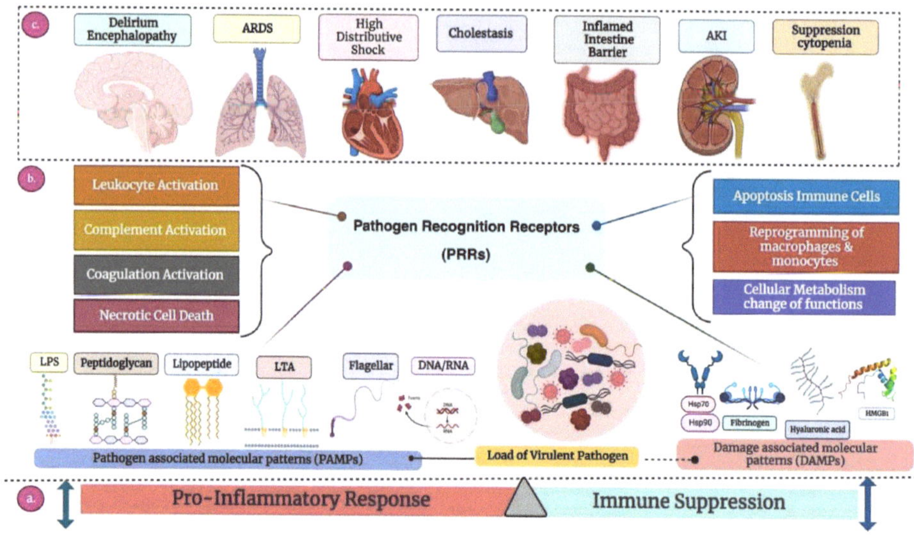

Figure 1. Overview of the pathogenesis of sepsis. (**a**) Immune Response: Sepsis occurs when the body responds to infection with an excessive immune system reaction, causing a disturbance in the usual equilibrium of the inflammatory response to maintain homeostasis. Activation of PRRs initiates both proinflammatory responses and immune suppression, ensuing hyperinflammation and immune suppression to the extent that is detrimental to the host. (**b**) Receptor Response: Once a pathogen successfully breaches the host's mucosal barrier, it can induce sepsis, depending on its quantity and virulence. The host's defense system identifies molecular components of invading pathogens (PAMPs) through specialized receptors called PRRs. This activation triggers the expression of target genes responsible for proinflammatory cytokines (resulting in leukocyte activation), inefficient utilization of the complement system, coagulation system activation, simultaneous downregulation of anticoagulant mechanisms, and necrotic cell death. This sets in motion a detrimental cycle, leading to the progression of sepsis, exacerbated by the release of endogenous molecules from injured cells (DAMPs or alarmins), further stimulating PRRs. Immune suppression manifests as extensive apoptosis, causing depletion of immune cells, reprogramming monocytes and macrophages into a state with reduced capacity to release proinflammatory cytokines, and an imbalance in cellular metabolic processes. (**c**) Organ Response: Organs respond to internal or external stimuli by initiating inflammation, undergoing changes in function, or activating compensatory mechanisms aimed at maintaining homeostasis and resolving disturbances. These responses are crucial for the body to cope with stress, injury, infection, or other challenges, ensuring proper functioning and survival. The main organs and their specific responses are described below. 1. Brain: (i) Delirium: Acute disturbance in attention and cognition, leading to confusion and altered perception. (ii) Encephalopathy: Brain dysfunction causing altered mental function, affecting cognition, consciousness, and behaviors. 2. Lungs: Acute Respiratory Distress Syndrome (ARDS) triggered by MDR bacteria is a severe and potentially life-threatening condition characterized by the rapid onset of widespread inflammation in the lungs. Infections, especially severe bacterial infections caused by multidrug-resistant bacteria, lead to direct lung injury, cytokine storms, secondary infections, and ventilator-associated pneumonia (VPA). 3. Heart: High distributive shock with MDR sepsis places immense strain on the

heart due to systemic vasodilation and reduced blood flow, leading to compromised cardiac function and potential myocardial damage. The combination of multidrug-resistant sepsis and shock increases the risk of cardiac dysfunction, contributing to the severity of the condition and complicating treatment. 4. Liver: Cholestasis during MDR sepsis involves a disruption in bile flow due to both the effects of severe infection and potential liver dysfunction from multidrug-resistant bacteria. This combination worsens jaundice, impairs detoxification processes, and contributes to the systemic complications of sepsis. 5. Gastrointestinal tract: An inflamed intestine barrier exacerbated by multidrug-resistant bacterial infections leads to severe inflammation and compromised intestinal integrity, increasing the risk of bacterial translocation. This can result in the systemic dissemination of pathogens, exacerbating MDR sepsis. 6. Kidney: In MDR sepsis, acute kidney injury is a combination of sepsis-induced circulatory changes, and the potential nephrotoxicity of the pathogens contributes to kidney dysfunction, increasing the risk of severe complications and mortality. 7. Suppression cytopenia: During MDR sepsis, suppression cytopenia leads to a significant reduction in blood cell counts. The combination of multidrug-resistant pathogens and the immunosuppressive effect of sepsis increases the risk of complications, including compromised immunity and susceptibility to bleeding or infections. Abbreviation: ARDS, acute respiratory distress syndrome; AKI, acute kidney injury; DAMPs, danger-associated molecular patterns; DNA, deoxyribonucleic acid; HMGB1, high-mobility group box-1 protein; HSPs, heat shock proteins; LPS, lipopolysaccharide; LTA, lipoteichoic acid; PAMPs, pathogen-associated molecular patterns; PPRs, pattern recognition receptors; RNA, ribonucleic acid. The dashed lines depict the disrupted immune response triggered by infection, which makes the body unable to restore its equilibrium and causes harm to the organs. This culminates in a severe and life-threatening state known as sepsis.

These concomitant secretions result in immunological paralysis, a significant reason for high mortality rates in patients who experience septic shock caused by MDR pathogens [65]. This could be attributed to previous exposure to an initial inadequate antimicrobial therapy, which cannot treat the infection; instead, it can affect the host defense system and may lead to altered immune function. Indeed, inadequate antimicrobial therapy can have detrimental ecological effects on the microenvironment as it can cause superinfection with MDR pathogens [66]. Similarly, weeks or months of continuous immune activation against pathogens, as in the case of sepsis patients, may lead to a chronic state, impairing the ability of cells to recognize antigens and creating a microenvironment where cells of innate and acquired immunity (neutrophils, macrophages, monocytes, T-cells, and B-cells) receive numerous stimuli that devastatingly affect their activity. The overall performance of receptors located on the cell surface and within the cell, which play a crucial role in the detection of microbial substances and internal warning signals, is crucially affected [67]. This concept is represented in Figure 2.

In the clinical management of sepsis, physicians strive to offer effective empirical antimicrobial treatment for hospitalized patients with sepsis, sometimes restoring to prescribing antibiotics without precise diagnostic confirmation. Unfortunately, while intended to save lives, this practice comes at the expense of potentially prescribing unnecessary antibiotics. This excessive treatment is associated with the emergence of MDR bacteria. Moreover, many patients use antibiotics without any prescription, whereas others take excessive doses of prescribed antibiotics, contributing to antibiotic resistance in bacteria. The higher incidence of multidrug resistance in sepsis patients could also be attributed to multiple patient-specific factors, including older age, comorbidities, immunosuppression or excessive use of immunosuppressive drugs, chemotherapy for cancer patients, and living in countries with lower and middle-income economies with deprived healthcare infrastructure and inaccessibility to healthcare facilities [68]. Polypharmacy, which involves the concurrent use of five or more drugs, is another significant factor contributing to the emergence of MDR bacteria in sepsis cases. Polypharmacy is often associated with the natural aging process, which, due to simultaneous biological and pathological changes, elevates the risk of multimorbidity and the necessity for multiple concurrent medications [69]. Since AMR in bacteria and fungi is complex and rooted in millions of years of evolution, these microorganisms have adopted different strategies to withstand antimicrobials, survive, and reproduce.

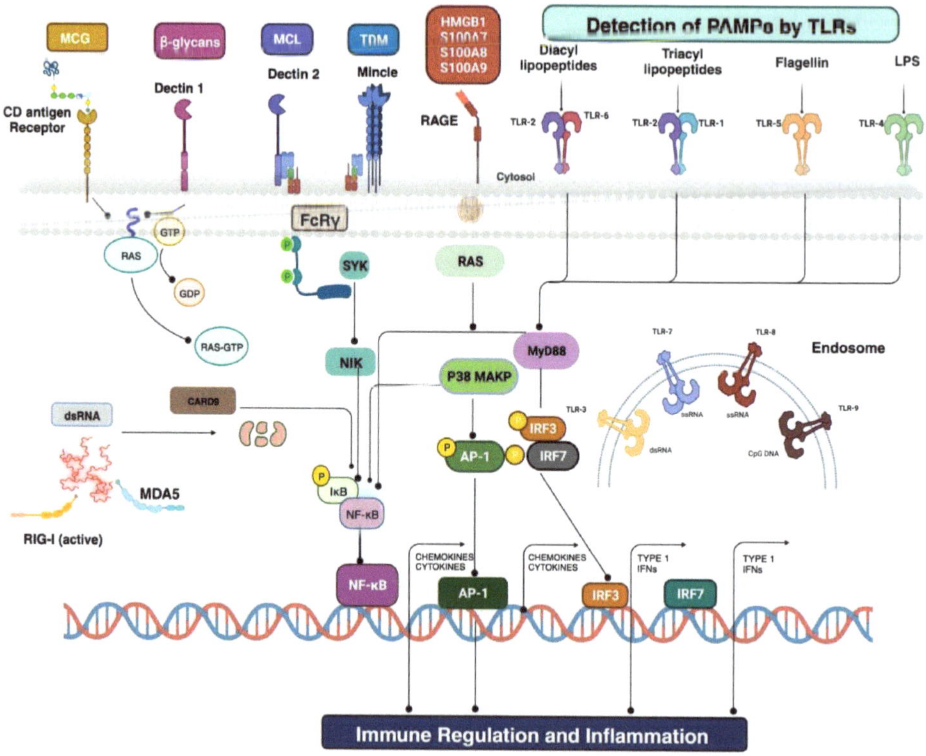

Figure 2. Cell surface and intercellular receptors amend for the recognition of PAMPs and DAMPs. The onset of sepsis is heralded by the host's detection, prompting the activation of inflammatory signaling pathways. An extensive array of cellular and intracellular receptors is used to identify PAMPs or DAMPs. Examples include microbial and host-originated glycoproteins, lipoproteins, and nucleic acids. The corresponding PRRs encompass Toll-like receptors, dectin 1 (a member of the C-type lectin domain family 7), and dectin 2 (a member of the C-type lectin domain family 6). At least ten distinct TLRs have been identified, usually forming homodimers or heterodimers. Upon activation, these signaling pathways typically integrate into interferon regulatory factor signaling and nuclear factor-κB. IRF is in charge of type I interferon production. NF-κB and activator protein 1 signaling predominantly oversee the early activation of genes involved in inflammation, such as TNF and IL1, as well as those encoding for endothelial cell surface molecules. Among the other notable components within this sepsis-related network are caspase recruitment domain-containing protein 9, lipopolysaccharide, myeloid differentiation primary response protein 88, and stimulator of interferon genes protein. Loss of lymphocytes is directly immunosuppressive, contributing to the lymphopenia observed in patients. The genetic mutation or pharmacological intervention that decreases sepsis-induced apoptosis improves survival in severe sepsis. The degree of lymphocyte apoptosis in animal models of sepsis correlates with the severity of sepsis, and persistent lymphopenia predicts sepsis mortality. The next generation of treatments evaluated for suppressing immune function through interaction with sepsis includes therapies targeting lymphocytes and leukocytes. Abbreviations: CARD9, caspase recruitment domain-containing protein 9; dsDNA, double-stranded DNA; dsRNA, double-stranded RNA; FcRγ, Fcγ receptor; HMGB1, high-mobility group box 1; iE-DAP, d-glutamyl-meso-diaminopimelic acid; LGP2, laboratory of genetics and physiology 2; LPL, lipoprotein lipase; LPS, lipopolysaccharide; LY96, lymphocyte antigen 96; MAPK, Mitogen-activated

protein kinase; MCG, mannose-containing glycoprotein; MDA5, melanoma differentiation-associated protein 5; DAMPs, damage-associated molecular patterns; MDP, muramyl dipeptide; MYD88, myeloid differentiation primary response 88; TLRs, Toll-like receptors; C-type lectin domain family 7 member A (dectin 1) and C-type lectin domain family 6; NIK, NF-κB-inducing kinase; NOD, nucleotide-binding oligomerization domain; RAF1, RAF proto-oncogene member A (dectin 2); RIG-I, retinoic acid-inducible gene 1 protein; ssRNA, single-stranded RNA; STING, stimulator of interferon genes; NF-κB, nuclear factor-κB; SYK, spleen tyrosine kinase; NF-κB and activator protein 1 (AP-1).

Consequently, most bacteria carry natural resistance to one or even multiple antibiotics. Contrarily, many bacteria can alter their antibiotic-targeting sites and become antibiotic-resistant. Similarly, self-medication is frequently non-specific to the target disease; hence, it may occasionally result in resistance development in opportunistic pathogens [70].

4. Common Pathogens Involved in MDR Sepsis

Sepsis can result in septic shock, multiple organ dysfunction, and ultimately death if it cannot be diagnosed timely and managed adequately. Sepsis can be infectious and caused by various microorganisms, including bacteria, viruses, and fungi. Bacteria are the most prevalent etiological pathogens, and *Streptococcus pneumoniae*, *S. aureus*, *E. coli*, *Hemophilus influenzae*, *Salmonella* spp., and *Neisseria meningitidis* are some of the most common bacterial pathogens involved in sepsis or sepsis-related comorbidities [71]. Fungi are responsible for around 15% of all infections, with aggressive fungal infections being the primary cause of sepsis, particularly in patients with immunosuppression or severe illnesses. For example, Candida species are the most prominent cause of fungal sepsis, responsible for around 5% of all sepsis cases. Invasive Candida infections are linked with a significantly increased sepsis-associated mortality risk. Various studies have linked inadequate antifungal therapy with higher mortality rates in patients with candidemia (a bloodstream infection-BSI caused by Candida species) or septic shock attributed to Candida [72]. Additionally, sepsis and septic shock indicators can be the lethal recurrent outcomes of infections caused by seasonal or periodic influenza, dengue viruses, and highly contagious pathogens of community health significance. Notable examples include swine and avian influenza viruses, the Middle East respiratory syndrome-related [MERS] coronavirus, the severe acute respiratory syndrome-related (SARS) coronavirus, and, most lately, the Ebola and yellow fever viruses [71].

Furthermore, anyone suffering from a severe infection, damage, or chronic disease can progress to sepsis; however, specific populations are more likely to develop the condition, including the elderly, pregnant or recently pregnant women, newborns, hospitalized patients, ICU patients, immunocompromised patients, and patients suffering from co-morbidities or chronic medical conditions (like kidney disease or cirrhosis) [68]. Some other studies have also confirmed that populations of underdeveloped countries, females, and older people, particularly those with comorbidities, are high-risk populations [11,60]. Similarly, sepsis, which may be acquired in healthcare settings, is among the most common adverse events during medical care establishment. This condition affects hundreds of millions of individuals worldwide each year. Infections contracted in healthcare settings and frequently brought on by MDR bacteria are described in Table 1, which can rapidly deteriorate the clinical condition of patients. This is the reason behind the higher risk of hospital-associated mortality among sepsis patients infected with MDR pathogens [73]. Several studies evaluated a relationship between gender, infection, and risk of sepsis and found that male patients with respiratory infections have higher chances of developing sepsis than females (36% versus 29%). Contrarily, female patients with genitourinary infection are more prone to develop sepsis than males (35% versus 27%) [74]. Accordingly, BSI caused by *P. aeruginosa* and *S. aureus* is more prevalent in males than females [75]. Conversely, approximately 60% of BSIs with *E. coli* occur in females, consistent with the higher risk of females developing sepsis due to urinary tract infections [76]. Similarly, various published manuscripts have confirmed that male patients with candidemia have a higher risk of developing sepsis than females [77,78].

Table 1. A systematic table covering reported mechanisms of multidrug-resistant bacteria in sepsis.

Bacterial Species	Mechanisms of Multidrug Resistance	Association with Sepsis
Gram-Positive Bacteria		
Staphylococcus aureus (including MRSA)	Altered penicillin-binding proteins (PBP2a)	Increased severity of infections, including skin and soft tissue infections, pneumonia, and bloodstream infections.
	Efflux pumps	MRSA is commonly associated with healthcare-associated infections.
	Biofilm formation	Virulence factors contribute to pathogenicity.
Enterococcus faecium/faecalis (including VRE)	Altered target site (D-Ala-D-Ala to D-Ala-D-Lac)	Frequent in healthcare-associated infections, especially in immunocompromised patients.
	Biofilm formation	High resistance to vancomycin, a crucial antibiotic.
Gram-Negative Bacteria		
Escherichia coli (Including ESBL-producing)	Production of extended-spectrum beta-lactamases	High resistance to beta-lactam antibiotics, leading to challenging treatment
	Plasmid-mediated resistance	Common in urinary tract, respiratory, and bloodstream infections.
	Porin mutations	Associated with nosocomial infections, which can progress to sepsis.
Klebsiella pneumoniae (Including CRE strains)	Production of carbapenemases	Limited treatment options due to resistance to broad-spectrum antibiotics.
	Plasmid-mediated resistance	High mortality rates associated with bloodstream infections.
	Reduced permeability of the outer membrane	Commonly found in healthcare settings.
Acinetobacter baumannii	Efflux pumps	Common cause of healthcare-associated infections, especially in ICUs.
	Biofilm formation	Associated with high mortality rates in bloodstream infections
	Intrinsic resistance mechanisms	Often involved in ventilator-associated pneumonia and septicemia.
Pseudomonas aeruginosa	Efflux pumps	Commonly implicated in hospital-acquired infections, including sepsis.
	Biofilm formation	Infections associated with a higher risk of treatment failure.

Additionally, some studies have identified a relationship between various factors affecting mortality rates among sepsis patients, including age, gender, comorbidities, disease severity [79], and the early initiation and appropriateness of antimicrobial and non-antimicrobial therapy [80]. One study identified a higher mortality rate of sepsis among individuals in the age group 15–50 years than older patients (58.5% versus 39.1%). The same study also revealed that most factors affecting mortality rates in sepsis patients are uncontrolled. They further found that 21.05% of sepsis patients were discharged from hospitals on medical advice. In general, the relationship between age and mortality among sepsis patients was controversial in this study [81]. Conversely, a study conducted by Carbajal-Guerrero and colleagues revealed that older patients (>65 years) with sepsis had a higher risk of comorbidities compared to the younger patients, and these comorbidities were found to be a potential factor contributing to the high mortality rate among the elderly [82]. Furthermore, the effect of gender on sepsis is still under debate among researchers. One study identified a higher incidence of sepsis among males than in females [83]. Another study that investigated the effect of gender on the survival of sepsis patients [83,84] showed that survival was better in females [85]. These differences in mortality rates for sepsis between male and female patients can be attributed to differences in their immune responses. For instance, estrogen production is higher in female patients than in males, which positively influences immune activity. This is because increasing body mass index and age in females increase the production of estrogen by elevating aromatase activity in adipose tissues, and high estrogen provides better protection to female patients with sepsis through immune activation [86].

5. Diagnostic Challenges and Innovations

Diagnosing sepsis at an early stage and promptly initiating treatment are essential for enhancing clinical outcomes and reducing the death rate of sepsis. Until a suitable alternative test is available, pathogen detection through conventional blood culturing has traditionally been the accepted method for diagnosing sepsis, as shown in Table 2. However, routine blood culturing takes 2–3 days to identify bacteria and even more time to test for antibiotic sensitivity, which is deemed inadequate in the case of sepsis. In such conditions, each hour of delay in treatment worsens patient conditions and increases morbidity and mortality [87]. Much research has been conducted on identifying the importance of blood culture for sepsis patients. One meta-analysis comprising 22,655 individuals with sepsis and septic shock from seven studies revealed only a positive blood culture result for 40.1% of patients [88]. Another study identified only 10–15% of positive blood culture results in neonates with sepsis [89]. Studies have confirmed multiple factors contributing to this poor diagnosis. For instance, most sepsis patients whose blood samples were taken for blood culturing had non-infectious inflammatory conditions caused by inflammatory, neurological, or metabolic disorders [90]. Conversely, sepsis patients with probably infectious inflammatory conditions receive antibiotics even before their sepsis worsens or before blood culturing, resulting in the inability of culture techniques to diagnose pathogens. Cheng and colleagues confirmed this phenomenon by demonstrating a 12% absolute difference in the count of positive blood culture outcomes before and after antimicrobial testing [91], which decreases the probability of detecting pathogens [92]. Finally, several microbial pathogens, such as fungi, bacteria, and some viruses, are undetectable through the traditional culturing approach and require alternative indicators for detection, including urinary antigens and non-specific markers for fungal presence. However, this may become increasingly challenging due to the rising incidence of sepsis caused by unusual pathogens [93].

These comparisons outline the key differences between the two approaches used in diagnosing antimicrobial resistance, highlighting the strengths and weaknesses of each method. Clinical diagnostic challenges and the need for immediate diagnosis and treatment have led to a dependence on identifying biomarkers in the blood, including procalcitonin (PCT), C-reactive protein (CRP), and white blood cell (WBC) count. Indeed, the early consensus definition of sepsis, Sepsis-1, incorporated a decreased ($<4 \times 10^9$/L) and an increased ($>1.2 \times 10^{10}$/L) WBC count into the criteria for systemic inflammatory response syndrome (SIRS) [94]. However, predicting infection in sepsis patients through serum biomarkers is debated due to the lack of sensitivity and specificity of many serological tests. A retrospective cohort study by Marik and Stephenson found a very poor predictive value (as low as 0.52 AUROC, an area under the receiver operating characteristic) of the WBC count for bacteremia in patients suspected of sepsis [95]. Similarly, Siegel and colleagues found a normal WBC count in 52% of patients with confirmed blood culture results showing bacteremia [96]. A meta-analysis study found that WBC count had minimal diagnostic significance in serious infections, with a negative probability ratio as low as 0.61 [97]. Similar results were obtained for CRP [98]. In contrast to the WBC count, the ratio of the neutrophil-to-lymphocyte count has constantly been found to be a far more accurate biomarker of physiological strain than absolute neutrophil or WBC counts [99]. An increase in neutrophil count and a decrease in lymphocyte count are frequently observed in systemic illnesses like sepsis, which may be attributed to the endogenous actions of hormones like cortisol and catecholamines. Moreover, sepsis prompts the migration of lymphocytes to inflammatory tissues, while increased lymphocyte apoptosis causes an increase in the ratio of neutrophils to lymphocytes [100]. A prospective study by Ljungström and a group comprising 1572 patients revealed a higher ratio of neutrophil-to-lymphocyte count compared to PCT and CRP (AUROC 0.68 versus 0.64 versus 0.57) or diagnosing bacterial sepsis [101]. However, a recent study predicting disease severity in COVID-19 patients confirmed that any kind of severe physiological strain can result in a rise in the ratio of neutrophils to lymphocytes, irrespective of the sepsis [102]. Additionally, studies confirmed that the neutrophil-to-lymphocyte ratio was invariably elevated even in non-infectious sepsis, making it significantly less precise to diagnose sepsis in critical care patients [103].

Table 2. A comparison table outlining the differences between conventional methods and molecular methods for diagnosing antimicrobial resistance (AMR).

Sl No	Aspect	Conventional Methods	Molecular Methods
1	Sample Type	Limited range of sample types	More adaptable to various sample types
2	Identification Speed	Relatively slow, it may take days to provide results.	Rapid results, often within hours.
3	Sensitivity and Specificity	It may have lower sensitivity and specificity.	Generally, higher sensitivity and specificity
4	Range of Pathogens Detected	Limited to certain pathogens (Genera of the Pathogen)	Broad range, capable of detecting various pathogens (exact Species of the Pathogen)
5	Type of Information	Phenotypic information (e.g., growth inhibition).	Genotypic information (specific genes or mutations).
6	Multiplexing Capability	Limited ability to test for multiple resistance genes	High multiplexing capability, detecting multiple targets in a single test
7	Equipment Required	Often requires specialized equipment and expertise	Requires specific equipment but can be more accessible
8	Ease of Use	It may require trained personnel and specialized equipment.	User-friendly protocols, less technical expertise needed
9	Accuracy	Subject to human handling error	Less prone to human error, higher accuracy
10	Resistance Detection Method	Culture-based methods, susceptibility testing	DNA sequencing, PCR, genotypic assays
11	Cost	Lower initial cost in some cases	Higher initial cost, but potentially cost-effective over time

Although CRP is a commonly used biomarker in critical illnesses, it is non-specific for bacterial infections; instead, CRP levels increased in most other causes of inflammation. A meta-analysis study evaluating the diagnostic performance of CRP in sepsis identified that CRP has a better-pooled sensitivity (80%) but only 61% specificity [104]. Studies have confirmed that CRP levels have a minimal association with the disease severity in sepsis, whereas they serve as the most commonly used biomarker for predicting the disease severity in patients with pancreatitis [105], with 100% and 81.4% sensitivity and specificity, respectively [106]. However, CRP cannot constantly differentiate sterile from infected pancreatic necrosis. Therefore, it is not a suggested biomarker to initiate antimicrobial therapy [107]. Similarly, the production of PCT is increased in response to sepsis [108], and it rises within 2–3 h of infection and gains a peak at 24 h, which is a much quicker rise than CRP (which reaches a peak at 72 h). A systematic review and meta-analysis conducted by Wacker and colleagues reported that PCT has an AUROC of 0.85; thus, it is an excellent biomarker for distinguishing sepsis from other non-inflammatory syndromes [109]. Another meta-analysis study comprising 12 articles found that PCT may exhibit limited effectiveness in differentiating viral and bacterial infections, with Kamat and group identifying a poor sensitivity of 55% and a moderate specificity of 76% [110]. Additionally, PCT has lower sensitivity and diagnostic AUROC to predict bacterial infection in individuals with autoimmune diseases [111], chronic renal failure [112], and immunosuppression [113].

Novel diagnostic approaches for pathogen detection can be helpful alternatives to conventional techniques. Surface-enhanced Raman spectroscopy (SERS) is becoming an increasingly significant method for detection due to its ability to amplify the Raman scattering of target particles on a superficial layer of metal-made or graphene-based surface [114]. Moreover, this method can easily detect label-free nucleic acids. Similarly, numerous studies have confirmed that matrix-assisted laser desorption ionization time-of-flight mass spectrometry (MALDI-TOF MS) is a rapid diagnostic method for accurate identification of various microscopic life, including yeast, bacteria, fungi, and even Nocardia and mycobacteria species within a very short time frame, thereby minimizing the amount of time needed for adequate and effective antimicrobial therapy in sepsis [115–117]. Studies also confirmed that MALDI-TOF MS significantly reduced the hospital stay of sepsis patients by 1.75 to 6 days [115,116] and enhanced overall survival by 4 to 9% [115,117], thereby highlighting the significance of early detection of pathogens. Unfortunately, MALDI-TOF MS cannot identify AMR mechanisms, and testing antibiotic susceptibility depends on conventional methods [118]. However, more sophisticated systems have been developed that use polymerase chain reaction (PCR) for microbial amplification before MS detection to rapidly identify clinically relevant bacterial and yeast species with a higher diagnostic strength than cultural techniques [119]. These systems can also detect microbial species that do not typically grow in blood cultures, including *Mycoplasma pneumoniae*, *Rickettsia typhi*, *Legionella pneumophila*, *Nocardia* spp., and various fungi [120]. Although these systems can rapidly and efficiently diagnose the microbiological cause of sepsis, only a few studies have confirmed their usefulness over conventional cultures [121]. Furthermore, these methods are currently limited to detecting only a few of the large diversity of antibiotic resistance markers, which are crucial for providing tailored treatment [122].

Much research has demonstrated that timely diagnosis of sepsis episodes and medical intervention improve clinical outcomes [123]. Many other studies have identified that timely antibiotic treatment yields a lesser impact than the patient control group, indicating the variability of the disease and the necessity for continued analysis and medication [124,125]. Therefore, the optimal point-of-care sensors make it possible to rapidly compile patient health data, increase healthcare coverage, and improve the efficiency of healthcare services while simultaneously reducing healthcare costs [126,127]. Furthermore, it is extensive and fast enough to provide researchers with sufficient information regarding pathogen and host–response virtually anywhere in a very short time, which enables the treatment of sepsis in two major streams: firstly, POCT-based devices can speed up the identification step where optimum care is delayed, thereby improving outcomes, and secondly, they can identify numerous things, including pathogens, cell-surface proteins, and plasma proteins, which are ascribed as representatives of the immune response of hosts and which, when coupled with complex data analytics, can assist in stratifying sepsis even at the patient bedside. This kind of information might accelerate the procedure for detecting patients who may benefit from supplementary therapy [123]. POCT may also see the evaluation of the development of various protein biomarkers (such as IL-6, IL-10, PCT, CRP, and TNF-α) linked with acute sepsis and septic shock in ICU patients and estimate the probability of all-cause mortality within 28 days [128], assisting in the decision-making process for the selection of antibiotics.

6. Clinical Management of MDR Sepsis

Despite substantial advancements in our knowledge of the pathophysiology of sepsis, numerous clinical trials have been unsuccessful in identifying novel therapies that can alter the course of the disease [129,130]. Recognizing sepsis as a medical emergency is essential since, in the absence of definitive treatment, therapeutic interventions involve timely management of infection and organ support [131]. The 2016 Surviving Sepsis Campaign

(SCC) guidelines strongly advise the prompt administration of intravenous broad-spectrum antibiotics, ideally within an hour following sepsis detection [132]. Several publications on sepsis and septic shock have found that delayed antibiotic administration is linked with adverse outcomes [133–135]. Beyond their apparent advantages, broad-spectrum antibiotics can cause substantial damage, such as antibiotic-associated adverse effects and potentially fatal AMR-related consequences [136,137]. Infections with MDRO have significantly increased worldwide, restricting our therapeutic options. The growing AMR is estimated to be responsible for approximately 10 million deaths each year by 2050. Therefore, treating patients with sepsis and septic shock by augmenting antimicrobial efficacy and avoiding the emergence of MDR strains is one of the primary concerns. Regarding this, antimicrobial stewardship (AS) is an important strategy for sepsis care since it focuses on multi-professional teamwork [for example, microbiologists, infectious disease specialists, and pharmacists] with appropriate, adequate, and optimized antimicrobial therapy [138].

Various studies have confirmed improved survival rates in sepsis patients with early and suitable antimicrobial administration and efficient source control [139], as validated by the inclusion of similar recommendations in 2016 SSC guidelines for early delivery of appropriate broad-spectrum antimicrobial drugs within one hour of hospital admission in patients afflicted by sepsis and septic shock [28,132]. Moreover, administering empiric antibiotic therapy directed at the most likely pathogens involved in infectious sepsis is crucial to improving patient outcomes. Numerous published manuscripts have discussed the adverse impact and consequences of inadequate empiric therapy in sepsis patients [138,140–144]. Notably, prescribing ineffective empiric therapy is prevalent in ICUs, occurring in 10–40% of sepsis cases, which varies depending on the frequency of MDR pathogens [144,145]. Recent studies have found that the patient group with higher disease severity scores is most likely to benefit from appropriate antibiotic treatment. In contrast, ineffective empiric antimicrobial therapy was linked with a 5-fold decrease in the survival of over 5000 individuals suffering from septic shock [133]. Another prospective study has found a significantly increased mortality rate among patients with septic shock and an average of three organ dysfunctions [143]. An appropriate empiric antimicrobial therapy means prescribing drugs that cover almost all potential pathogens responsible for the suspected infection. To achieve this, certain pathogen- and patient-related factors must be considered [138,146], including weight, age, allergies, comorbidities, chronic organ dysfunction, immunosuppressive therapy, and previous antibiotic or infection history. The risk of MDR pathogens should also be considered, including lengthy hospital stays, previous hospital admissions, the presence of invasive medical devices, and prior encounters with MDR pathogens [138]. Several investigations into the detrimental consequences and outcomes of delayed antimicrobial provision in patients with sepsis have concluded similar results [147–149]. These studies have confirmed that appropriate antibiotic therapy significantly decreased the mortality rate when it was given within ≤1 h [33], whereas each hour of delay in the treatment increased mortality [150] and dropped the overall survival rate by an average of 7.6% [87]. Besides delayed antibiotic administration, lengthy hospital stays [149,151], acute renal [152] and lung [153] diseases, and worsening organ dysfunction [154] have also been found to be common factors associated with increased mortality in sepsis patients.

Compared to these findings, various studies were unsuccessful in determining the usefulness of timely antimicrobial therapy [155–157]. A meta-analysis comprising over 16,000 individuals with sepsis and septic shock from 11 studies identified an insignificant difference between antibiotic administration (within 3 h) and mortality rate [158]. Another meta-analysis study comprising 11 studies on sepsis patients identified a 33% reduction in mortality among patients receiving early empiric antibiotic therapy (≤1 h) compared to those with delayed antibiotic administration (>1 h) [159]. A recent systematic review concluded that the mortality rate significantly decreased in patients with septic shock receiving early and adequate empiric antibiotic therapy [142]. Despite inconsistent outcomes, there is substantial agreement among international specialists on the need for prompt antimicrobial therapy in patients suffering from sepsis and septic shock, and novel ideas have recently been offered. A "door-to-needle" duration of 60 min has been advocated for antibiotic delivery, which indicates global concerns about launching a time window for successful therapy after sepsis detection [136]. Nonetheless, ensuring a competent application of institutional standards for antibiotic administration within 1 h after presentation remains difficult.

Given the rapidly growing prevalence of MDR infections, combined antibiotic therapy is commonly advised to warrant a larger antimicrobial spectrum and appropriate empiric coverage. The combined therapy is described as using antibiotics from two separate classes that have activity against a single infection, primarily to speed pathogen elimination and increase the susceptibility of pathogens to treatment [160]. To ensure the likelihood of having at least one active antibiotic against the possible pathogen involved, the Infectious Diseases Society of America (IDSA) endorses using two active medicines against Gram-negative bacilli for empiric treatment of septic shock [161]. Recognizing the need to encourage antibiotic judiciousness, the IDSA formed a committee to explore suggestions for prudent antibiotic usage in treating sepsis. The experts accepted ten antibiotic class combinations out of a total of 21. Concerns about rising resistance and proper pathogen coverage were stated as factors for selecting such combinations. The use of any combination involving macrolides or ciprofloxacin and specific pairings of aztreonam with cephalosporins and aminoglycosides with intravenous clindamycin were prohibited [162].

Studies on combination therapy have yielded conflicting findings, and there is a scarcity of well-powered randomized controlled trials examining this particular issue. Numerous observational studies, however, demonstrated that combination therapy outperformed monotherapy in individuals suffering from sepsis and septic shock [163,164]. For instance, a meta-regression analysis found a link between combination therapy and a high survival rate among severely ill sepsis patients with a higher mortality risk. Unexpectedly, this meta-analysis identified higher mortality among the patient group with a low risk of death [165]. Similar findings were reported in other studies where researchers linked higher mortality with nephrotoxic side effects leading to renal failure [166]. Based on these inconsistent findings, some specialists advocate employing a pair of antibiotics for the initial treatment of patients with septic shock and suspected MDR pathogen infections. Even with negative culture results, treatment can be cut down to personalized therapy at the minimum acceptable time after microbiological isolation or a satisfactory clinical response [167]. To assess the effectiveness of different antibiotic combinations, well-powered randomized controlled trials examining multiple antibiotic combinations in different situations should be conducted [168]. Additionally, individualized therapies tailored to patients' unique conditions, like diabetes, renal or hepatic failure, or immunosuppression, can yield favorable results instead of applying an uniform approach.

As sepsis is frequently accompanied by organ dysfunction, supportive care and management of organ dysfunction are critical in sepsis treatment to reduce complications and improve patient outcomes. Hemodynamic support and mechanical ventilation are the two fundamental pillars of supportive care. Hemodynamic support entails maintaining proper tissue perfusion and oxygen supply, fluid resuscitation to restore blood pressure, and adequate organ perfusion. Vasopressor medications may also be required to treat refractory hypotension and to sustain cardiac output. Similarly, mechanical ventilation techniques, such as low tidal volume ventilation and prone posture, benefit sepsis patients with acute respiratory distress syndrome induced by sepsis. Furthermore, renal and liver function should be constantly monitored to maintain optimal fluid and electrolyte balance and the fine balance of acids and bases. Some patients may need hemodialysis as a renal replacement therapy to prevent damage to other bodily organs caused by fluid imbalance and the presence of creatinine and urea in the blood, which hinder sepsis treatment [168–170].

7. Impact of MDR Sepsis on Critical Care

Studies have confirmed that sepsis and septic shock are highly prevalent among critically ill patients, which essentially require early and appropriate empiric antibiotic therapy within the first hour to manage these situations effectively [171]. However, MDR sepsis presents formidable challenges within ICUs, significantly impacting patients' well-being and straining healthcare resources. The complex nature of MDR microorganisms reduces antimicrobial treatment efficacy, often causing treatment failures and lengthy hospital stays. These MDR pathogens raise concerns about possible horizontal transmission within ICUs, highlighting the vital need for consistent infection prevention and control policies. Similarly, resource-restricted ICUs often lack essential equipment, laboratory assistance, and qualified physicians and nursing teams. Therefore, sepsis management guidelines in resource-limited ICUs, formulated by the Global Intensive Care Working Group of the European Society of Intensive Care Medicine (ESICM) [172], often differ in various aspects from the SSC recommendations, which were established in well-developed countries [173]. Notable instances include the meticulous management of glucose levels in the blood using insulin, a safe approach with consistent and accurate monitoring of blood glucose but risky when the effects of insulin on the blood are rarely or inadequately assessed. Furthermore, conventional culturing techniques cannot detect infectious sepsis due to empiric antibiotic administration to patients or take around 48–72 h to yield results. Therefore, early and precise identification of MDR pathogens is vital to support better infection control strategies [171].

Multiple infections, including ventilator-associated pneumonia (VAP) and hospital-acquired pneumonia (HAP), are widely prevalent in ICU settings and account for over half of all antibiotics provided in critical care situations. Despite attempts to enhance timely detection and therapy, the morbidity and mortality of sepsis and septic shock remain high, especially in patients with MDR sepsis [174]. Physicians in the ICU continue to have difficulty diagnosing VAP and HAP at the bedside. Routine CXR is no longer advised for ICU patients to evaluate disease progression and its response to treatment; instead, it is advisable to consider lung ultrasonography as a valuable diagnostic tool for VAP and HAP, especially when paired with the medical data of patients [171]. Previous studies have confirmed favorable effects and outcomes of β-lactam or β-lactamase inhibitors against VAP and HAP, especially for various Gram-negative bacilli that pose a significant concern in ICU settings. The β-lactam antibiotics are widely used in ICUs and are one of the safest antibiotics; however, they also have side effects. For example, neurotoxic symptoms have been identified in 10–15% of patients admitted to hospital ICUs. Similarly, there was an increased incidence of renal failure observed in ICU patients when they were administered β-lactam antibiotics in combination with nephrotoxic medicines like vancomycin [175].

Implementing strategies for controlling and preventing infection, including prudent antibiotic stewardship, strict adherence to hand hygiene protocols, comprehensive environmental disinfection regimens, and timely detection of MDR microorganisms, is critical to restrain the transmission of AMR pathogens. This necessitates a united effort and collaboration among healthcare practitioners, effective monitoring systems, and knowledgeable antimicrobial management teams to mount a staunch defense against the growing danger of MDR sepsis in critical care settings [176].

8. Frequency and Causes of Readmission in Sepsis Patients

Despite recent advancements in the medical field, the mortality rates associated with sepsis are significantly high, affecting almost 42% of sepsis patients [31]. However, alarmingly, even patients who survive are not immune to the effects of sepsis, as nearly one-third of sepsis survivors were readmitted within 180 days. Readmissions following sepsis-related hospital stays are frequent and expensive, with severe physical and financial implications. The relationship between surviving sepsis and subsequent readmissions is a relatively new area of research, with prior studies focusing solely on short-term and immediate outcomes. Consequently, we could only find a few studies for comparison, all from well-developed countries. The national study by Norman and group [177] found a 30-day readmission rate of 28% in the United States. Another study comprising patients from 21 community-based hospitals [178] found a readmission rate of 17.9%. Similarly, a 90-day readmission rate ranged between 30 and 42% [179]. Research conducted by Goodwin and colleagues on 43,452 sepsis survivors admitted to non-governmental hospitals in California, New York, and Florida found a significantly high 180-day readmission rate of 48% [180]. This high readmission rate may be attributed to a greater risk of depression [180], sleep deprivation, encephalopathy [177], mental illnesses, and cognitive and organ failure, all ultimately leading to death among sepsis survivors, as identified by various studies [181]. Besides high morbidity, the high readmission rate of sepsis survivors also comes with a significant financial burden, as recent research quoted an annual cost of over USD 38 billion spent on sepsis in the United States. On average, a solitary readmission may result in expenses ranging from USD 25,000 to USD 30,000. These horrifying figures can be attributed to the fact that sepsis is generally treated in the ICUs, which is extremely costly due to the cost of lengthy hospital stays, medications, laboratory tests, use of medical equipment, invasive devices, procedures, nursing staff, and taxes [182,183].

The financial burden on sepsis patients is exacerbated in developing countries, including India and Pakistan, where patients typically have to pay for healthcare-associated expenses out-of-pocket. Additionally, patients do not have medical insurance or loan facilities. A brief look at the per capita figures in developing countries puts these findings into proper perspective. With a per capita income of USD 1500 in a developing country compared to a substantial USD 53,000 in the United States, it is easy to assume how a single readmission could be overwhelming for patients and their families. Most households in developing countries have only one worker; therefore, an illness leading to prolonged hospitalization for that individual could be disastrous for the entire family. Moreover, the majority of employees live paycheck-to-paycheck and have few savings or investments. There is no choice for sick leave, and each day spent in the hospital results in no revenue for that day. Furthermore, it would not be easy to find a suitable substitute for the primary wage earner due to cultural factors in most patriarchal families. Consequently, families find themselves compelled to liquidate all of their assets or borrow money from relatives and friends, which might take years to repay. Other family members commonly offer nursing care in the home, resulting in reduced focus on childcare and diminished earning potential [184].

9. Preventive Measures and Infection Control

The Antimicrobial Stewardship Program (ASP) is a multifaceted, collaborative approach that engages various healthcare professionals, including clinicians, microbiologists, pharmacists, and nursing staff, to enhance treatment outcomes and prevention by minimizing AMR among microbial pathogens [185]. The ASP is one of three important principles of a comprehensive strategy for strengthening healthcare systems. Although infection prevention and control (IPC) and medicine followed by patient safety are the other two principles of ASP, ASP cannot be successful without including IPC [186] because healthcare epidemiologists and infection preventionists play a pivotal role in the implementation and success of ASP [187]. Notably, following the WHO essential medications list "AwaRe16" classification [Access, Watch, and Reserve], optimizing antibiotic usage, and surveillance are important aspects of ASP that are directly linked with reduced AMR [188]. This multifaceted strategy eliminates the need for antimicrobial therapy by preventing infection transmission, which reduces the emergence of resistance. The Centers for Disease Control and Prevention (CDC) developed seven fundamentals for ASP implementation in 2019. Notably, leadership and accountability are the first two concepts or principles responsible for the program's goals and outcomes, followed by education and local antibiogram deployment. The latter two are administrative components, which are based on the idea that common infections receive appropriate empiric therapy. Prescription preauthorization and resistance surveillance performed by pharmacists and laboratorians, respectively, are the two actionable tasks where the necessary interventions can be carried out as mandated by institutional standards and policies [189].

The WHO has designated AMR as a global threat because it is a well-established fact that threatens public health and national security [190]. Therefore, the association between healthcare providers (HCPs) and public health organizations is critical. It makes it easier to develop prevention initiatives, promote education, and conduct surveillance, all aimed at slowing down the spread of AMR [191]. Patients who exhibit resistance to currently available antimicrobials force physicians to employ reserved antibiotics like carbapenems and polymyxins. These reserved antibiotics are expensive, may not be readily accessible in some countries, and may have potentially unintended consequences [for example, colistin administration is linked with acute kidney injury [192]. Currently, healthcare professionals are facing a worldwide challenge of MDR-ESKAPE pathogens, which are infamously branded as "bugs without borders" [193,194]. These are nosocomial pathogens with the ability to escape the biocidal effect of antimicrobials [195]. Hence, tackling AMR is a crucial aspect of ensuring safe and successful healthcare delivery, as highlighted by the implementation of ASP [196]. Since its start, ASP has been extremely effective in reducing antibiotic usage. Notably, the four Ds, which are the key facets of ideal antimicrobial therapy, encompass selecting the right drug, dose, de-escalation to pathogen-directed therapy, and the right duration of therapy and infection control. These are the guiding principles of ASP [195]. These approaches align closely with public health objectives and encompass the promotion of ASP by monitoring, ensuring data transparency, developing infrastructure, and increasing patient and healthcare professional knowledge and awareness [191].

The lack of novel antimicrobials necessitates the preservation of existing ones. To ensure the judicious use of novel antimicrobials, the Infectious Diseases Society of America [IDSA] and other public health bodies recommend the implementation of ASP to preserve the efficacy of these medications [83]. Moreover, a set of systematic ASP initiatives have been introduced globally in clinical settings to lessen selected pressures that favor highly resistant organisms [197]. The ASP is critical in preventing the AMR spread [198]; however, a meta-analysis study found significant variability in the included studies, and collaborations between the IPC department and the ASP team were found to be more effective in limiting the AMR spread. Nonetheless, it is recommended that all fundamental features, including education programs and antimicrobial limitation through prospective audits and feedback, be employed in conjunction to improve outcomes. Notably, ASP efforts may not produce results without hospital leadership commitment. ASP has been found effective in reducing AMR and hospital costs in various regions worldwide, and a few of the safety measures and prevention controls are illustrated in Figure 3 [199,200].

Everyday self-care routines that incorporate cleansing and sanitizing both your body and hands are paramount to maintaining good health. Regular sterilization of surfaces prone to high contact is also significant in curbing the spread of harmful microorganisms. Face masks should be worn consistently, particularly when maintaining safe distances from others is difficult. Self-medication is a practice to avoid, especially in cases where the correct dosage and timing of intake are not known. To prevent potential contamination risks, hospital waste should be correctly deposited into the designated trash receptacles. Travel plans should be put on hold when one is unwell as a preventive measure against spreading the disease.

MDR microbial pathogens cause a significant proportion of infections in ICUs, with around 23,000 deaths annually in healthcare settings alone in the USA [201]. Besides host susceptibility, the complexity and logistics of critical care medications put patients at risk of contracting infectious pathogens. Invasive procedures and implantable devices, which are frequently used to provide supportive care to critically ill patients, also serve as entry points for pathogens. Similarly, the concurrent involvement of numerous medical team members and the utilization of numerous patient care devices for lifesaving critical care treatments may increase the chances of infection transmission from staff or fomites to patients. Generally, infection control precautions may not be prioritized in emergency conditions like sepsis and cardiac arrest, in which even seconds matter. Pathogenic microorganisms in the ICU are more prevalent on or in the human body [skin, respiratory epithelium, and gastrointestinal tract] and in the hospital environment and serve as transmission reservoirs. Additionally, antibiotics, chemotherapy, or acquiring nosocomial pathogens, among other things, might disrupt a patient's flora. Therefore, patients colonized with resistant bacteria can serve as potential reservoirs for the transmission and spread of infection. The proportion of patients in a given unit colonized with resistant bacteria, or colonization pressure, is an independent risk factor for transmission [202,203]. Moreover, person-to-person transmission of resistant pathogens mainly occurs through contaminated patient care equipment, the hands of healthcare providers, and contaminated surfaces.

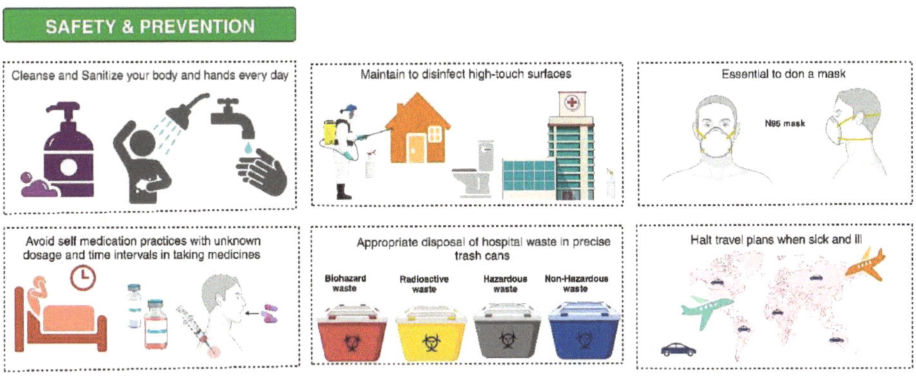

Figure 3. Prevention and control of the rise in multidrug-resistant microorganisms. Everyday self-care routines that involve cleaning and sanitizing your body and hands are paramount to maintaining good health. Regular sterilization of surfaces prone to high contact is also significant in curbing the spread of harmful microorganisms. Face masks should be worn consistently, particularly when maintaining safe distances from others is difficult. Self-medication is a practice to avoid, especially in cases where the correct dosage and timing of intake are not known. Hospital waste should be correctly deposited into the designated trash receptacles to prevent contamination risks. Travel plans should be put on hold when one is unwell as a preventive measure against spreading the disease.

A recent study found environmental contamination with MDROs in 40% of patient rooms in the hospital, including Vancomycin-resistant Enterobacterales [VRE] [204]. Studies also found the viability of difficult-to-treat MDROs like MRSA, VRE, and *A. baumannii* on fomites in the hospital environment, including dry surfaces, steel, and plastic materials. Other pathogens of high concern were also found prevalent under dry conditions, like carbapenemase-producing Enterobacteriaceae, including *bla*$_{KPC}$-carrying *Klebsiella pneumoniae* [205]. Studies have shown the high efficacy of approved hospital disinfectants against these pathogens. Using disposable patient care equipment, especially those known to be used for patients harboring MDROs, has been reported to minimize the risk of cross-transmission. Additionally, sharing items, including cooling blankets, blood pressure measuring devices, and portable radiology cassettes, should be thoroughly disinfected. Other items, like fabric privacy curtains, should be replaced with disposable curtains [206]. Numerous studies have documented the benefits of supplementary methods of disinfection, like hydrogen peroxide vapors and ultraviolet lights, to reduce the burden of bacterial pathogens and their spores. Hydrogen peroxide is effective in decontaminating hospital wards experiencing outbreaks [207] or environments where high-concern pathogens are present [208]. Contrarily, other studies have identified that although ultraviolet lights are less labor-intensive, less time-consuming, and do not require technical expertise for operation, they are less effective in eliminating all pathogens. Similarly, some pathogens can reside in damp environments and may form biofilms from which they can be transmitted to patients. For example, waterborne bacteria, including *Stenotrophomonas*, *Pseudomonas*, *Aeromonas*, and *Sphingomonas*, can colonize plumbing fittings such as sink drains, faucets, and aerators.

Preventing transmission via contaminated plumbing is a significant concern in hospital infection control and is currently being researched [209]. Some basic methods include ensuring that hospital water has an adequate amount of free chlorine, choosing sinks with low-splash designs, and keeping patient care items away from handwashing sinks, where they could be polluted by pathogen-contaminated drain splash-back. Plumbing fixtures may require disassembly, special cleaning and disinfection measures, or even replacement in an epidemic environment in which plumbing fittings are implicated [210]. Therefore, ASP, hand hygiene, and adequate disinfection of hospital surfaces and equipment are essential in preventing the spread of MDROs. Further, hospital administrations comprising infection control specialists, microbiologists, and critical care experts should collaboratively constitute policies and procedures for infection control in critical care units and emergency rooms, the necessary training and education of ICU staff for infection control, and other relevant outcome measures. Additionally, infection control protocols and procedures must be followed by having adequate nursing personnel, setting up infrastructure like handwashing stations, and providing hospital supplies like masks, gloves, and alcohol-based hand gels.

Furthermore, vaccines are commonly administered as a preventative measure and are applicable before the bacteria grow and spread following the initial infection (during low pathogen burden) and before various tissues and organs are affected. This significantly lowers the probability of mutations that confer resistance arising and spreading. Antibiotics often only have one mode of action or one target, like the cell wall of bacteria or bacterial translation machinery. This is because antibiotics are designed to be highly specific in killing pathogens. Bacteria can naturally resist antibiotics or acquire or develop this resistance over time (like avoiding access to antibiotic targets, drug efflux, modifications of drug targeting sites, or even inactivation of the antibiotics themselves). Therefore, changes in the drug target site caused by a single mutation render the antibiotic useless. Additionally, the selective pressure from antibiotic usage encourages the development of drug-resistant clones. Conversely, vaccines reduce the likelihood of resistant clones being selected for further development since they have a preventative effect. Moreover, because vaccines frequently target several antigens or various epitopes of the same antigen, for instance, polyclonal antibodies, the development of vaccine-evasion variations would require many mutations that would each have an impact on a distinct epitope, making the emergence of resistance in bacteria challenging [211].

10. Global Efforts and Collaborations

To underscore the serious threats posed by AMR, the CDC has published a study to characterize the important challenges associated with AMR and threat level classifications for MDROs [212]. The report classified pathogens into three distinct types: urgent, serious, and concerning. With ESKAPE pathogens being the most urgent threat to sepsis patients, policymakers and stakeholders have initiated numerous programs in this area. For instance, the National Action Plan for Combating Antibiotic-Resistant Bacteria (CARB) was launched to confront the escalating challenge of AMR through a well-coordinated and collaborative effort as part of the US government's national response focused on addressing AMR. Five areas were focused on by the action plan, including (i) reducing and stopping the emergence of resistant bacteria, (ii) strengthening One Health monitoring efforts, (iii) promoting the development and use of rapid and novel diagnostics for detecting resistant organisms, (iv) expediting research for new antibiotics, alternative therapeutics, and vaccines, and (v) improving global collaboration [213]. Additionally, the WHO has approved an action plan focusing on AMR with five goals, including (1) increased awareness of AMR through efficient communication and education, (2) strengthened knowledge and evidence base for monitoring and scientific research, (3) decrease in the frequency of infection through infection prevention and hygiene measures, (4) optimization of the judicious use of antimicrobials in both human and animal health, and (5) creation of an economic rationale for long-term investment that considers the demands of all countries [214]. From a public health perspective, the CDC has led a multidimensional effort involving activities aimed at detecting and treating resistance on time and investing in prevention measures. The establishment of an Antibiotic Resistance Solutions Initiative, the Antibiotic Resistance Lab Network, and fundamental advice for ASP in various healthcare settings [215] are specific CDC activities.

The word "stewardship" was first coined in 1970, when an international initiative for optimal antibiotic administration, dosage, and duration was taken. In 2012, the Global Sepsis Alliance (GSA), an organization committed to decreasing the influence of sepsis and coordinating national and global initiatives against sepsis, introduced World Sepsis Day. Before that, numerous national public health organizations were unfamiliar with sepsis knowledge; even the Global Burden of Disease Report did not mention sepsis. Later, the White House issued the National Action Plan to demand the implementation of ASPs by 2020 in all hospitals providing acute care to patients. In this regard, by 2016, 64.2% of the critical care hospitals in the US had satisfied the essential criteria of the ASP proposed by the CDC [216]. The CDC has focused on ASP by releasing recommendations called "the Core Elements of Hospital Antibiotic Stewardship Programs". The basic features are designed to help hospitals of all sizes and complexity confront the dangers of AMR while also promoting patient safety through the deployment of effective ASPs. The guidelines recognize the dynamic nature of ASP and the need for greater flexibility in project and program implementation. Key components include leadership dedication, responsibility, pharmacy knowledge and expertise, action, monitoring, reporting, and education [215]. The Centers for Medicare and Medicaid Services provided additional support and engagement in 2019 by mandating the establishment and advancement of an ASP as a prerequisite for participation for all acute care hospitals participating in Medicare and Medicaid programs [217]. Over the years, these policies and activities have aided in the formulation and execution of ASPs across various settings, encompassing both integrated and non-integrated healthcare systems. Indeed, forward-thinking healthcare systems have initiated efforts to encourage and subsidize ASPs. Similarly, Europe has taken numerous initiatives to implement ASPs at regional and national levels [218]. In this regard, ESGAP, the ESCMID Study Group on ASP, has played an especially prominent role in these activities.

Besides these initiatives, public awareness of the AMR problem is critical. A survey analysis using the Amazon Mechanical Turk Crowdsourcing platform to recruit respondents found that, despite a substantial majority of respondents (93%) agreeing that unsuitable antibiotic usage contributes to antibiotic resistance, 70% of the survey respondents expressed a neutral stance or disagreed with the assertion that antibiotic resistance is a problem [219]. Another poll found that 65% of the American populace perceives antibiotic resistance as a matter of public health concern, and 81% are concerned that diseases may become progressively more difficult to treat as a result of antibiotic resistance. An annual observance to raise awareness about AMR was held as part of the CDC's initiatives to combat AMR and involve the public [220]. Increased public education regarding the substantial strain imposed by antibiotic-resistant infections on healthcare resources and the communal issues involved in a holistic approach to countering AMR will remain critical in the future. Regarding sepsis, the WHO took significant measures to address the pressing global health threat of sepsis, resulting in the publication of the WHO Secretariat Report and the adoption of Resolution WHA70.7 by the 70th World Health Assembly (WHA) in May 2017 on "Improving the prevention, diagnosis, and clinical management of sepsis". The first progress report on implementing the resolution was issued in 2020 for WHA 73. Among the significant accomplishments were identifying sepsis treatment gaps and developing global guidelines for the clinical management of sepsis [221].

Sepsis Alliance, founded in 2007, is another prominent sepsis organization working in all 50 states of the US to "save lives and reduce suffering from sepsis." Sepsis surveillance is dedicated to saving lives and improving suffering by enhancing sepsis awareness and treatment. Its goal is to make this world free of sepsis. Sepsis Alliance is also a proud co-founder of the GSA, founded in 2010, and currently represents over one million caregivers in over 70 countries [222]. GSA initiated World Sepsis Day [WSD] in 2012. Since then, events have occurred worldwide every September 13th to promote awareness about sepsis. Various events are also organized for medical personnel, including sports activities, pink picnics, photo exhibitions, dinners, grand galas, multiple possibilities for public gatherings, including hospital open houses and community healthcare events, and online campaigns, including the "World Sepsis Congress", and movements across various social media websites like Facebook, Twitter, and WhatsApp [221]. Similarly, the International Surviving Sepsis Campaign (SSC) is a collaborative project of the Society of Critical Care Medicine (SCCM) and the European Society of Intensive Care Medicine (ESICM), which are dedicated to lowering morbidity and mortality occurring globally due to sepsis and septic shock. SCCM is also committed to enhancing the prognosis for sepsis survivors, particularly those with post-sepsis syndrome. The SSC campaign was initiated in 2002 during the annual meeting at ESICM and has established guidelines and bundles for managing sepsis [168].

Several challenges to lowering the massive global burden of sepsis include difficulties in identifying related morbidity and mortality, insufficient knowledge, poverty, health inequities, resource-limited public health, and a fragile acute healthcare delivery system. Context-specific solutions to this serious problem are essential due to considerable disparities in susceptible populations, the infecting microorganisms, and the healthcare ability to manage sepsis globally, particularly in low and middle-income countries (LMIC) [223]. The high variability of typical critical care syndromes, including sepsis, has hampered developments in finding therapy targets; consequently, the demands of severely ill patients in LMIC are frequently unmet, and some patients are even subjected to therapies that might be harmful. Given the substantial resource variance, it may be impossible to anticipate identical goals and worldwide agreement in all management areas. Therefore, regional critical care management teams nationwide must customize diagnostic and treatment methods for various problems in their respective environments. Similarly, investments in the acute care of sepsis patients should be appropriate and effective compared to expensive and technology-concentrated ones. Such assets can provide substantial returns across several clinical specialties and positively affect population health outcomes [224,225].

11. Discussion

Sepsis is a life-threatening emergency condition of global public health concern with substantial mortality and financial costs. Sepsis definition has significantly evolved in recent years, with the currently acceptable Sepsis-3 definition, which emphasizes the role of the immune system in sepsis development. This review comprehensively evaluated various research and reports on MDR-sepsis and its associated healthcare challenges. Epidemiological data from various studies highlighted a high prevalence and incidence of sepsis, with significant disparities at regional and global levels. One study reported 48.9 million cases of sepsis, with 11 million deaths occurring annually worldwide. Another study found that the annual healthcare costs of sepsis reached USD 38 billion alone in the USA. Studies conducted in India revealed a higher incidence of sepsis among elderly ICU patients and with Gram-negative bacterial pathogens, particularly *E. coli* and *A. baumanni* [34,35]; however, one Indian study also identified Gram-positive *S. aureus* as the prevalent cause of sepsis [34]. Similarly, studies identified a significant increase in the number of MDR sepsis among hospitalized neonates and the elderly, notably with Gram-negative bacterial pathogens. Studies have shown consistency regarding the causative pathogens of sepsis. For instance, most studies highlighted the presence of ESKAPE pathogens in infectious sepsis [226–228], whereas only a few studies identified *K. aerogenes* and *Enterobacter cloacae* as being responsible for sepsis [229].

The worldwide escalating incidence of sepsis and sepsis-associated healthcare costs may be attributed to the growing incidence of AMR in MDR pathogens, which significantly challenge sepsis treatment. Antibiotic resistance development in MDR pathogens, particularly against the commonly prescribed antimicrobials, results in substantial delays in providing effective antimicrobial treatment. These delays in treatment even worsen the health conditions of susceptible populations like children, the elderly, those with a previous history of infection, and patients with comorbidities. These delays are also correlated with increased mortality rates, prolonged hospital stays, and increased healthcare expenses. Consequently, in the face of MDR infections and delayed microbiological results, the widespread use of broad-spectrum antibiotics in empirical antimicrobial therapy has become a common practice. However, this reliance on broad-spectrum antibiotics further risks individual health by continuing the overuse and misuse of these drugs, consequently exacerbating the development of antimicrobial resistance [230].

Infections with *Enterobacteriaceae*, including *E. coli* and *K. pneumoniae*, are a significant concern in ICU patients with MDR sepsis. Studies have identified an almost 51% incidence of infection in ICU patients, with infection incidence density ranging from 13 to 20.3 episodes per thousand patient days [231]. A study conducted from June 2009 to December 2013 identified a 14.9% mortality rate among patients infected with *Enterobacteriaceae* bacteremia. The authors identified that increasing sepsis severity was significantly correlated with higher mortality, with 3.5%, 9.9%, and 28.6% mortality for sepsis, severe sepsis, and septic shock, respectively. They further identified that time to antimicrobial therapy was not significantly associated with mortality; however, prolonged ICU and hospital stays were found to be significantly associated with increased severity of sepsis, ultimately increasing the death rates among sepsis patients [232]. Furthermore, researchers from another study found that 48% of *Enterobacteriaceae*-infected patients developed recurrent infections within a 12-month follow-up period. Over half of these recurrent infections were caused by the same bacterial species and at the same culture site. Studies also identified that patients harboring MDR Gram-negative bacteria were independent predictors of subsequent mortality after discharge from the index hospitalization. Furthermore, researchers found that the chances of recurrent infection were high within the first three months of hospital discharge [233]. This is a significant finding, as it indicates a critical post-hospitalization period for monitoring and intervention. Therefore, timely and precise prognosis and outcomes for ICU sepsis patients with MDR *Enterobacteriaceae* infections are critical in managing MDR sepsis.

The AMR emergence in bacteria is a complex phenomenon attributed to diverse molecular mechanisms that rely on both the antimicrobial agent in question and the specific pathogen. These AMR mechanisms encompass a spectrum of genetic events, including constitutive or inducible expression of resistance genes and upregulation of these resistance genes [234]. Additionally, certain bacteria inherently possess resistance to specific types or entire classes of antimicrobial agents. Notably, in infectious diseases, bacteria can produce biofilms, which are implicated in over 65% of human infectious diseases [235]. Composed of structured entities like extracellular polymeric substances—polysaccharides, proteins, and extracellular DNA [236]—bacterial biofilms confer resistance through multiple mechanisms, including impeding the cell cycle, facilitating horizontal gene transfer, secreting enzymes that alter or bind antibiotics, and limiting antibiotic diffusion [237].

Studies have identified varied patterns of resistance among the pathogens involved in sepsis. As ESKAPE pathogens were found to be the major cause of MDR sepsis, pathogens have shown noticeable resistance to β-lactam antibiotics [58]. Most ESKAPE pathogens were found to be involved in the production of all four classes of β-lactamase enzymes [59]. Other pathogen resistance mechanisms involved in MDR sepsis were the production of AMEs by *P. aeruginosa* [60] and various resistance proteins and genes by *S. aureus*, including penicillin-binding proteins (PBP and PBP2a), *mecA*, *mecC*, *VanA*, *gyrA*, *gyrB*, and *erm* (*ermA*, *ermB*, *ermC*, and *ermF*) genes [62,63]. Additionally, studies have confirmed that most resistance cases were also attributable to patient-specific factors like older age, inadequate or excessive empiric antibiotic therapy, antibiotic usage without any prescription, immunosuppression [68], and in some cases, concomitant secretions of inflammatory and non-inflammatory molecules, leading to immunological paralysis [65]. Moreover, ESBL production is seen only in one-third of *E. coli* in early-onset sepsis, compared to a far higher ESBL production by *K. pneumonia* in late-onset sepsis [238]. Similarly, a Chinese study identified a significantly higher proportion of MDROs among patients with late-onset sepsis [239].

A comprehensive analysis of multiple AMR mechanisms reveals the diverse nature of antibiotic resistance patterns in sepsis-related pathogens. For instance, ESKAPE pathogens dominate in cases of MDR sepsis, which is attributed to their ability to produce multiple β-lactamase enzymes that target β-lactam antibiotics. Specific resistance mechanisms in *P. aeruginosa* and *S. aureus*, involving AMEs and varied resistance proteins and genes, highlight the complexity of resistance, which indicates that although sepsis can be treated through an appropriate antimicrobial regime, treating MDR sepsis is challenging due to the development of resistance in sepsis pathogens, which not only restricts the treatment options but also reduces the survival rate of patients, where each hour delay in treatment significantly increases the patient's mortality rate. Moreover, patient-related factors like age, inappropriate antibiotic usage, immunosuppression, and the release of inflammatory molecules were also found to contribute significantly to the emergence of MDR sepsis. Variations in ESBL production between *E. coli* and *K. pneumoniae* in early and late-onset sepsis, along with a higher prevalence of MDROs in late-onset sepsis, highlight temporal and pathogen-specific resistance dynamics. These insights emphasize the multifaceted nature of antibiotic resistance in sepsis, demanding tailored treatment strategies that account for microbial and patient-specific complexities.

Timely diagnosis and management of sepsis are crucial for minimizing mortality rates. Previous studies have focused on conventional blood culturing techniques for appropriate empiric antibiotic therapy; however, diagnosing sepsis is still under debate among researchers due to the infectious as well as non-infectious nature of sepsis. For instance, antibiotic therapy can be applicable only to patients with infectious causes of sepsis. However, culturing techniques take a longer time to detect pathogens, thereby increasing the mortality rates of sepsis patients [90]. In this regard, various researchers mentioned CRP, WBC count, and PCT tests as significant biomarkers for diagnosing sepsis and initiating prompt antibiotic therapy [94]. Unfortunately, studies have found variable results in serological tests. They identified a lower to normal WBC count even in patients with culture-positive bacteremia [96]. Similarly, a higher neutrophil-to-lymphocyte count was considered a positive biomarker for infectious sepsis; however, many patients with even non-infectious sepsis were identified to have higher neutrophil-to-lymphocyte counts [102]. In addition to having low sensitivity and specificity, inconsistent findings were observed in the case of CRP [106] and PCT levels, making them less valuable for diagnosing sepsis [110]. Considering these issues, researchers have shifted their focus to identifying novel approaches for pathogen detection to diagnose sepsis. For instance, SERS [114] and MALDI-TOF MS techniques [115,116] have shown promising results in diagnosing sepsis. However, unfortunately, AMR mechanisms cannot be detected by MALDI-TOF MS [118], thus requiring an advanced system capable of identifying microbial AMR genes. In this regard, researchers have designed MALDI-TOF MS systems that incorporate PCR for microbial amplification before MS detection. These systems have a higher diagnostic potential than conventional culture techniques [119], and they can even detect microorganisms that do not normally grow in blood cultures, including *Mycoplasma pneumoniae*, *Rickettsia typhi*, *Legionella pneumophila*, *Nocardia* spp., and various fungi [120]. However, only a few studies have confirmed their usefulness over conventional cultures [121]. Therefore, further research is needed to develop technologically advanced systems to rapidly and accurately identify microbial genes associated with multidrug resistance in sepsis patients.

Additionally, the optimal POCT tests are rapid and extensive in providing sufficient information regarding pathogen and host–response virtually anywhere in a very short time [126,127]. Although these novel diagnostic techniques are promising, they have certain limitations. They can currently detect a limited number of resistance biomarkers. Hence, additional research is required to use these tests as a standard for diagnosing sepsis.

12. Strengths and Limitations

Focusing on MDR sepsis in relation to healthcare settings is one of the key strengths of this review. While previous reviews have examined aspects of MDR sepsis, none have focused on the increasing problem of MDROs or the healthcare burden of MDR sepsis. Another strength of this review was its rigorous approach to finding studies focusing on multiple aspects of MDR sepsis, from sepsis epidemiology and healthcare burden and cost to the evaluation of different pathogens involved and their resistance mechanisms. This review aimed to address a comprehensive range of pertinent topics related to MDR sepsis.

The type of evidence obtained determines the limitations of this review. These were geographically distributed and the result of various methodological approaches. None of the included studies utilized an in-depth qualitative approach to investigate the complete spectrum of factors that may affect the patient's experience of care. Moreover, there was a lack of studies describing the significance of MDR sepsis and its impact on healthcare in the subcontinent region, particularly in India and Pakistan. Finally, only peer-reviewed and published research was included because it was deemed to be of the highest quality [237–240]. Gray literature was not included; this may be regarded as a further limitation because a deeper search in this direction may have generated additional material that could have contributed to this review and expanded its scope.

13. Future Directions in Research and Therapeutics

The field of sepsis care has seen a significant transformation over the past few decades, notably in therapeutic approaches. There have been various advancements, ranging from more precise therapies and the creation of new drugs to the discovery of novel alternative antibiotics. These recent advancements indicate an ever-changing landscape on the cusp of redefining sepsis care. Developing cutting-edge novel therapies and medicines is a step in the right direction, holding promising outcomes. Researchers have investigated various immunomodulatory drugs and targeted therapies to disrupt the complicated mechanisms that drive the advancement of sepsis [241]. These therapies have the potential to tip the scales in favor of patient recovery since they focus on reducing excessive inflammatory responses. However, as the case of drotrecogin-α demonstrates, turning scientific promise into clinical success is complicated and requires careful inspection [130].

The escalation of AMR necessitates the development of novel antibiotic resistance mitigation techniques. The failure of traditional therapy can be because the pathophysiology of sepsis is the consequence of a highly complex series of mechanisms in which a dysregulated host response produces cellular damage, tissue damage, and, eventually, organ failure. Therefore, to enhance the effectiveness of antibiotics, it is advisable to employ adjunct therapies that complement antibiotic treatment, such as improving supportive care, targeting bacterial virulence factors, and targeting host response factors. Supportive care involves employing oxygenation or ventilation strategies or optimizing fluid or vasopressor use based on patient-specific characteristics. Bacterial virulence factors can be targeted by using anti-endotoxin antibodies or endotoxin removal columns [242,243]. Hemadsorption methods, such as polymyxin B adsorption, are an example of an endotoxin removal column that has shown potential for filtering out endotoxins and creating new pathways to neutralize the detrimental effects of septic shock [244].

Similarly, host response factors can be targeted using anticoagulants or anti-cytokine drugs [243]. Contrarily, researchers have identified various other options that can serve as substitutes for antibiotic therapy. For instance, phage therapy may cause distinct forms of immunomodulation in successive phases of sepsis. Interestingly, the possible applicability of phages [and their enzymes, lysins] in treating sepsis has previously been supported by animal and clinical experiments [245]. However, these techniques require rigorous validation and incorporation into comprehensive care paradigms.

Immune-based therapies to alleviate the sepsis burden have consistently failed to improve patient outcomes. Recent advancements in immune medicine against cancer and the realization that extended immunosuppression in sepsis patients can leave them susceptible to secondary infection and mortality have prompted a renewal of sepsis immune therapy research. Earlier immune treatments were based on targeting a single mediator and were administered to varied patient groups with complicated and dynamic immune responses. In this regard, personalized immune therapy is on the rise due to advances in genomics, proteomics, metabolomics, and point-of-care technologies, together with an enhanced knowledge of sepsis pathophysiology [246]. During the past decade, changing preferences from immunosuppressive medications to immunostimulatory treatments have displayed promising effects in preclinical studies, case series, and small clinical investigations [247]. For instance, immunostimulatory agents such as interferon-gamma [IFNγ] and granulocyte–macrophage colony-stimulating factor (GM-CSF) have been the most widely researched in sepsis. These compounds exhibit robust potential for stimulating myeloid cell activity: they improve the antigen presentation capabilities by enhancing the monocyte human leukocyte antigen-DR [mHLA-DR] gene and the production of pro-inflammatory cytokines by monocytes [248]. A randomized controlled trial, guided by biomarkers that are associated with GM-CSF, compared its effects against a placebo and found it to be safe and effective in restoring monocytic immunocompetence. The treatment positively impacted sepsis patients, with preliminary results indicating that the mechanical ventilation time was reduced and disease severity declined swiftly in treated individuals [249]. The advent of precision medicine is ushering in an era of personalized therapies tailored to the individual [250]. Immune-based treatments, such as monoclonal and polyclonal antibodies and other immunomodulators tailored to the specific patient profile, represent a paradigm shift [251]. Utilizing the body's natural defense, these treatments aim to retune immunological responses, minimizing the damage caused by sepsis. As we decipher the complex interplay between individual genetics, immunological state, and treatment results, the potential of precision medicine emerges as a beacon of hope. Compared to traditional immunosuppressive therapies, precision medicine is pivotal for the success of upcoming immunostimulatory drug studies. Consequently, it is crucial to identify individuals with a highly repressed or overactive immune system who are expected to benefit from immunosuppressive and immunostimulatory medication and to evaluate the immunological and therapeutic response accurately [252].

14. Conclusions

Sepsis is a complicated disruption of immunologic equilibrium, highlighting its complexity and the intricate link between immune function and clinical symptoms. The mortality, morbidity, and economic impact of sepsis are global concerns. Non-target-specific antibiotic therapy, misuse or abuse of antibiotic therapy, polypharmacy, and inadequate empiric antibiotic therapy may favor the emergence of MDROs, thereby having substantially adverse ecological side effects and economic burdens. That is why patients with infectious sepsis, particularly those harboring MDROs, have a higher risk of hospital-associated mortality. Antimicrobial resistance, including MDR pathogens, challenges treatment efficacy, increases the risk of adverse effects, and hampers treatment success. Antimicrobial resistance determines treatment ineffectiveness in clinical settings, leading to rapid advancement to sepsis and septic shock. Multidisciplinary strategies for timely diagnosis and application of appropriate antimicrobial treatment are critical in managing septic patients and limiting sepsis-related complications. Therefore, there is an urgent need for early administration of antimicrobials and organ support due to the time-dependent nature and severity of sepsis. Further, researchers should focus on developing diagnostic methods such as POCT that would detect sepsis early in the infection to avoid critical damage to organs and identifying better and more effective alternatives to antibiotics, such as phage therapy, immune-based therapies involving monoclonal and polyclonal antibodies, and precision medicine. Healthcare stakeholders must prioritize early and adequate administration of antimicrobials, preferably within the first hour of diagnosis, along with organ support. Public health organizations like the WHO collaborate with worldwide organizations and stakeholders to improve the treatment of sepsis and infection prevention control, including vaccinations, which should yield maximum outputs. With technological advancements, the POCT's role in bedside detection of sepsis is substantially increasing. The use of nanoparticles and immune-based therapeutics, in combination with precision medicine, is an important field of research for healthcare providers, including physicians, pharmacists, and microbiologists. Furthermore, addressing the challenges associated with AMR is essential to ensure effective treatment and minimize adverse effects.

Author Contributions: N.R.K. and A.P. conceived the idea for the manuscript and wrote the initial and final drafts. N.R.K. and S.N.K. undertook the literature search; T.A.B. provided important intellectual content and critically revised the manuscript. N.R.K., A.P., S.N.K. and T.A.B., participated in at least one round of the process and contributed to the preparation of the manuscript. All authors have read and agreed to the published version of the manuscript.

Funding: This research received no external funding.

Conflicts of Interest: The authors declare that they have no competing interests.

List of Abbreviations

MDROs	Multidrug resistance organisms
MDR	Multidrug-resistant
POCT	Point-of-care testing
MRSA	Methicillin-resistant Staphylococcus aureus
CRE	Carbapenem-resistant Enterobacteriaceae
ICU	Intensive care unit
WHO	World Health Organization
NMSS	National Mortality Surveillance System
ESBLs	Extended spectrum-beta-lactamase
COVID-19	Coronavirus disease 2019
AMR	Antimicrobial resistance
ESKAPE	*Enterococcus faecium*, *S. aureus*, *Acinetobacter baumannii*, *Klebsiella pneumonia*, *Enterobacter species*, and *Pseudomonas aeruginosa*
AMEs	Aminoglycoside modifying enzymes
PBPs	penicillin-binding proteins
ARDS	Acute respiratory distress syndrome
AKI	Acute kidney injury
DAMPs	Danger-associated molecular patterns
DNA	Deoxyribonucleic
RNA	Ribonucleic acid
HMGB1	High-mobility group box-1 protein
HSPs	Heat shock proteins
LPS	Lipopolysaccharide
LTA	Lipoteichoic acid
PAMPs	Pathogen-associated molecular patterns
APCs	Antigen-presenting cells
PPR	Pattern recognition receptors
TLRs	Toll-like receptors
IFNs	Interferons
NF-κB	Nuclear factor-κB
IRF	Interferon regulatory factor
TNF	Tumor necrosis factor
IL	Interleukins
BSI	Bloodstream infection
MERS	Middle East respiratory syndrome-related
SARS	Severe acute respiratory syndrome
PCT	Procalcitonin
CRP	C-reactive protein
WBC	White blood cells
SIRS	Systemic Inflammatory Response Syndrome

AUROC	Area under the receiver operating characteristic
SERS	Surface-enhanced Raman spectroscopy
MALDI-TOF MS	Matrix-assisted laser desorption ionization time-of-flight mass spectrometry
PCR	Polymerase chain reaction
MS	Mass spectrometry
SCC	Surviving Sepsis Campaign
AS	Antimicrobial stewardship
IDSA	Infectious Diseases Society of America
ESICM	European Society of Intensive Care Medicine
VAP	Ventilator-associated pneumonia
HAP	hospital-acquired pneumonia
ASP	Antimicrobial Stewardship Program
IPC	Infection prevention and control
CDC	Centers for Disease Control and Prevention
AwaRe	Access, Watch, and Reserve
HCPs	Healthcare providers
VRE	Vancomycin-resistant Enterobacterales
IFNγ	interferon-gamma
GM-CSF	Granulocyte–macrophage colony-stimulating factor
mHLA-DR	monocyte human leukocyte antigen-DR
CARB	Combating Antibiotic-Resistant Bacteria
GSA	Global Sepsis Alliance
WHA	World Health Assembly
WSD	World Sepsis Day
SSCM	Society of Critical Care Medicine
LMIC	Low and middle-income countries

References

1. Jarczak, D.; Kluge, S.; Nierhaus, A. Sepsis-Pathophysiology and Therapeutic Concepts. *Front. Med.* **2021**, *8*, 628302. [CrossRef] [PubMed]
2. Singer, M.; Deutschman, C.S.; Seymour, C.W.; Shankar-Hari, M.; Annane, D.; Bauer, M.; Bellomo, R.; Bernard, G.R.; Chiche, J.D.; Coopersmith, C.M.; et al. The Third International Consensus Definitions for Sepsis and Septic Shock (Sepsis-3). *JAMA* **2016**, *315*, 801–810. [CrossRef] [PubMed]
3. Chousterman, B.G.; Swirski, F.K.; Weber, G.F. Cytokine Storm and Sepsis Disease Pathogenesis. *Semin. Immunopathol.* **2017**, *39*, 517–528. [CrossRef] [PubMed]
4. Rubio, I.; Osuchowski, M.F.; Shankar-Hari, M.; Skirecki, T.; Winkler, M.S.; Lachmann, G.; La Rosée, P.; Monneret, G.; Venet, F.; Bauer, M.; et al. Current Gaps in Sepsis Immunology: New Opportunities for Translational Research. *Lancet. Infect. Dis.* **2019**, *19*, e422–e436. [CrossRef] [PubMed]
5. Hotchkiss, R.S.; Moldawer, L.L.; Opal, S.M.; Reinhart, K.; Turnbull, I.R.; Vincent, J.L. Sepsis and Septic Shock. *Nat. Rev. Dis. Primers* **2016**, *2*, 16045. [CrossRef] [PubMed]
6. Tamayo, E.; Fernández, A.; Almansa, R.; Carrasco, E.; Heredia, M.; Lajo, C.; Goncalves, L.; Gómez-Herreras, J.I.; de Lejarazu, R.O.; Bermejo-Martin, J.F. Pro- and Anti-Inflammatory Responses Are Regulated Simultaneously from the First Moments of Septic Shock. *Eur. Cytokine Netw.* **2011**, *22*, 82–87. [CrossRef] [PubMed]
7. Andaluz-Ojeda, D.; Bobillo, F.; Iglesias, V.; Almansa, R.; Rico, L.; Gandía, F.; Resino, S.; Tamayo, E.; de Lejarazu, R.O.; Bermejo-Martin, J.F. A Combined Score of Pro- and Anti-Inflammatory Interleukins Improves Mortality Prediction in Severe Sepsis. *Cytokine* **2012**, *57*, 332–336. [CrossRef]
8. Chaudhry, H.; Zhou, J.; Zhong, Y.; Ali, M.M.; McGuire, F.; Nagarkatti, P.S.; Nagarkatti, M. Role Of Cytokines As A Double-Edged Sword In Sepsis. *In Vivo* **2013**, *27*, 669–684.
9. Tang, B.M.; Huang, S.J.; McLean, A.S. Genome-Wide Transcription Profiling of Human Sepsis: A Systematic Review. *Crit. Care* **2010**, *14*, R237. [CrossRef]
10. Reinhart, K.; Daniels, R.; Kissoon, N.; Machado, F.R.; Schachter, R.D.; Finfer, S. Recognizing Sepsis as a Global Health Priority—A WHO Resolution. *N. Engl. J. Med.* **2017**, *377*, 414–417. [CrossRef]
11. Rudd, K.E.; Johnson, S.C.; Agesa, K.M.; Shackelford, K.A.; Tsoi, D.; Kievlan, D.R.; Colombara, D.V.; Ikuta, K.S.; Kissoon, N.; Finfer, S. Global, Regional, and National Sepsis Incidence and Mortality, 1990–2017: Analysis for the Global Burden of Disease Study. *J. Lancet* **2020**, *395*, 200–211. [CrossRef]
12. Torio, C.M.; Moore, B.J. *National Inpatient Hospital Costs: The Most Expensive Conditions by Payer*; Agency for Healthcare Research and Quality (US): Rockville, MD, USA, 2016.
13. Iwashyna, T.J.; Ely, E.W.; Smith, D.M.; Langa, K.M. Long-Term Cognitive Impairment and Functional Disability among Survivors of Severe Sepsis. *JAMA* **2010**, *304*, 1787–1794. [CrossRef] [PubMed]

14. Nasa, P.; Juneja, D.; Singh, O.; Dang, R.; Arora, V. Severe Sepsis and Its Impact on Outcome in Elderly and Very Elderly Patients Admitted in Intensive Care Unit. *J. Intensive Care Med.* **2012**, *27*, 179–183. [CrossRef] [PubMed]
15. Wan Muhd Shukeri, W.F.; Mat Nor, M.B.; Md Ralib, A. Sepsis and Its Impact on Outcomes in Elderly Patients Admitted to a Malaysian Intensive Care Unit. *Malays. J. Med. Sci. MJMS* **2022**, *29*, 145–150. [CrossRef] [PubMed]
16. Plata-Menchaca, E.P.; Ferrer, R.; Ruiz Rodríguez, J.C.; Morais, R.; Póvoa, P. Antibiotic Treatment in Patients with Sepsis: A Narrative Review. *Hosp. Pract.* **2022**, *50*, 203–213. [CrossRef] [PubMed]
17. Zhu, Y.; Huang, W.E.; Yang, Q. Clinical Perspective of Antimicrobial Resistance in Bacteria. *Infect. Drug Resist.* **2022**, *15*, 735–746. [CrossRef] [PubMed]
18. Ugwu, M.C.; Shariff, M.; Nnajide, C.M.; Beri, K.; Okezie, U.M.; Iroha, I.R.; Esimone, C.O. Phenotypic and Molecular Characterization of β-Lactamases among Enterobacterial Uropathogens in Southeastern Nigeria. *Can. J. Infect. Dis. Med. Microbiol.* **2020**, *2020*, 5843904. [CrossRef] [PubMed]
19. Tacconelli, E.; Carrara, E.; Savoldi, A.; Harbarth, S.; Mendelson, M.; Monnet, D.L.; Pulcini, C.; Kahlmeter, G.; Kluytmans, J.; Carmeli, Y.; et al. Discovery, Research, and Development of New Antibiotics: The WHO Priority List of Antibiotic-Resistant Bacteria and Tuberculosis. *Lancet Infect. Dis.* **2018**, *18*, 318–327. [CrossRef] [PubMed]
20. Dutescu, I.A.; Hillier, S.A. Encouraging the Development of New Antibiotics: Are Financial Incentives the Right Way Forward? A Systematic Review and Case Study. *Infect. Drug Resist.* **2021**, *14*, 415–434. [CrossRef]
21. Saleem, N. Antibiotics Modulate Variable Immunological Responses in Sepsis—A Narrative Review. *Preprints* **2022**, 2022100218.
22. Efunshile, A.M.; Ezeanosike, O.; Nwangwu, C.C.; König, B.; Jokelainen, P.; Robertson, L.J. Apparent Overuse of Antibiotics in the Management of Watery Diarrhoea in Children in Abakaliki, Nigeria. *BMC Infect. Dis.* **2019**, *19*, 275. [CrossRef] [PubMed]
23. Murray, C.J.L.; Ikuta, K.S.; Sharara, F.; Swetschinski, L.; Robles Aguilar, G.; Gray, A.; Han, C.; Bisignano, C.; Rao, P.; Wool, E.; et al. Global Burden of Bacterial Antimicrobial Resistance in 2019: A Systematic Analysis. *Lancet* **2022**, *399*, 629–655. [CrossRef] [PubMed]
24. Belachew, S.A.; Hall, L.; Selvey, L.A. Non-Prescription Dispensing of Antibiotic Agents among Community Drug Retail Outlets in Sub-Saharan African Countries: A Systematic Review and Meta-Analysis. *Antimicrob. Resist. Infect. Control* **2021**, *10*, 13. [CrossRef] [PubMed]
25. Suy, S.; Rego, S.; Bory, S.; Chhorn, S.; Phou, S.; Prien, C.; Heng, S.; Wu, S.; Legido-Quigley, H.; Hanefeld, J.; et al. Invisible Medicine Sellers and Their Use of Antibiotics: A Qualitative Study in Cambodia. *BMJ Glob. Health* **2019**, *4*, e001787. [CrossRef] [PubMed]
26. Tangcharoensathien, V.; Chanvatik, S.; Sommanustweechai, A. Complex Determinants of Inappropriate Use of Antibiotics. *Bull. World Health Organ.* **2018**, *96*, 141–144. [CrossRef] [PubMed]
27. Garau, J. Impact of Antibiotic Restrictions: The Ethical Perspective. *Clin. Microbiol. Infect.* **2006**, *12*, 16–24. [CrossRef] [PubMed]
28. Evans, L.; Rhodes, A.; Alhazzani, W.; Antonelli, M.; Coopersmith, C.M.; French, C.; Machado, F.R.; McIntyre, L.; Ostermann, M.; Prescott, H.C.; et al. Surviving Sepsis Campaign: International Guidelines for Management of Sepsis and Septic Shock 2021. *Intensive Care Med.* **2021**, *47*, 1181–1247. [CrossRef] [PubMed]
29. Chiu, C.; Legrand, M. Epidemiology of Sepsis and Septic Shock. *Curr. Opin. Anaesthesiol.* **2021**, *34*, 71–76. [CrossRef]
30. World Health Organization. *Global Report on the Epidemiology and Burden of Sepsis: Current Evidence, Identifying Gaps and Future Directions*; World Health Organization: Geneva, Switzerland, 2020.
31. Fleischmann, C.; Scherag, A.; Adhikari, N.K.J.; Hartog, C.S.; Tsaganos, T.; Schlattmann, P.; Angus, D.C.; Reinhart, K. Assessment of Global Incidence and Mortality of Hospital-Treated Sepsis. Current Estimates and Limitations. *Am. J. Respir. Crit. Care Med.* **2015**, *193*, 259–272. [CrossRef]
32. Sakr, Y.; Jaschinski, U.; Wittebole, X.; Szakmany, T.; Lipman, J.; Ñamendys-Silva, S.A.; Martin-Loeches, I.; Leone, M.; Lupu, M.-N.; Vincent, J.-L.; et al. Sepsis in Intensive Care Unit Patients: Worldwide Data from the Intensive Care over Nations Audit. *Open Forum Infect. Dis.* **2018**, *5*, ofy313. [CrossRef]
33. Todi, S.; Chatterjee, S.; Bhattacharyya, M. Epidemiology of Severe Sepsis in India. *Crit. Care* **2007**, *11*, P65. [CrossRef]
34. Anand, A.; Kumar, N.; Gambhir, I. Clinicomicrobiological Profile of the Indian Elderly with Sepsis. *Ann. Trop. Med. Public Health* **2016**, *9*, 316–320. [CrossRef]
35. Chatterjee, S.; Bhattacharya, M.; Todi, S.K. Epidemiology of Adult-Population Sepsis in India: A Single Center 5 Year Experience. *Indian J. Crit. Care Med.* **2017**, *21*, 573–577. [CrossRef] [PubMed]
36. Prabhudev, P.; Ramamoorthi, K.; Acharya, R.V. A Clinical and Demographic Profile of Elderly (>65 Years) in the Medical Intensive Care Units of a Tertiary Care Center. *Indian J. Crit. Care Med.* **2023**, *27*, 166–175. [CrossRef] [PubMed]
37. Weng, L.; Zeng, X.Y.; Yin, P.; Wang, L.J.; Wang, C.Y.; Jiang, W.; Zhou, M.G.; Du, B. Sepsis-Related Mortality in China: A Descriptive Analysis. *Intensive Care Med.* **2018**, *44*, 1071–1080. [CrossRef] [PubMed]
38. Weng, L.; Xu, Y.; Yin, P.; Wang, Y.; Chen, Y.; Liu, W.; Li, S.; Peng, J.-M.; Dong, R.; Hu, X.-Y.; et al. National Incidence and Mortality of Hospitalized Sepsis in China. *Crit. Care* **2023**, *27*, 84. [CrossRef] [PubMed]
39. Fleischmann-Struzek, C.; Mellhammar, L.; Rose, N.; Cassini, A.; Rudd, K.E.; Schlattmann, P.; Allegranzi, B.; Reinhart, K. Incidence and Mortality of Hospital- and ICU-Treated Sepsis: Results from an Updated and Expanded Systematic Review and Meta-Analysis. *Intensive Care Med.* **2020**, *46*, 1552–1562. [CrossRef] [PubMed]

40. Marchaim, D.; Gottesman, T.; Schwartz, O.; Korem, M.; Maor, Y.; Rahav, G.; Karplus, R.; Lazarovitch, T.; Braun, E.; Sprecher, H.; et al. National Multicenter Study of Predictors and Outcomes of Bacteremia upon Hospital Admission Caused by Enterobacteriaceae Producing Extended-Spectrum Beta-Lactamases. *Antimicrob. Agents Chemother.* **2010**, *54*, 5099–5104. [CrossRef]
41. Trecarichi, E.M.; Cauda, R.; Tumbarello, M. Detecting Risk and Predicting Patient Mortality in Patients with Extended-Spectrum β-Lactamase-Producing Enterobacteriaceae Bloodstream Infections. *Future Microbiol.* **2012**, *7*, 1173–1189. [CrossRef]
42. Sibila, O.; Rodrigo-Troyano, A.; Shindo, Y.; Aliberti, S.; Restrepo, M.I. Multidrug-Resistant Pathogens in Patients with Pneumonia Coming from the Community. *Curr. Opin. Pulm. Med.* **2016**, *22*, 219–226. [CrossRef]
43. Paul, M.; Shani, V.; Muchtar, E.; Kariv, G.; Robenshtok, E.; Leibovici, L. Systematic Review and Meta-Analysis of the Efficacy of Appropriate Empiric Antibiotic Therapy for Sepsis. *Antimicrob. Agents Chemother.* **2010**, *54*, 4851–4863. [CrossRef] [PubMed]
44. Kollef, M.H.; Zilberberg, M.D.; Shorr, A.F.; Vo, L.; Schein, J.; Micek, S.T.; Kim, M. Epidemiology, Microbiology and Outcomes of Healthcare-Associated and Community-Acquired Bacteremia: A Multicenter Cohort Study. *J. Infect.* **2011**, *62*, 130–135. [CrossRef] [PubMed]
45. European Centre for Disease Prevention and Control; World Health Organization Regional Office for Europe. *Antimicrobial Resistance Surveillance in Europe 2022–2020 Data*; World Health Organization Regional Office for Europe: Copenhagen, Denmark, 2022.
46. Capsoni, N.; Bellone, P.; Aliberti, S.; Sotgiu, G.; Pavanello, D.; Visintin, B.; Callisto, E.; Saderi, L.; Soldini, D.; Lardera, L.; et al. Prevalence, Risk Factors and Outcomes of Patients Coming from the Community with Sepsis Due to Multidrug Resistant Bacteria. *Multidiscip. Respir. Med.* **2019**, *14*, 23. [CrossRef] [PubMed]
47. Liu, V.; Escobar, G.J.; Greene, J.D.; Soule, J.; Whippy, A.; Angus, D.C.; Iwashyna, T.J. Hospital Deaths in Patients with Sepsis from 2 Independent Cohorts. *JAMA* **2014**, *312*, 90–92. [CrossRef] [PubMed]
48. Rubens, M.; Saxena, A.; Ramamoorthy, V.; Das, S.; Khera, R.; Hong, J.; Armaignac, D.; Veledar, E.; Nasir, K.; Gidel, L. Increasing Sepsis Rates in the United States: Results From National Inpatient Sample, 2005 to 2014. *J. Intensive Care Med.* **2020**, *35*, 858–868. [CrossRef] [PubMed]
49. Garg, P.; Krishak, R.; Shukla, D.K. NICU in a Community Level Hospital. *Indian J. Pediatr.* **2005**, *72*, 27–30. [CrossRef] [PubMed]
50. Jayaram, R.; Ramakrishnan, N. Cost of Intensive Care in India. *Indian J. Crit. Care Med.* **2008**, *12*, 55–61. [CrossRef] [PubMed]
51. Farrah, K.; McIntyre, L.; Doig, C.J.; Talarico, R.; Taljaard, M.; Krahn, M.; Fergusson, D.; Forster, A.J.; Coyle, D.; Thavorn, K. Sepsis-associated Mortality, Resource Use, and Healthcare Costs: A Propensity-Matched Cohort Study. *J. Crit. Care Med.* **2021**, *49*, 215–227. [CrossRef]
52. Oami, T.; Imaeda, T.; Nakada, T.a.; Abe, T.; Takahashi, N.; Yamao, Y.; Nakagawa, S.; Ogura, H.; Shime, N.; Umemura, Y.; et al. Temporal Trends of Medical Cost and Cost-Effectiveness in Sepsis Patients: A Japanese Nationwide Medical Claims Database. *J. Intensive Care* **2022**, *10*, 33. [CrossRef]
53. Lewis, K. The science of antibiotic discovery. *Cell* **2020**, *181*, 29–45. [CrossRef]
54. Lebreton, F.; Manson, A.L.; Saavedra, J.T.; Straub, T.J.; Earl, A.M.; Gilmore, M.S. Tracing the Enterococci from Paleozoic Origins to the Hospital. *Cell* **2017**, *169*, 849–861.e813. [CrossRef] [PubMed]
55. Du, D.; Wang-Kan, X.; Neuberger, A.; Van Veen, H.W.; Pos, K.M.; Piddock, L.J.; Luisi, B.F. Multidrug Efflux Pumps: Structure, Function and Regulation. *J Nat. Rev. Microbiol.* **2018**, *16*, 523–539. [CrossRef] [PubMed]
56. Reygaert, W.C. An Overview of the Antimicrobial Resistance Mechanisms of Bacteria. *AIMS Microbiol.* **2018**, *4*, 482–501. [CrossRef] [PubMed]
57. Friedrich, A.W. Control of Hospital Acquired Infections and Antimicrobial Resistance in Europe: The Way to Go. *Wien. Med. Wochenschr.* **2019**, *169*, 25–30. [CrossRef] [PubMed]
58. Pandey, R.; Mishra, S.K.; Shrestha, A. Characterisation of ESKAPE Pathogens with Special Reference to Multidrug Resistance and Biofilm Production in a Nepalese Hospital. *Infect. Drug Resist.* **2021**, *14*, 2201–2212. [CrossRef] [PubMed]
59. Kyriakidis, I.; Vasileiou, E.; Pana, Z.D.; Tragiannidis, A. Acinetobacter baumannii Antibiotic Resistance Mechanisms. *Pathogens* **2021**, *10*, 373. [CrossRef] [PubMed]
60. Recio, R.; Mancheño, M.; Viedma, E.; Villa, J.; Orellana, M.; Lora-Tamayo, J.; Chaves, F. Predictors of Mortality in Bloodstream Infections Caused by *Pseudomonas aeruginosa* and Impact of Antimicrobial Resistance and Bacterial Virulence. *Antimicrob. Agents Chemother.* **2020**, *64*, e01759-19. [CrossRef]
61. Tong, S.Y.; Davis, J.S.; Eichenberger, E.; Holland, T.L.; Fowler, V.G., Jr. *Staphylococcus aureus* Infections: Epidemiology, Pathophysiology, Clinical Manifestations, and Management. *Clin. Microbiol. Rev.* **2015**, *28*, 603–661. [CrossRef]
62. Turner, N.A.; Sharma-Kuinkel, B.K.; Maskarinec, S.A.; Eichenberger, E.M.; Shah, P.P.; Carugati, M.; Holland, T.L.; Fowler, V.G., Jr. Methicillin-Resistant *Staphylococcus aureus*: An Overview of Basic and Clinical Research. *Nat. Rev. Microbiol.* **2019**, *17*, 203–218. [CrossRef]
63. Timsina, R.; Shrestha, U.; Singh, A.; Timalsina, B. Inducible Clindamycin Resistance and Erm Genes in *Staphylococcus aureus* in School Children in Kathmandu, Nepal. *Future Sci. OA* **2020**, *7*, Fso361. [CrossRef]
64. Boomer, J.S.; Green, J.M.; Hotchkiss, R.S. The Changing Immune System in Sepsis: Is Individualized Immuno-Modulatory Therapy the Answer? *Virulence* **2014**, *5*, 45–56. [CrossRef] [PubMed]
65. Venet, F.; Monneret, G. Advances in the Understanding and Treatment of Sepsis-Induced Immunosuppression. *Nat. Rev. Nephrol.* **2018**, *14*, 121–137. [CrossRef] [PubMed]
66. Becattini, S.; Taur, Y.; Pamer, E.G. Antibiotic-Induced Changes in the Intestinal Microbiota and Disease. *Trends Mol. Med.* **2016**, *22*, 458–478. [CrossRef] [PubMed]

67. Girardis, M.; Cossarizza, A. Early Alterations of B Cells in Patients with Septic Shock: Another Piece in the Complex Puzzle of the Immune Response in Sepsis. *Crit. Care* **2013**, *17*, 162. [CrossRef] [PubMed]
68. Vakkalanka, J.P.; Harland, K.K.; Swanson, M.B.; Mohr, N.M. Clinical and Epidemiological Variability in Severe Sepsis: An Ecological Study. *J. Epidemiol. Community Health* **2018**, *72*, 741–745. [CrossRef] [PubMed]
69. Varghese, D.; Ishida, C.; Haseer Koya, H. *Polypharmacy*; StatPearls Publishing: Treasure Island, FL, USA, 2022.
70. Macias, J.; Kahly, O.; Edward, R.; Khan, S.; Qureshi, A.; Shaik, A.; Shala, A.; Shah, D. Sepsis: A Systematic Review of Antibiotic Resistance and Antimicrobial Therapies. *J. Mod. Res. Inflamm.* **2022**, *11*, 9–23. [CrossRef]
71. Reinhart, K. How Antimicrobial Resistance Amplifies the Burden of Sepsis. Available online: https://www.news-medical.net/health/How-Antimicrobial-Resistance-Amplifies-the-Burden-of-Sepsis.aspx (accessed on 15 August 2023).
72. Hurst, J. Invasive Fungal Infections Can Lead to Sepsis—And Have a High Mortality Rate. Available online: https://www.biomerieuxconnection.com/2019/09/24/invasive-fungal-infections-can-lead-to-sepsis-and-have-a-high-mortality-rate/ (accessed on 15 August 2023).
73. Gotts, J.E.; Matthay, M.A. Sepsis: Pathophysiology and Clinical Management. *BMJ* **2016**, *353*, i1585. [CrossRef]
74. Esper, A.M.; Moss, M.; Lewis, C.A.; Nisbet, R.; Mannino, D.M.; Martin, G.S. The Role of Infection and Comorbidity: Factors that Influence Disparities in Sepsis. *Crit. Care Med.* **2006**, *34*, 2576–2582. [CrossRef]
75. Al-Hasan, M.N.; Wilson, J.W.; Lahr, B.D.; Eckel-Passow, J.E.; Baddour, L.M. Incidence of *Pseudomonas aeruginosa* bacteremia: A Population-Based Study. *Am. J. Med.* **2008**, *121*, 702–708. [CrossRef]
76. Laupland, K.B.; Gregson, D.B.; Church, D.L.; Ross, T.; Pitout, J.D. Incidence, Risk Factors and Outcomes of *Escherichia coli* Bloodstream Infections in a Large Canadian Region. *Clin. Microbiol. Infect.* **2008**, *14*, 1041–1047. [CrossRef]
77. Ulu Kilic, A.; Alp, E.; Cevahir, F.; Ture, Z.; Yozgat, N. Epidemiology and Cost Implications of Candidemia, a 6-Year Analysis from a Developing Country. *Mycoses* **2017**, *60*, 198–203. [CrossRef] [PubMed]
78. Horn, D.L.; Neofytos, D.; Anaissie, E.J.; Fishman, J.A.; Steinbach, W.J.; Olyaei, A.J.; Marr, K.A.; Pfaller, M.A.; Chang, C.H.; Webster, K.M. Epidemiology and Outcomes of Candidemia in 2019 Patients: Data from the Prospective Antifungal Therapy Alliance Registry. *Clin. Infect. Dis.* **2009**, *48*, 1695–1703. [CrossRef] [PubMed]
79. Degoricija, V.; Sharma, M.; Legac, A.; Gradiser, M.; Sefer, S.; Vucicević, Z. Survival Analysis of 314 Episodes of Sepsis in Medical Intensive Care Unit in University Hospital: Impact of Intensive Care Unit Performance and Antimicrobial Therapy. *Croat. Med. J.* **2006**, *47*, 385–397. [PubMed]
80. Garnacho-Montero, J.; Ortiz-Leyba, C.; Herrera-Melero, I.; Aldabó-Pallás, T.; Cayuela-Dominguez, A.; Marquez-Vacaro, J.A.; Carbajal-Guerrero, J.; Garcia-Garmendia, J.L. Mortality and Morbidity Attributable to Inadequate Empirical Antimicrobial Therapy in Patients Admitted to the ICU with Sepsis: A Matched Cohort Study. *J. Antimicrob. Chemother.* **2008**, *61*, 436–441. [CrossRef] [PubMed]
81. Pradipta, I.S.; Sodik, D.C.; Lestari, K.; Parwati, I.; Halimah, E.; Diantini, A.; Abdulah, R. Antibiotic Resistance in Sepsis Patients: Evaluation and Recommendation of Antibiotic Use. *North Am. J. Med. Sci.* **2013**, *5*, 344–352. [CrossRef] [PubMed]
82. Carbajal-Guerrero, J.; Cayuela-Domínguez, A.; Fernández-García, E.; Aldabó-Pallás, T.; Márquez-Vácaro, J.A.; Ortiz-Leyba, C.; Garnacho-Montero, J. Epidemiology and Long-Term Outcome of Sepsis in Elderly Patients. *Med. Intensiv.* **2014**, *38*, 21–32. [CrossRef] [PubMed]
83. Martin, G.S.; Mannino, D.M.; Eaton, S.; Moss, M. The Epidemiology of Sepsis in the United States from 1979 through 2000. *N. Engl. J. Med.* **2003**, *348*, 1546–1554. [CrossRef] [PubMed]
84. Wichmann, M.W.; Inthorn, D.; Andress, H.J.; Schildberg, F.W. Incidence and Mortality of Severe Sepsis in Surgical Intensive Care Patients: The Influence of Patient Gender on Disease Process and Outcome. *Intensive Care Med.* **2000**, *26*, 167–172. [CrossRef]
85. Adrie, C.; Azoulay, E.; Francais, A.; Clec'h, C.; Darques, L.; Schwebel, C.; Nakache, D.; Jamali, S.; Goldgran-Toledano, D.; Garrouste-Orgeas, M.; et al. Influence of Gender on the Outcome of Severe Sepsis: A Reappraisal. *Chest* **2007**, *132*, 1786–1793. [CrossRef]
86. Berkowitz, D.M.; Martin, G.S. Sepsis and Sex: Can We Look beyond Our Hormones? *Chest* **2007**, *132*, 1725–1727. [CrossRef]
87. Kumar, A.; Roberts, D.; Wood, K.E.; Light, B.; Parrillo, J.E.; Sharma, S.; Suppes, R.; Feinstein, D.; Zanotti, S.; Taiberg, L.; et al. Duration of Hypotension before Initiation of Effective Antimicrobial Therapy Is the Critical Determinant of Survival in Human Septic Shock. *Crit. Care Med.* **2006**, *34*, 1589–1596. [CrossRef]
88. Li, Y.; Guo, J.; Yang, H.; Li, H.; Shen, Y.; Zhang, D. Comparison of Culture-Negative and Culture-Positive Sepsis or Septic Shock: A Systematic Review and Meta-Analysis. *Crit. Care* **2021**, *25*, 167. [CrossRef] [PubMed]
89. Blackburn, R.M.; Muller-Pebody, B.; Planche, T.; Johnson, A.; Hopkins, S.; Sharland, M.; Kennea, N.; Heath, P.T. Neonatal Sepsis--Many Blood Samples, Few Positive Cultures: Implications for Improving Antibiotic Prescribing. *Arch. Dis. Child. Fetal Neonatal Ed.* **2012**, *97*, F487–F488. [CrossRef] [PubMed]
90. Cohen, J.; Brun-Buisson, C.; Torres, A.; Jorgensen, J. Diagnosis of Infection in Sepsis: An Evidence-Based Review. *Crit. Care Med.* **2004**, *32*, S466–S494. [CrossRef] [PubMed]
91. Cheng, M.P.; Stenstrom, R.; Paquette, K.; Stabler, S.N.; Akhter, M.; Davidson, A.C.; Gavric, M.; Lawandi, A.; Jinah, R.; Saeed, Z.; et al. Blood Culture Results Before and After Antimicrobial Administration in Patients With Severe Manifestations of Sepsis: A Diagnostic Study. *Ann. Intern. Med.* **2019**, *171*, 547–554. [CrossRef] [PubMed]

92. Scheer, C.S.; Fuchs, C.; Gründling, M.; Vollmer, M.; Bast, J.; Bohnert, J.A.; Zimmermann, K.; Hahnenkamp, K.; Rehberg, S.; Kuhn, S.O. Impact of Antibiotic Administration on Blood Culture Positivity at the Beginning of Sepsis: A Prospective Clinical Cohort Study. *Clin. Microbiol. Infect.* **2019**, *25*, 326–331. [CrossRef] [PubMed]
93. Lin, G.L.; McGinley, J.P.; Drysdale, S.B.; Pollard, A.J. Epidemiology and Immune Pathogenesis of Viral Sepsis. *Front. Immunol.* **2018**, *9*, 2147. [CrossRef] [PubMed]
94. Bone, R.C.; Balk, R.A.; Cerra, F.B.; Dellinger, R.P.; Fein, A.M.; Knaus, W.A.; Schein, R.M.; Sibbald, W.J. Definitions for sepsis and organ failure and guidelines for the use of innovative therapies in sepsis. The ACCP/SCCM Consensus Conference Committee. American College of Chest Physicians/Society of Critical Care Medicine. *Chest* **1992**, *101*, 1644–1655. [CrossRef]
95. Marik, P.E.; Stephenson, E. The Ability of Procalcitonin, Lactate, White Blood Cell Count and Neutrophil-Lymphocyte Count Ratio to Predict Blood Stream Infection. Analysis of a Large Database. *J. Crit. Care* **2020**, *60*, 135–139. [CrossRef]
96. Seigel, T.A.; Cocchi, M.N.; Salciccioli, J.; Shapiro, N.I.; Howell, M.; Tang, A.; Donnino, M.W. Inadequacy of temperature and white blood cell count in predicting bacteremia in patients with suspected infection. *J. Emerg. Med.* **2012**, *42*, 254–259. [CrossRef]
97. Van den Bruel, A.; Thompson, M.J.; Haj-Hassan, T.; Stevens, R.; Moll, H.; Lakhanpaul, M.; Mant, D. Diagnostic value of laboratory tests in identifying serious infections in febrile children: Systematic review. *BMJ* **2011**, *342*, d3082. [CrossRef] [PubMed]
98. Panagiotopoulou, I.G.; Parashar, D.; Lin, R.; Antonowicz, S.; Wells, A.D.; Bajwa, F.M.; Krijgsman, B. The diagnostic value of white cell count, C-reactive protein and bilirubin in acute appendicitis and its complications. *Ann. R. Coll. Surg. Engl.* **2013**, *95*, 215–221. [CrossRef] [PubMed]
99. Farkas, J.D. The complete blood count to diagnose septic shock. *J. Thorac. Dis.* **2020**, *12*, S16–S21. [CrossRef] [PubMed]
100. Azab, B.; Camacho-Rivera, M.; Taioli, E. Average values and racial differences of neutrophil lymphocyte ratio among a nationally representative sample of United States subjects. *PLoS ONE* **2014**, *9*, e112361. [CrossRef] [PubMed]
101. Ljungström, L.; Pernestig, A.K.; Jacobsson, G.; Andersson, R.; Usener, B.; Tilevik, D. Diagnostic accuracy of procalcitonin, neutrophil-lymphocyte count ratio, C-reactive protein, and lactate in patients with suspected bacterial sepsis. *PLoS ONE* **2017**, *12*, e0181704. [CrossRef] [PubMed]
102. Liu, J.; Liu, Y.; Xiang, P.; Pu, L.; Xiong, H.; Li, C.; Zhang, M.; Tan, J.; Xu, Y.; Song, R.; et al. Neutrophil-to-lymphocyte ratio predicts critical illness patients with 2019 coronavirus disease in the early stage. *J. Transl. Med.* **2020**, *18*, 206. [CrossRef] [PubMed]
103. Westerdijk, K.; Simons, K.S.; Zegers, M.; Wever, P.C.; Pickkers, P.; de Jager, C.P.C. The value of the neutrophil-lymphocyte count ratio in the diagnosis of sepsis in patients admitted to the Intensive Care Unit: A retrospective cohort study. *PLoS ONE* **2019**, *14*, e0212861. [CrossRef]
104. Tan, M.; Lu, Y.; Jiang, H.; Zhang, L. The diagnostic accuracy of procalcitonin and C-reactive protein for sepsis: A systematic review and meta-analysis. *J. Cell. Biochem.* **2019**, *120*, 5852–5859. [CrossRef]
105. Stirling, A.D.; Moran, N.R.; Kelly, M.E.; Ridgway, P.F.; Conlon, K.C. The predictive value of C-reactive protein (CRP) in acute pancreatitis—Is interval change in CRP an additional indicator of severity? *HPB* **2017**, *19*, 874–880. [CrossRef]
106. Khanna, A.K.; Meher, S.; Prakash, S.; Tiwary, S.K.; Singh, U.; Srivastava, A.; Dixit, V.K. Comparison of Ranson, Glasgow, MOSS, SIRS, BISAP, APACHE-II, CTSI Scores, IL-6, CRP, and Procalcitonin in Predicting Severity, Organ Failure, Pancreatic Necrosis, and Mortality in Acute Pancreatitis. *HPB Surg.* **2013**, *2013*, 367581. [CrossRef]
107. Póvoa, P.; Coelho, L.; Almeida, E.; Fernandes, A.; Mealha, R.; Moreira, P.; Sabino, H. Early identification of intensive care unit-acquired infections with daily monitoring of C-reactive protein: A prospective observational study. *Crit. Care* **2006**, *10*, R63. [CrossRef] [PubMed]
108. Assicot, M.; Gendrel, D.; Carsin, H.; Raymond, J.; Guilbaud, J.; Bohuon, C. High serum procalcitonin concentrations in patients with sepsis and infection. *Lancet* **1993**, *341*, 515–518. [CrossRef] [PubMed]
109. Wacker, C.; Prkno, A.; Brunkhorst, F.M.; Schlattmann, P. Procalcitonin as a diagnostic marker for sepsis: A systematic review and meta-analysis. *Lancet Infect. Dis.* **2013**, *13*, 426–435. [CrossRef] [PubMed]
110. Kamat, I.S.; Ramachandran, V.; Eswaran, H.; Guffey, D.; Musher, D.M. Procalcitonin to Distinguish Viral From Bacterial Pneumonia: A Systematic Review and Meta-analysis. *Clin. Infect. Dis.* **2020**, *70*, 538–542. [CrossRef]
111. Joo, K.; Park, W.; Lim, M.J.; Kwon, S.R.; Yoon, J. Serum procalcitonin for differentiating bacterial infection from disease flares in patients with autoimmune diseases. *J. Korean Med. Sci.* **2011**, *26*, 1147–1151. [CrossRef]
112. El-Sayed, D.; Grotts, J.; Golgert, W.A.; Sugar, A.M. Sensitivity and specificity of procalcitonin in predicting bacterial infections in patients with renal impairment. *Open Forum Infect. Dis.* **2014**, *1*, ofu068. [CrossRef]
113. Hoeboer, S.H.; van der Geest, P.J.; Nieboer, D.; Groeneveld, A.B. The diagnostic accuracy of procalcitonin for bacteraemia: A systematic review and meta-analysis. *Clin. Microbiol. Infect.* **2015**, *21*, 474–481. [CrossRef]
114. Jeong, K.; Stanwix, P.L.; May, E.F.; Aman, Z.M. Surface-Enhanced Raman Scattering Imaging of Cetylpyridinium Chloride Adsorption to a Solid Surface. *J Anal. Chem.* **2022**, *94*, 14169–14176. [CrossRef]
115. Delport, J.A.; Strikwerda, A.; Armstrong, A.; Schaus, D.; John, M. MALDI-ToF short incubation identification from blood cultures is associated with reduced length of hospitalization and a decrease in bacteremia associated mortality. *Eur. J. Clin. Microbiol. Infect. Dis.* **2017**, *36*, 1181–1186. [CrossRef]
116. Beganovic, M.; Costello, M.; Wieczorkiewicz, S.M. Effect of Matrix-Assisted Laser Desorption Ionization-Time of Flight Mass Spectrometry (MALDI-TOF MS) Alone versus MALDI-TOF MS Combined with Real-Time Antimicrobial Stewardship Interventions on Time to Optimal Antimicrobial Therapy in Patients with Positive Blood Cultures. *J. Clin. Microbiol.* **2017**, *55*, 1437–1445. [CrossRef]

117. Patel, T.S.; Kaakeh, R.; Nagel, J.L.; Newton, D.W.; Stevenson, J.G. Cost Analysis of Implementing Matrix-Assisted Laser Desorption Ionization-Time of Flight Mass Spectrometry Plus Real-Time Antimicrobial Stewardship Intervention for Bloodstream Infections. *J. Clin. Microbiol.* **2017**, *55*, 60–67. [CrossRef] [PubMed]
118. Luethy, P.M.; Johnson, J.K. The Use of Matrix-Assisted Laser Desorption/Ionization Time-of-Flight Mass Spectrometry (MALDI-TOF MS) for the Identification of Pathogens Causing Sepsis. *J. Appl. Lab. Med.* **2019**, *3*, 675–685. [CrossRef] [PubMed]
119. Westh, H.; Lisby, G.; Breysse, F.; Böddinghaus, B.; Chomarat, M.; Gant, V.; Goglio, A.; Raglio, A.; Schuster, H.; Stuber, F.; et al. Multiplex real-time PCR and blood culture for identification of bloodstream pathogens in patients with suspected sepsis. *Clin. Microbiol. Infect.* **2009**, *15*, 544–551. [CrossRef] [PubMed]
120. Mustafa, M.I.; Al-Marzooq, F.; How, S.H.; Kuan, Y.C.; Ng, T.H. The use of multiplex real-time PCR improves the detection of the bacterial etiology of community acquired pneumonia. *Trop. Biomed.* **2011**, *28*, 531–544. [PubMed]
121. van de Groep, K.; Bos, M.P.; Varkila, M.R.J.; Savelkoul, P.H.M.; Ong, D.S.Y.; Derde, L.P.G.; Juffermans, N.P.; van der Poll, T.; Bonten, M.J.M.; Cremer, O.L. Moderate positive predictive value of a multiplex real-time PCR on whole blood for pathogen detection in critically ill patients with sepsis. *Eur. J. Clin. Microbiol. Infect. Dis.* **2019**, *38*, 1829–1836. [CrossRef] [PubMed]
122. Sinha, M.; Jupe, J.; Mack, H.; Coleman, T.P.; Lawrence, S.M.; Fraley, S.I. Emerging Technologies for Molecular Diagnosis of Sepsis. *Clin. Microbiol. Rev.* **2018**, *31*, e00089-17. [CrossRef] [PubMed]
123. Levy, M.M.; Gesten, F.C.; Phillips, G.S.; Terry, K.M.; Seymour, C.W.; Prescott, H.C.; Friedrich, M.; Iwashyna, T.J.; Osborn, T.; Lemeshow, S. Mortality Changes Associated with Mandated Public Reporting for Sepsis. The Results of the New York State Initiative. *Am. J. Respir. Crit. Care Med.* **2018**, *198*, 1406–1412. [CrossRef] [PubMed]
124. Kumar, V. Targeting macrophage immunometabolism: Dawn in the darkness of sepsis. *Int. Immunopharmacol.* **2018**, *58*, 173–185. [CrossRef]
125. Rello, J.; van Engelen, T.S.R.; Alp, E.; Calandra, T.; Cattoir, V.; Kern, W.V.; Netea, M.G.; Nseir, S.; Opal, S.M.; van de Veerdonk, F.L.; et al. Towards precision medicine in sepsis: A position paper from the European Society of Clinical Microbiology and Infectious Diseases. *Clin. Microbiol. Infect.* **2018**, *24*, 1264–1272. [CrossRef]
126. Antonioli, L.; Blandizzi, C.; Fornai, M.; Pacher, P.; Lee, H.T.; Haskó, G. P2X4 receptors, immunity, and sepsis. *Curr. Opin. Pharmacol.* **2019**, *47*, 65–74. [CrossRef]
127. Mirasoli, M.; Bonvicini, F.; Lovecchio, N.; Petrucci, G.; Zangheri, M.; Calabria, D.; Costantini, F.; Roda, A.; Gallinella, G.; Caputo, D.; et al. On-chip LAMP-BART reaction for viral DNA real-time bioluminescence detection. *Sens. Actuators B Chem.* **2018**, *262*, 1024–1033. [CrossRef]
128. Nasseri, B.; Soleimani, N.; Rabiee, N.; Kalbasi, A.; Karimi, M.; Hamblin, M.R. Point-of-care microfluidic devices for pathogen detection. *Biosens. Bioelectron.* **2018**, *117*, 112–128. [CrossRef] [PubMed]
129. Sprung, C.L.; Annane, D.; Keh, D.; Moreno, R.; Singer, M.; Freivogel, K.; Weiss, Y.G.; Benbenishty, J.; Kalenka, A.; Forst, H.; et al. Hydrocortisone therapy for patients with septic shock. *N. Engl. J. Med.* **2008**, *358*, 111–124. [CrossRef] [PubMed]
130. Ranieri, V.M.; Thompson, B.T.; Barie, P.S.; Dhainaut, J.F.; Douglas, I.S.; Finfer, S.; Gårdlund, B.; Marshall, J.C.; Rhodes, A.; Artigas, A.; et al. Drotrecogin alfa (activated) in adults with septic shock. *N. Engl. J. Med.* **2012**, *366*, 2055–2064. [CrossRef] [PubMed]
131. Plata-Menchaca, E.P.; Ferrer, R. Life-support tools for improving performance of the Surviving Sepsis Campaign Hour-1 bundle. *Med. Intensiv.* **2018**, *42*, 547–550. [CrossRef]
132. Rhodes, A.; Evans, L.E.; Alhazzani, W.; Levy, M.M.; Antonelli, M.; Ferrer, R.; Kumar, A.; Sevransky, J.E.; Sprung, C.L.; Nunnally, M.E.; et al. Surviving Sepsis Campaign: International Guidelines for Management of Sepsis and Septic Shock: 2016. *Intensive Care Med.* **2017**, *43*, 304–377. [CrossRef]
133. Kumar, A.; Ellis, P.; Arabi, Y.; Roberts, D.; Light, B.; Parrillo, J.E.; Dodek, P.; Wood, G.; Kumar, A.; Simon, D.; et al. Initiation of inappropriate antimicrobial therapy results in a fivefold reduction of survival in human septic shock. *Chest* **2009**, *136*, 1237–1248. [CrossRef]
134. Ferrer, R.; Martin-Loeches, I.; Phillips, G.; Osborn, T.M.; Townsend, S.; Dellinger, R.P.; Artigas, A.; Schorr, C.; Levy, M.M. Empiric antibiotic treatment reduces mortality in severe sepsis and septic shock from the first hour: Results from a guideline-based performance improvement program. *Crit. Care Med.* **2014**, *42*, 1749–1755. [CrossRef]
135. Liu, V.X.; Fielding-Singh, V.; Greene, J.D.; Baker, J.M.; Iwashyna, T.J.; Bhattacharya, J.; Escobar, G.J. The Timing of Early Antibiotics and Hospital Mortality in Sepsis. *Am. J. Respir. Crit. Care Med.* **2017**, *196*, 856–863. [CrossRef]
136. Singer, M. Antibiotics for Sepsis: Does Each Hour Really Count, or Is It Incestuous Amplification? *Am. J. Respir. Crit. Care Med.* **2017**, *196*, 800–802. [CrossRef]
137. Mi, M.Y.; Klompas, M.; Evans, L. Early Administration of Antibiotics for Suspected Sepsis. *N. Engl. J. Med.* **2019**, *380*, 593–596. [CrossRef] [PubMed]
138. Ulldemolins, M.; Nuvials, X.; Palomar, M.; Masclans, J.R.; Rello, J. Appropriateness is critical. *Crit. Care Clin.* **2011**, *27*, 35–51. [CrossRef] [PubMed]
139. Levy, M.M.; Rhodes, A.; Phillips, G.S.; Townsend, S.R.; Schorr, C.A.; Beale, R.; Osborn, T.; Lemeshow, S.; Chiche, J.D.; Artigas, A.; et al. Surviving Sepsis Campaign: Association between performance metrics and outcomes in a 7.5-year study. *Crit. Care Med.* **2015**, *43*, 3–12. [CrossRef]
140. Coz Yataco, A.; Jaehne, A.K.; Rivers, E.P. Protocolized Early Sepsis Care Is Not Only Helpful for Patients: It Prevents Medical Errors. *Crit. Care Med.* **2017**, *45*, 464–472. [CrossRef] [PubMed]

141. Garnacho-Montero, J.; Gutiérrez-Pizarraya, A.; Escoresca-Ortega, A.; Fernández-Delgado, E.; López-Sánchez, J.M. Adequate antibiotic therapy prior to ICU admission in patients with severe sepsis and septic shock reduces hospital mortality. *Crit. Care* **2015**, *19*, 302. [CrossRef] [PubMed]
142. Sherwin, R.; Winters, M.E.; Vilke, G.M.; Wardi, G. Does Early and Appropriate Antibiotic Administration Improve Mortality in Emergency Department Patients with Severe Sepsis or Septic Shock? *J. Emerg. Med.* **2017**, *53*, 588–595. [CrossRef] [PubMed]
143. Suberviola Cañas, B.; Jáuregui, R.; Ballesteros, M.; Leizaola, O.; González-Castro, A.; Castellanos-Ortega, Á. Effects of antibiotic administration delay and inadequacy upon the survival of septic shock patients. *Med. Intensiv.* **2015**, *39*, 459–466. [CrossRef]
144. Zilberberg, M.D.; Shorr, A.F.; Micek, S.T.; Vazquez-Guillamet, C.; Kollef, M.H. Multi-drug resistance, inappropriate initial antibiotic therapy and mortality in Gram-negative severe sepsis and septic shock: A retrospective cohort study. *Crit. Care* **2014**, *18*, 596. [CrossRef]
145. Ferrer, R.; Martínez, M.L.; Gomà, G.; Suárez, D.; Álvarez-Rocha, L.; de la Torre, M.V.; González, G.; Zaragoza, R.; Borges, M.; Blanco, J.; et al. Improved empirical antibiotic treatment of sepsis after an educational intervention: The ABISS-Edusepsis study. *Crit. Care* **2018**, *22*, 167. [CrossRef]
146. Green, R.S.; Gorman, S.K. Emergency department antimicrobial considerations in severe sepsis. *Emerg. Med. Clin. N. Am.* **2014**, *32*, 835–849. [CrossRef]
147. Hamandi, B.; Holbrook, A.M.; Humar, A.; Brunton, J.; Papadimitropoulos, E.A.; Wong, G.G.; Thabane, L. Delay of adequate empiric antibiotic therapy is associated with increased mortality among solid-organ transplant patients. *Am. J. Transplant.* **2009**, *9*, 1657–1665. [CrossRef]
148. Garnacho-Montero, J.; García-Cabrera, E.; Diaz-Martín, A.; Lepe-Jiménez, J.A.; Iraurgi-Arcarazo, P.; Jiménez-Alvarez, R.; Revuelto-Rey, J.; Aznar-Martín, J. Determinants of outcome in patients with bacteraemic pneumococcal pneumonia: Importance of early adequate treatment. *Scand. J. Infect. Dis.* **2010**, *42*, 185–192. [CrossRef] [PubMed]
149. Harmankaya, M.; Oreskov, J.O.; Burcharth, J.; Gögenur, I. The impact of timing of antibiotics on in-hospital outcomes after major emergency abdominal surgery. *Eur. J. Trauma Emerg. Surg.* **2020**, *46*, 221–227. [CrossRef] [PubMed]
150. Puskarich, M.A.; Trzeciak, S.; Shapiro, N.I.; Arnold, R.C.; Horton, J.M.; Studnek, J.R.; Kline, J.A.; Jones, A.E. Association between timing of antibiotic administration and mortality from septic shock in patients treated with a quantitative resuscitation protocol. *Crit. Care Med.* **2011**, *39*, 2066–2071. [CrossRef] [PubMed]
151. Zhang, D.; Micek, S.T.; Kollef, M.H. Time to Appropriate Antibiotic Therapy Is an Independent Determinant of Postinfection ICU and Hospital Lengths of Stay in Patients With Sepsis. *Crit. Care Med.* **2015**, *43*, 2133–2140. [CrossRef] [PubMed]
152. Bagshaw, S.M.; Lapinsky, S.; Dial, S.; Arabi, Y.; Dodek, P.; Wood, G.; Ellis, P.; Guzman, J.; Marshall, J.; Parrillo, J.E.; et al. Acute kidney injury in septic shock: Clinical outcomes and impact of duration of hypotension prior to initiation of antimicrobial therapy. *Intensive Care Med.* **2009**, *35*, 871–881. [CrossRef] [PubMed]
153. Iscimen, R.; Cartin-Ceba, R.; Yilmaz, M.; Khan, H.; Hubmayr, R.D.; Afessa, B.; Gajic, O. Risk factors for the development of acute lung injury in patients with septic shock: An observational cohort study. *Crit. Care Med.* **2008**, *36*, 1518–1522. [CrossRef] [PubMed]
154. Hwang, S.Y.; Shin, J.; Jo, I.J.; Park, J.E.; Yoon, H.; Cha, W.C.; Sim, M.S.; Shin, T.G. Delayed Antibiotic Therapy and Organ Dysfunction in Critically Ill Septic Patients in the Emergency Department. *J. Clin. Med.* **2019**, *8*, 222. [CrossRef]
155. Labelle, A.; Juang, P.; Reichley, R.; Micek, S.; Hoffmann, J.; Hoban, A.; Hampton, N.; Kollef, M. The determinants of hospital mortality among patients with septic shock receiving appropriate initial antibiotic treatment. *Crit. Care Med.* **2012**, *40*, 2016–2021. [CrossRef]
156. van Paridon, B.M.; Sheppard, C.; Joffe, A.R.; Alberta Sepsis Network. Timing of antibiotics, volume, and vasoactive infusions in children with sepsis admitted to intensive care. *Crit. Care* **2015**, *19*, 293. [CrossRef]
157. Alam, N.; Oskam, E.; Stassen, P.M.; Exter, P.V.; van de Ven, P.M.; Haak, H.R.; Holleman, F.; Zanten, A.V.; Leeuwen-Nguyen, H.V.; Bon, V.; et al. Prehospital antibiotics in the ambulance for sepsis: A multicentre, open label, randomised trial. *Lancet Respir. Med.* **2018**, *6*, 40–50. [CrossRef] [PubMed]
158. Sterling, S.A.; Miller, W.R.; Pryor, J.; Puskarich, M.A.; Jones, A.E. The Impact of Timing of Antibiotics on Outcomes in Severe Sepsis and Septic Shock: A Systematic Review and Meta-Analysis. *Crit. Care Med.* **2015**, *43*, 1907–1915. [CrossRef] [PubMed]
159. Johnston, A.N.B.; Park, J.; Doi, S.A.; Sharman, V.; Clark, J.; Robinson, J.; Crilly, J. Effect of Immediate Administration of Antibiotics in Patients With Sepsis in Tertiary Care: A Systematic Review and Meta-analysis. *Clin. Ther.* **2017**, *39*, 190–202.e196. [CrossRef] [PubMed]
160. Vincent, J.L.; Bassetti, M.; François, B.; Karam, G.; Chastre, J.; Torres, A.; Roberts, J.A.; Taccone, F.S.; Rello, J.; Calandra, T.; et al. Advances in antibiotic therapy in the critically ill. *Crit. Care* **2016**, *20*, 133. [CrossRef] [PubMed]
161. Force, I.S.T. Infectious Diseases Society of America (IDSA) POSITION STATEMENT: Why IDSA Did Not Endorse the Surviving Sepsis Campaign Guidelines. *Clin. Infect. Dis.* **2017**, *66*, 1631–1635. [CrossRef] [PubMed]
162. Septimus, E.J.; Coopersmith, C.M.; Whittle, J.; Hale, C.P.; Fishman, N.O.; Kim, T.J. Sepsis National Hospital Inpatient Quality Measure (SEP-1): Multistakeholder Work Group Recommendations for Appropriate Antibiotics for the Treatment of Sepsis. *Clin. Infect. Dis.* **2017**, *65*, 1565–1569. [CrossRef] [PubMed]
163. Delannoy, P.Y.; Boussekey, N.; Devos, P.; Alfandari, S.; Turbelin, C.; Chiche, A.; Meybeck, A.; Georges, H.; Leroy, O. Impact of combination therapy with aminoglycosides on the outcome of ICU-acquired bacteraemias. *Eur. J. Clin. Microbiol. Infect. Dis.* **2012**, *31*, 2293–2299. [CrossRef]

164. Micek, S.T.; Welch, E.C.; Khan, J.; Pervez, M.; Doherty, J.A.; Reichley, R.M.; Kollef, M.H. Empiric combination antibiotic therapy is associated with improved outcome against sepsis due to Gram-negative bacteria: A retrospective analysis. *Antimicrob. Agents Chemother.* **2010**, *54*, 1742–1748. [CrossRef]
165. Kumar, A.; Safdar, N.; Kethireddy, S.; Chateau, D. A survival benefit of combination antibiotic therapy for serious infections associated with sepsis and septic shock is contingent only on the risk of death: A meta-analytic/meta-regression study. *Crit. Care Med.* **2010**, *38*, 1651–1664. [CrossRef]
166. Ong, D.S.Y.; Frencken, J.F.; Klein Klouwenberg, P.M.C.; Juffermans, N.; van der Poll, T.; Bonten, M.J.M.; Cremer, O.L. Short-Course Adjunctive Gentamicin as Empirical Therapy in Patients With Severe Sepsis and Septic Shock: A Prospective Observational Cohort Study. *Clin. Infect. Dis.* **2017**, *64*, 1731–1736. [CrossRef]
167. Klompas, M. Monotherapy Is Adequate for Septic Shock Due to Gram-Negative Organisms. *Crit. Care Med.* **2017**, *45*, 1930–1932. [CrossRef]
168. Coopersmith, C.M.; De Backer, D.; Deutschman, C.S.; Ferrer, R.; Lat, I.; Machado, F.R.; Martin, G.S.; Martin-Loeches, I.; Nunnally, M.E.; Antonelli, M.; et al. Surviving Sepsis Campaign: Research Priorities for Sepsis and Septic Shock. *Crit. Care Med.* **2018**, *46*, 1334–1356. [CrossRef]
169. Bellomo, R.; Kellum, J.A.; Ronco, C.; Wald, R.; Martensson, J.; Maiden, M.; Bagshaw, S.M.; Glassford, N.J.; Lankadeva, Y.; Vaara, S.T.; et al. Acute kidney injury in sepsis. *Intensive Care Med.* **2017**, *43*, 816–828. [CrossRef] [PubMed]
170. Pearce, A.K.; McGuire, W.C.; Malhotra, A. Prone Positioning in Acute Respiratory Distress Syndrome. *NEJM Evid.* **2022**, *1*, EVIDra2100046. [CrossRef]
171. Spadaro, S. Multidrug Resistance in Critically Ill Patients: An Unresolved Issue. *Microorganisms* **2023**, *11*, 946. [CrossRef] [PubMed]
172. Dünser, M.W.; Festic, E.; Dondorp, A.; Kissoon, N.; Ganbat, T.; Kwizera, A.; Haniffa, R.; Baker, T.; Schultz, M.J. Recommendations for sepsis management in resource-limited settings. *Intensive Care Med.* **2012**, *38*, 557–574. [CrossRef] [PubMed]
173. Dellinger, R.P.; Levy, M.M.; Rhodes, A.; Annane, D.; Gerlach, H.; Opal, S.M.; Sevransky, J.E.; Sprung, C.L.; Douglas, I.S.; Jaeschke, R.; et al. Surviving sepsis campaign: International guidelines for management of severe sepsis and septic shock: 2012. *Crit. Care Med.* **2013**, *41*, 580–637. [CrossRef] [PubMed]
174. Klompas, M.; Branson, R.; Eichenwald, E.C.; Greene, L.R.; Howell, M.D.; Lee, G.; Magill, S.S.; Maragakis, L.L.; Priebe, G.P.; Speck, K.; et al. Strategies to prevent ventilator-associated pneumonia in acute care hospitals: 2014 update. *Infect. Control Hosp. Epidemiol.* **2014**, *35*, 915–936. [CrossRef] [PubMed]
175. Spadaro, S.; Berselli, A.; Fogagnolo, A.; Capuzzo, M.; Ragazzi, R.; Marangoni, E.; Bertacchini, S.; Volta, C.A. Evaluation of a protocol for vancomycin administration in critically patients with and without kidney dysfunction. *BMC Anesthesiol.* **2015**, *15*, 95. [CrossRef] [PubMed]
176. Schinas, G.; Polyzou, E.; Spernovasilis, N.; Gogos, C.; Dimopoulos, G.; Akinosoglou, K. Preventing Multidrug-Resistant Bacterial Transmission in the Intensive Care Unit with a Comprehensive Approach: A Policymaking Manual. *Antibiotics* **2023**, *12*, 1255. [CrossRef]
177. Norman, B.C.; Cooke, C.R.; Ely, E.W.; Graves, J.A. Sepsis-Associated 30-Day Risk-Standardized Readmissions: Analysis of a Nationwide Medicare Sample. *Crit. Care Med.* **2017**, *45*, 1130–1137. [CrossRef]
178. Liu, V.; Lei, X.; Prescott, H.C.; Kipnis, P.; Iwashyna, T.J.; Escobar, G.J. Hospital readmission and healthcare utilization following sepsis in community settings. *J. Hosp. Med.* **2014**, *9*, 502–507. [CrossRef]
179. DeMerle, K.M.; Royer, S.C.; Mikkelsen, M.E.; Prescott, H.C. Readmissions for Recurrent Sepsis: New or Relapsed Infection? *Crit. Care Med.* **2017**, *45*, 1702–1708. [CrossRef]
180. Goodwin, A.J.; Rice, D.A.; Simpson, K.N.; Ford, D.W. Frequency, cost, and risk factors of readmissions among severe sepsis survivors. *Crit. Care Med.* **2015**, *43*, 738–746. [CrossRef] [PubMed]
181. Prescott, H.C. Variation in Postsepsis Readmission Patterns: A Cohort Study of Veterans Affairs Beneficiaries. *Ann. Am. Thorac. Soc.* **2017**, *14*, 230–237. [CrossRef] [PubMed]
182. Ortego, A.; Gaieski, D.F.; Fuchs, B.D.; Jones, T.; Halpern, S.D.; Small, D.S.; Sante, S.C.; Drumheller, B.; Christie, J.D.; Mikkelsen, M.E. Hospital-based acute care use in survivors of septic shock. *Crit. Care Med.* **2015**, *43*, 729–737. [CrossRef] [PubMed]
183. Sun, A.; Netzer, G.; Small, D.S.; Hanish, A.; Fuchs, B.D.; Gaieski, D.F.; Mikkelsen, M.E. Association Between Index Hospitalization and Hospital Readmission in Sepsis Survivors. *Crit. Care Med.* **2016**, *44*, 478–487. [CrossRef]
184. Arshad, A.; Ayaz, A.; Haroon, M.A.; Jamil, B.; Hussain, E. Frequency and Cause of Readmissions in Sepsis Patients Presenting to a Tertiary Care Hospital in a Low Middle Income Country. *Crit. Care Explor.* **2020**, *2*, e0080. [CrossRef] [PubMed]
185. Liaskou, M.; Duggan, C.; Joynes, R.; Rosado, H. Pharmacy's role in antimicrobial resistance and stewardship. *Pharm. J.* **2018**, *10*. [CrossRef]
186. Dyar, O.J.; Huttner, B.; Schouten, J.; Pulcini, C. What is antimicrobial stewardship? *Clin. Microbiol. Infect.* **2017**, *23*, 793–798. [CrossRef]
187. Bondarenka, C.M.; Bosso, J.A. Successful Implementation of an Antimicrobial Stewardship Program at an Academic Medical Center. *Hosp. Pharm.* **2017**, *52*, 508–513. [CrossRef]
188. Zanichelli, V.; Sharland, M.; Cappello, B.; Moja, L.; Getahun, H.; Pessoa-Silva, C.; Sati, H.; van Weezenbeek, C.; Balkhy, H.; Simão, M.; et al. The WHO AWaRe (Access, Watch, Reserve) antibiotic book and prevention of antimicrobial resistance. *Bull. World Health Organ.* **2023**, *101*, 290–296. [CrossRef]

189. Majumder, M.A.A.; Rahman, S.; Cohall, D.; Bharatha, A.; Singh, K.; Haque, M.; Gittens-St Hilaire, M. Antimicrobial Stewardship: Fighting Antimicrobial Resistance and Protecting Global Public Health. *Infect. Drug Resist.* **2020**, *13*, 4713–4738. [CrossRef] [PubMed]
190. Ostrowsky, B.; Banerjee, R.; Bonomo, R.A.; Cosgrove, S.E.; Davidson, L.; Doron, S.; Gilbert, D.N.; Jezek, A.; Lynch, J.B., 3rd; Septimus, E.J.; et al. Infectious Diseases Physicians: Leading the Way in Antimicrobial Stewardship. *Clin. Infect. Dis.* **2018**, *66*, 995–1003. [CrossRef] [PubMed]
191. Trivedi, K.K.; Pollack, L.A. The role of public health in antimicrobial stewardship in healthcare. *Clin. Infect. Dis.* **2014**, *59* (Suppl. S3), S101–S103. [CrossRef] [PubMed]
192. Gai, Z.; Samodelov, S.L.; Kullak-Ublick, G.A.; Visentin, M. Molecular Mechanisms of Colistin-Induced Nephrotoxicity. *Molecules* **2019**, *24*, 653. [CrossRef] [PubMed]
193. Mulani, M.S.; Kamble, E.E.; Kumkar, S.N.; Tawre, M.S.; Pardesi, K.R. Emerging Strategies to Combat ESKAPE Pathogens in the Era of Antimicrobial Resistance: A Review. *Front. Microbiol.* **2019**, *10*, 539. [CrossRef] [PubMed]
194. Lloyd, D.H.; Page, S.W. Antimicrobial Stewardship in Veterinary Medicine. *Microbiol. Spectr.* **2018**, *6*. [CrossRef] [PubMed]
195. Doron, S.; Davidson, L.E. Antimicrobial stewardship. *Mayo Clin. Proc.* **2011**, *86*, 1113–1123. [CrossRef]
196. Dadgostar, P. Antimicrobial Resistance: Implications and Costs. *Infect. Drug Resist.* **2019**, *12*, 3903–3910. [CrossRef]
197. Bertollo, L.G.; Lutkemeyer, D.S.; Levin, A.S. Are antimicrobial stewardship programs effective strategies for preventing antibiotic resistance? A systematic review. *Am. J. Infect. Control* **2018**, *46*, 824–836. [CrossRef] [PubMed]
198. García-Rodríguez, J.F.; Bardán-García, B.; Peña-Rodríguez, M.F.; Álvarez-Díaz, H.; Mariño-Callejo, A. Meropenem antimicrobial stewardship program: Clinical, economic, and antibiotic resistance impact. *Eur. J. Clin. Microbiol. Infect. Dis.* **2019**, *38*, 161–170. [CrossRef] [PubMed]
199. Zequinao, T.; Gasparetto, J.; Oliveira, D.D.S.; Silva, G.T.; Telles, J.P.; Tuon, F.F. A broad-spectrum beta-lactam-sparing stewardship program in a middle-income country public hospital: Antibiotic use and expenditure outcomes and antimicrobial susceptibility profiles. *Braz. J. Infect. Dis.* **2020**, *24*, 221–230. [CrossRef] [PubMed]
200. Kolar, M.; Htoutou Sedlakova, M.; Urbanek, K.; Mlynarcik, P.; Roderova, M.; Hricova, K.; Mezerova, K.; Kucova, P.; Zapletalova, J.; Fiserova, K.; et al. Implementation of Antibiotic Stewardship in a University Hospital Setting. *Antibiotics* **2021**, *10*, 93. [CrossRef] [PubMed]
201. Strich, J.R.; Palmore, T.N. Preventing Transmission of Multidrug-Resistant Pathogens in the Intensive Care Unit. *Infect. Dis. Clin. N. Am.* **2017**, *31*, 535–550. [CrossRef] [PubMed]
202. Williams, V.R.; Callery, S.; Vearncombe, M.; Simor, A.E. The role of colonization pressure in nosocomial transmission of methicillin-resistant *Staphylococcus aureus*. *Am. J. Infect. Control* **2009**, *37*, 106–110. [CrossRef] [PubMed]
203. Masse, J.; Elkalioubie, A.; Blazejewski, C.; Ledoux, G.; Wallet, F.; Poissy, J.; Preau, S.; Nseir, S. Colonization pressure as a risk factor of ICU-acquired multidrug resistant bacteria: A prospective observational study. *Eur. J. Clin. Microbiol. Infect. Dis.* **2017**, *36*, 797–805. [CrossRef]
204. Shams, A.M.; Rose, L.J.; Edwards, J.R.; Cali, S.; Harris, A.D.; Jacob, J.T.; LaFae, A.; Pineles, L.L.; Thom, K.A.; McDonald, L.C.; et al. Assessment of the Overall and Multidrug-Resistant Organism Bioburden on Environmental Surfaces in Healthcare Facilities. *Infect. Control Hosp. Epidemiol.* **2016**, *37*, 1426–1432. [CrossRef]
205. Lyons, A.; Rose, L.; Noble-Wang, J.A. *Survival of Healthcare Pathogens on Hospital Surfaces*; SHEA: St. Louis, MO, USA, 2017.
206. Ohl, M.; Schweizer, M.; Graham, M.; Heilmann, K.; Boyken, L.; Diekema, D. Hospital privacy curtains are frequently and rapidly contaminated with potentially pathogenic bacteria. *Am. J. Infect. Control* **2012**, *40*, 904–906. [CrossRef]
207. Snitkin, E.S.; Zelazny, A.M.; Thomas, P.J.; Stock, F.; Henderson, D.K.; Palmore, T.N.; Segre, J.A. Tracking a hospital outbreak of carbapenem-resistant Klebsiella pneumoniae with whole-genome sequencing. *Sci. Transl. Med.* **2012**, *4*, 148ra116. [CrossRef]
208. Otter, J.A.; Mepham, S.; Athan, B.; Mack, D.; Smith, R.; Jacobs, M.; Hopkins, S. Terminal decontamination of the Royal Free London's high-level isolation unit after a case of Ebola virus disease using hydrogen peroxide vapor. *Am. J. Infect. Control* **2016**, *44*, 233–235. [CrossRef] [PubMed]
209. Kotay, S.; Chai, W.; Guilford, W.; Barry, K.; Mathers, A.J. Spread from the Sink to the Patient: In Situ Study Using Green Fluorescent Protein (GFP)-Expressing *Escherichia coli* To Model Bacterial Dispersion from Hand-Washing Sink-Trap Reservoirs. *Appl. Environ. Microbiol.* **2017**, *83*, e03327-16. [CrossRef] [PubMed]
210. Williams, M.M.; Armbruster, C.R.; Arduino, M.J. Plumbing of hospital premises is a reservoir for opportunistically pathogenic microorganisms: A review. *Biofouling* **2013**, *29*, 147–162. [CrossRef] [PubMed]
211. Bagnoli, F.; Payne, D.J. Reaction: Alternative Modalities to Address Antibiotic-Resistant Pathogens. *Chem* **2017**, *3*, 369–372. [CrossRef]
212. CDC. *Antibiotic Resistance Threats in the United States, 2019*; Control and Prevention Centers for Disease and Division of Healthcare Quality Promotion; Antibiotic Resistance and Coordination Strategy Unit National Center for Emerging Zoonotic and Infectious Diseases: Atlanta, GA, USA, 2019.
213. The White House. *National Action Plan for Combating Antibiotic-Resistant Bacteria*; The White House: Washington, DC, USA, 2015.
214. McEwen, S.A.; Collignon, P.J. Antimicrobial Resistance: A One Health Perspective. *Microbiol. Spectr.* **2018**, *6*. [CrossRef] [PubMed]
215. Bosserman, R.E.; Kwon, J.H. Know your Microbe Foes: The Role of Surveillance in Combatting Antimicrobial Resistance. *Yale J. Biol. Med.* **2022**, *95*, 517–523. [PubMed]

216. CDC. Core Elements of Hospital Antibiotic Stewardship Programs. US Department of Health and Human Services, CDC. Available online: https://www.cdc.gov/antibiotic-use/core-elements/hospital.html (accessed on 14 August 2023).
217. Turcotte, J.; Sanford, Z.; Broda, A.; Patton, C. Centers for Medicare & Medicaid Services Hierarchical Condition Category score as a predictor of readmission and reoperation following elective inpatient spine surgery. *J. Neurosurg. Spine* **2019**, *31*, 600–606. [CrossRef]
218. Pulcini, C. Antibiotic stewardship: A European perspective. *FEMS Microbiol. Lett.* **2017**, *364*, fnx230. [CrossRef]
219. Carter, R.R.; Sun, J.; Jump, R.L. A Survey and Analysis of the American Public's Perceptions and Knowledge About Antibiotic Resistance. *Open Forum Infect. Dis.* **2016**, *3*, ofw112. [CrossRef]
220. Hayes, J.F. Fighting Back against Antimicrobial Resistance with Comprehensive Policy and Education: A Narrative Review. *Antibiotics* **2022**, *11*, 644. [CrossRef]
221. WHO. Sepsis, Infection Prevention and Control. Available online: https://www.who.int/teams/integrated-health-services/infection-prevention-control/sepsis (accessed on 14 August 2023).
222. Stehr, S.N.; Reinhart, K. Sepsis as a Global Health Problem—Why We Need a Global Sepsis Alliance. *Shock* **2013**, *39*, 3–4. [CrossRef]
223. Rudd, K.E.; Kissoon, N.; Limmathurotsakul, D.; Bory, S.; Mutahunga, B.; Seymour, C.W.; Angus, D.C.; West, T.E. The global burden of sepsis: Barriers and potential solutions. *Crit. Care* **2018**, *22*, 232. [CrossRef] [PubMed]
224. Ranjit, S.; Kissoon, N. Challenges and Solutions in translating sepsis guidelines into practice in resource-limited settings. *Transl. Pediatr.* **2021**, *10*, 2646–2665. [CrossRef]
225. The Lancet Respiratory, M. The future of critical care: Lessons from the COVID-19 crisis. *Lancet Respir. Med.* **2020**, *8*, 527. [CrossRef]
226. Tiskumara, R.; Fakharee, S.H.; Liu, C.Q.; Nuntnarumit, P.; Lui, K.M.; Hammoud, M.; Lee, J.K.; Chow, C.B.; Shenoi, A.; Halliday, R.; et al. Neonatal infections in Asia. *Arch. Dis. Child. Fetal Neonatal Ed.* **2009**, *94*, F144–F148. [CrossRef] [PubMed]
227. Shim, G.H.; Kim, S.D.; Kim, H.S.; Kim, E.S.; Lee, H.J.; Lee, J.A.; Choi, C.W.; Kim, E.K.; Choi, E.H.; Kim, B.I.; et al. Trends in epidemiology of neonatal sepsis in a tertiary center in Korea: A 26-year longitudinal analysis, 1980–2005. *J. Korean Med. Sci.* **2011**, *26*, 284–289. [CrossRef] [PubMed]
228. Mintz, A.; Mor, M.; Klinger, G.; Scheuerman, O.; Pirogovsky, A.; Sokolover, N.; Bromiker, R. Changing epidemiology and resistance patterns of pathogens causing neonatal bacteremia. *Eur. J. Clin. Microbiol. Infect. Dis.* **2020**, *39*, 1879–1884. [CrossRef] [PubMed]
229. Investigators of the Delhi Neonatal Infection Study (DeNIS) Collaboration. Characterisation and antimicrobial resistance of sepsis pathogens in neonates born in tertiary care centres in Delhi, India: A cohort study. *Lancet Glob. Health* **2016**, *4*, e752–e760. [CrossRef] [PubMed]
230. Kalın, G.; Alp, E.; Chouaikhi, A.; Roger, C. Antimicrobial Multidrug Resistance: Clinical Implications for Infection Management in Critically Ill Patients. *Microorganisms* **2023**, *11*, 2575. [CrossRef]
231. Vincent, J.-L.; Rello, J.; Marshall, J.; Silva, E.; Anzueto, A.; Martin, C.D.; Moreno, R.; Lipman, J.; Gomersall, C.; Sakr, Y.; et al. International Study of the Prevalence and Outcomes of Infection in Intensive Care Units. *JAMA* **2009**, *302*, 2323–2329. [CrossRef]
232. Burnham, J.P.; Lane, M.A.; Kollef, M.H. Impact of Sepsis Classification and Multidrug-Resistance Status on Outcome Among Patients Treated With Appropriate Therapy. *Crit. Care Med.* **2015**, *43*, 1580–1586. [CrossRef] [PubMed]
233. Tseng, W.P.; Chen, Y.C.; Chen, S.Y.; Chen, S.Y.; Chang, S.C. Risk for subsequent infection and mortality after hospitalization among patients with multidrug-resistant gram-negative bacteria colonization or infection. *Antimicrob. Resist. Infect. Control* **2018**, *7*, 93. [CrossRef] [PubMed]
234. Chan, K.-G. Whole-genome sequencing in the prediction of antimicrobial resistance. *Expert Rev. Anti-Infect. Ther.* **2016**, *14*, 617–619. [CrossRef] [PubMed]
235. Potera, C. Forging a Link Between Biofilms and Disease. *Science* **1999**, *283*, 1837–1839. [CrossRef] [PubMed]
236. Di Martino, P. Extracellular polymeric substances, a key element in understanding biofilm phenotype. *AIMS Microbiol.* **2018**, *4*, 274–288. [CrossRef] [PubMed]
237. Hall, C.W.; Mah, T.-F. Molecular mechanisms of biofilm-based antibiotic resistance and tolerance in pathogenic bacteria. *FEMS Microbiol. Rev.* **2017**, *41*, 276–301. [CrossRef] [PubMed]
238. Sehgal, R.; Gaind, R.; Chellani, H.; Agarwal, P. Extended-spectrum beta lactamase-producing gram-negative bacteria: Clinical profile and outcome in a neonatal intensive care unit. *Ann. Trop. Paediatr.* **2007**, *27*, 45–54. [CrossRef]
239. Xie, J.; Li, S.; Xue, M.; Yang, C.; Huang, Y.; Chihade, D.B.; Liu, L.; Yang, Y.; Qiu, H. Early- and Late-Onset Bloodstream Infections in the Intensive Care Unit: A Retrospective 5-Year Study of Patients at a University Hospital in China. *J. Infect. Dis.* **2020**, *221*, S184–S192. [CrossRef]
240. Higgins, J.P.T.; Thomas, J.; Chandler, J.; Cumpston, M.; Li, T.; Page, M.J.; Welch, V.A. (Eds.) *Cochrane Handbook for Systematic Reviews of Interventions*; Cochrane: Oxford, UK, 2023.
241. Seymour, C.W.; Kennedy, J.N.; Wang, S.; Chang, C.H.; Elliott, C.F.; Xu, Z.; Berry, S.; Clermont, G.; Cooper, G.; Gomez, H.; et al. Derivation, Validation, and Potential Treatment Implications of Novel Clinical Phenotypes for Sepsis. *JAMA* **2019**, *321*, 2003–2017. [CrossRef]
242. Cohen, J. Non-antibiotic strategies for sepsis. *Clin. Microbiol. Infect.* **2009**, *15*, 302–307. [CrossRef]

243. Polat, G.; Ugan, R.A.; Cadirci, E.; Halici, Z. Sepsis and Septic Shock: Current Treatment Strategies and New Approaches. *Eurasian J. Med.* **2017**, *49*, 53–58. [CrossRef] [PubMed]
244. Sergio, L.; Guido, B.; Carlotta, R.; Fiorenza, F.; Michele, G.; Marco, P.; Giuseppe, R.; GiVi, T.I.G.I.p.l.V.d.I.i.T.I.i.a.i.c. Efficacy of coupled plasma filtration adsorption (CPFA) in patients with septic shock: A multicenter randomised controlled clinical trial. *BMJ Open* **2014**, *4*, e003536. [CrossRef]
245. Górski, A.; Jończyk-Matysiak, E.; Łusiak-Szelachowska, M.; Międzybrodzki, R.; Weber-Dąbrowska, B.; Borysowski, J. The Potential of Phage Therapy in Sepsis. *Front. Immunol.* **2017**, *8*, 1783. [CrossRef] [PubMed]
246. Davies, R.; O'Dea, K.; Gordon, A. Immune therapy in sepsis: Are we ready to try again? *J. Intensive Care Soc.* **2018**, *19*, 326–344. [CrossRef] [PubMed]
247. Delsing, C.E.; Gresnigt, M.S.; Leentjens, J.; Preijers, F.; Frager, F.A.; Kox, M.; Monneret, G.; Venet, F.; Bleeker-Rovers, C.P.; van de Veerdonk, F.L.; et al. Interferon-gamma as adjunctive immunotherapy for invasive fungal infections: A case series. *BMC Infect. Dis.* **2014**, *14*, 166. [CrossRef] [PubMed]
248. Flohé, S.; Lendemans, S.; Selbach, C.; Waydhas, C.; Ackermann, M.; Schade, F.U.; Kreuzfelder, E. Effect of granulocyte-macrophage colony-stimulating factor on the immune response of circulating monocytes after severe trauma. *Crit. Care Med.* **2003**, *31*, 2462–2469. [CrossRef] [PubMed]
249. Meisel, C.; Schefold, J.C.; Pschowski, R.; Baumann, T.; Hetzger, K.; Gregor, J.; Weber-Carstens, S.; Hasper, D.; Keh, D.; Zuckermann, H.; et al. Granulocyte-macrophage colony-stimulating factor to reverse sepsis-associated immunosuppression: A double-blind, randomized, placebo-controlled multicenter trial. *Am. J. Respir. Crit. Care Med.* **2009**, *180*, 640–648. [CrossRef] [PubMed]
250. Gyawali, B.; Ramakrishna, K.; Dhamoon, A.S. Sepsis: The evolution in definition, pathophysiology, and management. *SAGE Open Med.* **2019**, *7*, 2050312119835043. [CrossRef]
251. Sparrow, E.; Friede, M.; Sheikh, M.; Torvaldsen, S. Therapeutic antibodies for infectious diseases. *Bull. World Health Organ.* **2017**, *95*, 235–237. [CrossRef]
252. Peters van Ton, A.M.; Kox, M.; Abdo, W.F.; Pickkers, P. Precision Immunotherapy for Sepsis. *Front. Immunol.* **2018**, *9*, 1926. [CrossRef]

antibiotics

MDPI

Article

Teicoplanin-Resistant Coagulase-Negative Staphylococci: Do the Current Susceptibility Testing Methods Reliably Detect This Elusive Phenotype?

Adriana D. Balasiu * and Colin R. MacKenzie

Institute of Medical Microbiology and Hospital Hygiene, Heinrich-Heine University Hospital, 40225 Dusseldorf, Germany
* Correspondence: adriana.balasiu@med.uni-duesseldorf.de

Abstract: Coagulase-negative staphylococci (CoNS), members of the skin commensal microbiota, are increasingly associated with local or systemic infections due to a shift in patient populations in recent decades. Subsequently, more CoNS strains have been subjected to antibiotic susceptibility testing (AST), thus leading to the increased detection of teicoplanin resistance. However, data concerning teicoplanin resistance among CoNS strains remain limited, heterogeneous, and inconclusive. We collected 162 consecutive CoNS strains identified using Vitek-2 as teicoplanin-resistant and tested them with a range of AST methods. The results of standard and high inoculum broth microdilution (sBMD; hBMD), agar dilution (AD) after 24 h and 48 h incubation, standard and macrogradient diffusion strip (sGDT, MET), screening agar, and disc diffusion were compared to assess their robustness and to establish a diagnostic algorithm to detect teicoplanin resistance. sBMD was used as the reference method, and the lowest number of strains were teicoplanin-resistant using this method. sGDT and disc diffusion generated similar results to sBMD. Compared with sBMD, AD-24 h generated the lowest number of false teicoplanin-resistant strains, followed by hBMD, AD-48 h, and Vitek-2. sGDT, a fast, easy, affordable method in diagnostic settings, generated the highest rate of false teicoplanin-susceptible strains. Vitek-2 testing produced the highest number of teicoplanin-resistant strains. Only in two strains was the initial Vitek-2 teicoplanin resistance confirmed using five other AST methods. In conclusion, the different antibiotic susceptibility testing methods generated inconsistent, inconclusive, and discrepant results, thus making it difficult to establish a diagnostic algorithm for suspected teicoplanin resistance. Teicoplanin testing proved to be challenging and easily influenced by technical factors. This study aimed not only to raise awareness of teicoplanin resistance testing but also of the need for future studies focusing on the clinical efficacy of teicoplanin in relation to its susceptibility results.

Keywords: CoNS; teicoplanin; therapy; resistance; susceptibility testing

1. Introduction

The coagulase-negative staphylococci (CoNS) include a large number of different *Staphylococcus* species and are part of the skin and mucous membrane commensal microbiota. In certain circumstances (interference with skin health, ecology, and structure, or the immune system), they may cause opportunistic local or systemic infections. Advances in modern medicine have led to an increased role of CoNS among patients who are immunocompromised, critically ill, long-term hospitalized, or have implanted medical devices [1–6].

CoNS strains have been reported to play a significant role not only among device-associated infections (intravascular catheters, cerebrospinal fluid shunts, prosthetic joint, vascular grafts, and peritoneal dialysis catheters) but also in osteomyelitis, infective endocarditis [3], surgical site infections [5], and infections in neonates [7]. Van Epps et al. showed [1] that 50–70% of healthcare-associated infections in the USA are a consequence of a broad spectrum of available implantable medical devices, from the easily replaceable peripheral cannula to long-term devices, including extracorporeal life support, left ventricular assist devices, neurological devices, and joint prostheses.

CoNS strains cause 20–30%, and in some studies even up to 45% [2,8] of central-line-associated bloodstream infections (CLABSIs) in intensive care units and 35–55% of cardiovascular implantable electronic device (CIED) infections [9]. Furthermore, the 2018 ECDC report showed that [5], overall, 50% of surgical site infections (SSIs) are due to Gram-positive cocci. CoNS strains were found in 26.4% of SSI after coronary artery bypass graft and 18.9% after hip prosthesis surgery. Amat-Santos et al. found that 24.5% of prosthetic valve endocarditis cases after transcatheter aortic valve replacement were caused by CoNS [3]. In addition, CoNS is a major cause of late-onset sepsis among neonates [7].

Different AST methods, depending on the setting, can be performed: semi-automated or manually, using microdilution or agar dilution, disc diffusion, or gradient test. Most CoNS strains display resistance to beta-lactam agents; therefore, glycopeptide antibiotics (GAs) are often the therapy of choice for these infections. Vancomycin and teicoplanin are naturally occurring actinomycete-derived glycopeptide antibiotics [10]. GAs share the same mechanism of action (inhibition of the cell synthesis), structure, and spectrum of activity (mainly aerobic Gram-positive bacteria). GAs bind to the N-Acyl-D-Ala-D-Ala subunit of peptidoglycan, thus inhibiting cell-wall biosynthesis and inducing cell death [11]. Teicoplanin has similar efficacy to vancomycin but has been associated with fewer side effects and less nephrotoxicity than vancomycin [12,13]. Therefore, teicoplanin has become a therapeutic alternative to vancomycin for certain patients (e.g., those with neutropenia [14], or renal dysfunction).

Teicoplanin resistance has been increasingly reported over the years, but the published results are disparate. In our laboratory, we have made the same observation, and thus the main concern as to whether teicoplanin resistance is increasing remains unanswered. This leads to the question of which method is the most reliable to detect resistance to ensure that patients receive the appropriate therapy. Therefore, the aims of this study were to (i) assess the robustness of the routinely employed susceptibility testing by comparing it with other available methods and (ii) propose a diagnostic algorithm to detect the teicoplanin resistance and heteroresistance, thus avoiding labor-intensive population analysis methods.

2. Results

2.1. Patients and Included Isolates

Of the 162 tested isolates, 157 (96.9%) were *Staphylococcus epidermidis*, followed by *S. hominis* (3 isolates, 1.9%) and *S. haemolyticus* (2 isolates, 1.2%). In total, 96 (59.2%) strains were recovered from blood cultures, 76 (46.9%) of which were peripheral, and 20 (12.3%) were from central lines. The remaining 66 strains (40.8%) were isolated from tissue, intraoperative swabs, catheter tips, cerebrospinal fluid (CSF) from external ventricular drains (EVD), aspirates, respiratory samples, urine (from immunocompromised patients), and cell culture media (cell therapy products).

2.2. Vitek-2

The number of teicoplanin-resistant strains detected using Vitek-2 in our laboratory varied over 6 years between 20% and 32%, as shown by the annual resistance statistics listed in Table 1.

On retesting the 162 isolates using Vitek-2, 88 (54.3%) strains were susceptible to teicoplanin, and 74/162 (45.7%) were resistant. Most of the teicoplanin-resistant strains found with Vitek-2, i.e., 46/74 (62.2%), had a MIC of 8, while 27/74 (36.5%) had a MIC of 16, and 1 strain (1.3%) an MIC of 32. Moreover, Vitek-2 MIC distribution shows that the MICs are within close range of EUCAST defined teicoplanin breakpoint (Tables 2 and 3). Vitek-2 found 63 teicoplanin-resistant strains not confirmed by sBMD, and in 14 strains, sBMD testing correlated with the Vitek-2 results (Table 2).

Table 1. Annual resistance statistics for CoNS for the entire clinic between 2015 and 2020, susceptible strains in percentage (%).

Year	Total	OXA	GEN	LEV	SXT	ERN	CLI	VAN	TEI	LIN	TIG	FOS	FUS	RIF	TET	DAP
2015	650	36	57	44	72	30	47	100	**68**	100	99	56	61	93	55	100
2016	669	31	54	42	72	28	44	100	**71**	100	100	57	-	92	-	99
2017	759	32	58	45	72	31	43	100	**74**	100	100	51	-	93	-	99
2018	619	39	64	49	73	33	50	100	**69**	100	100	59	-	92	-	100
2019	562	36	63	54	71	34	50	100	**84**	99	100	56	-	92	-	99
2020	497	37	66	54	70	36	52	100	**80**	99	100	61	-	94	-	98

OXA (oxacillin), GEN (gentamicin), LEV (levofloxacin), SXT (trimethoprim–sulfamethoxazole), ERN (erythromycin), CLI (clindamycin), VAN (vancomycin), TEI (teicoplanin), LIN (linezolid), TIG (tigecycline), FOS (fosfomycin), FUS (fusidic acid), RIF (rifampicin), TET (tetracycline), DAP (daptomycin).

Table 2. Antimicrobial resistance routinely performed using Vitek-2 *.

Vitek-2	EUCAST							
	Susceptible (S) ≤ 4 mg/L (%)				Resistant > 4 mg/L (%)			Total
MIC Teicoplanin mg/L	≤0.5	1	2	4	8	16 *	32 *	
S. epidermidis	15 (9.3)	3 (1.8)	24 (14.8)	46 (28.4)	44 (27.2)	24 (14.8)	1 (0.6)	157
S. haemolyticus	-	-	-	-	1 (0.6)	1 (0.6)	-	2
S. hominis	-	-	-	-	1 (0.6)	2 (1.2)	-	3
Total %	88 (54.3)				74 (45.7)			162

* According to CLSI: strains with a MIC of 16 mg/L would be assigned to intermediate strains and the strain with a MIC of 32 mg/L would be resistant.

Table 3. Teicoplanin MIC distribution with Vitek-2.

Tei	AB	OXA	GEN	LEV	SXT	ERN	CLI	VAN	LIN	TIG	FOS	FUS	RIF	TET	DAP
S (88)	R	62	37	42	23	59	41	-	-	-	17	42	5	57	1
	S	26	51	46	65	29	57	88	88	88	71	46	83	31	87
R (74)	R	66	38	58	25	57	53	-	-	-	17	35	3	29	-
	S	8	36	16	49	17	21	74	74	73 **	56 **	28 **	71	45	73 **
Total															
	R	128	75	100	48	116	94	-	-	-	34	77	8	86	1
	%	79	46.3	61.7	29.6	71.6	58.0	-	-	-	21.0	47.5	4.9	53.1	0.6
	S	34	87	62	114	46	68	162	161	161	127	84	154	76	160
	%	21	53.7	38.3	70.4	28.4	42.0	100	99.4	99.4	78.4	51.8	95.1	46.9	98.7

AB (antibiotic), OXA (oxacillin), GEN (gentamicin), LEV (levofloxacin), SXT (trimethoprim–sulfamethoxazole), ERN (erythromycin), CLI (clindamycin), VAN (vancomycin), TEI (teicoplanin), LIN (linezolid), TIG (tigecycline), FOS (fosfomycin), FUS (fusidic acid), RIF (rifampicin), TET (tetracycline), DAP (daptomycin). ** For one strain, TIG, FOS, FUS, and DAP were not tested.

Using Vitek-2, 79% of the strains were oxacillin-resistant, and none of the strains displayed resistance to vancomycin or linezolid. All the antibiotics tested using Vitek-2 are listed in Table 3.

2.3. Standard and High-Broth Microdilution (sBMD and hBMD)

Using sBMD, 146/162 (90.1%) were teicoplanin-susceptible, and 16/162 (9.9%) were resistant. With hBMD, only 109/162 strains (67.3%) were susceptible, and 53/162 (32.7%)

were resistant. With sBMD, most of the strains (62.9%) had a MIC of 2 or 4, whereas using hBMD, the majority (58.0%) had a MIC of 4 or 8. The 39 (24.1%) teicoplanin-resistant strains in the hBMD assay had a MIC of 4 (28 strains), 2 (10 strains), and 0.5 (1 strain) in sBMD. These results are summarized in Tables 4–6.

Table 4. sBMD MICs.

sBMD	**EUCAST**							
	Susceptible (S) ≤ 4 mg/L (%)				**Resistant > 4 mg/L (%)**			**Total**
MIC Teicoplanin mg/L	≤0.5	1	2	4	8	16	>16	
S. epidermidis	15	29	48	53	12	-	-	157
S. haemolyticus	-	-	-	-	-	1 *	1 *	2
S. hominis	-	-	-	-	2	-	-	3
	9.3	17.9	29.6	33.3	8.6	0.6	0.6	162

* According to CLSI: the strain with a MIC of 16 mg/L would be assigned to intermediate and the strain with a MIC of >16 would be resistant.

Table 5. The minimum inhibitory concentration of the staphylococci strains using hBMD.

hBMD	**EUCAST**							
	Susceptible (S) ≤ 4 mg/L (%)				**Resistant > 4 mg/L (%)**			**Total**
MIC Teicoplanin mg/L	≤0.5	1	2	4	8	16	>16	
S. epidermidis	11	16	35	45	44	5 *	1 *	157
S. haemolyticus	-	-	-	-	2	-	-	2
S. hominis	-	-	-	-	1	-	-	3
	6.8	9.9	21.6	29.0	29.0	3.1	0.6	162

* According to CLSI: the strain with a MIC of 16 mg/L would be assigned to intermediate and the strain with a MIC of > 16 would be resistant.

Table 6. MIC distribution using sBMD vs. Vitek-2 and the respective EA, CA, and ME.

EUCAST Category	**sBMD Teicoplanin mg/L**	**sBMD No.**	**MIC by Vitek-2, (No.)**							
			Susceptible				**Resistant**			
			0.5	**1**	**2**	**4**	**8**	**16**	**32**	
Resistant (R)	32	-	-	-	-	-	-	-	-	EA 81 (50) CA 106 (65.4) ME 59 (36.4)
	16	2	-	-	-	-	2	-	-	
	8	11	-	-	-	-	3	7	1	
Total R		13 (8%)	-				13			
Susceptible (S)	4	40	-	-	4	6	18	12	-	
	2	49	2	2	4	20	18	3	-	
	1	22	2	1	6	7	3	3	-	
	0.5	25	5	-	8	11	-	1	-	
	<0.5	11	6	-	3	1	1	-	-	
Total S		147 (90.7%)	88 (54.3)				59 (36.4)			

Notably, 2 strains of 162 were not tested due to lack of growth. EA, essential agreement; CA, categorical agreement; ME, major error.

Vancomycin MIC was measured by means of sBMD and Vitek-2. All the samples were vancomycin-susceptible using both methods. While with sBMD, the majority

144/162 (88.9%) of the strains had a MIC of 2 mg/L, with Vitek-2, 70 (43.2%) had a MIC of 1 mg/L and 80 (49.4%) had a MIC of 2 mg/L.

2.4. *Agar-Diffusion 24 h and 48 h Incubation (AD-24 h and AD-48 h)*

In the AD-24 h assay, 128 (79%) strains were teicoplanin-susceptible, 33 (20.4%) were resistant, and 1 strain displayed no growth after 20–24 h incubation. Among the 33 teicoplanin-resistant strains in hBMD, 20 (12.3%) strains were susceptible, and 13 were resistant using sBMD, while with Vitek-2, 7 were susceptible, and 26 were resistant. AD-24 h, on the one hand, failed to recognize accurately 3 teicoplanin-resistant strains from sBMD, but on the other hand, generated 20 more resistant strains than sBMD. AD-24 h and the other AST results are summarized in Tables 7 and 8.

Table 7. MICs using agar dilution after 24 h and 48 h incubation.

Agar Dilution (AD) *	EUCAST									
	Susceptible (S) ≤ 4 mg/L								Resistant > 4 mg/L	
Incubation	24 h	48 h	24 h	48 h	24 h	48 h	24 h	48 h	24 h	48 h
Teicoplanin (mg/L)	0.5		1		2		4		≥8	≥8
S. epidermidis	1	1	16	6	61	46	48	44	31	59
S. haemolyticus	-	-	-	-	-	-	-	-	-	2
S. hominis	-	-	-	-	-	-	-	-	2	3

* One strain remained without growth.

Table 8. Results of AD-24 h vs. other AST assay methods.

AD-24 h	sBMD	AD-48 h	Vitek-2	sGDT	Screening McF 0.5	Screening McF 2	Disc Diffusion (CLSI)
128 S	125 S	97 S	80 S	125 S	80 pos	114 pos	126 S
	3 R	31 R	48 R	1 R	44 neg	12 neg	-
	-	-	-	2 NE	4 NE	2 NE	2 NE
33 R	20 S	-	7 S	31 S	33 pos	33 pos	31 S
	13 R	33 R	26 R	2 R	-	-	2 I
	-	-	-	-	-	-	
1 NG	1 S	NG	1 S	NE	-	-	NG

S, susceptible; R, resistant; NE, not evaluable; NG, no growth; pos, positive; neg, negative; I, intermediate according to CLSI.

In AD after 48 h incubation, only 98 (60.5%) strains remained susceptible, 63 (38.9%) were resistant, and 1 strain displayed no growth. Notably, 31 strains, initially tested in AD-24 h as susceptible with a MIC of 4 (30) and 2 (1), were resistant after 48 h incubation. Only 14 of the 16 teicoplanin-resistant strains in sBMD were among the 63 teicoplanin-resistant strains in AD-48 h. Testing with sBMD and AD-48 h found the highest number of susceptible strains, whereas using Vitek-2 and AD-48 h, most strains were teicoplanin-resistant (51). Further results are depicted in Table 9.

Table 9. Results of AD-48 h vs. other AST methods.

AD-24 h	sBMD	AD-48 h	Vitek-2	sGDT	Screening McF 0.5	Screening McF 2	Disc Diffusion (CLSI)
97 S	95 S	97 S	74 S	96 S	50 pos	84 pos	96 S
	2 R	-	23 R		44 neg	12 neg	-
	-	-	-	1 NE *	3 NE	1 NE	1 NE
64 R	50 S	31 S	13 S	60 S	63 pos	63 pos	60 S
	14 R	33 R	51 R	3 R	-	-	2 I
	-	-	-	1 NG	1 NE-	NE	2 NE
1 NG	1 S	NG	1 S	NE	-	-	NG

* S, susceptible; R, resistant; NE, not evaluable; NG, no growth; pos, positive; neg, negative; I, intermediate according to CLSI.

2.5. Standard Gradient Diffusion Test (sGDT) and Macrodilution Gradient Test (MET)

All but three strains tested teicoplanin-susceptible by means of sGDT. Most of the strains displayed a MIC of 1 mg/L (81/162) or 2 mg/L (44/162). The assay recognized only 3 of the 16 teicoplanin-resistant strains from sBMD, thus generating the highest rate not only of CA but also of vME (Table 10).

Table 10. Susceptibility results using sGDT.

sGDT	EUCAST					
	Susceptible (S) ≤ 4 mg/L (%)				Resistant > 4 mg/L (%)	
MIC Teicoplanin mg/L	≤0.5	1	2	4	8	16
S. epidermidis	23	81	42	8	-	-
S. haemolyticus	-	-	-	-	2	-
S. hominis	-	-	2	-	-	1
% *	14.2	50	27.2	4.9	1.2	0.6

* Two strains display no growth, and for one strain, the MIC could not be read (1.8%).

The values obtained using MET are not strictly speaking MICs. After 48 h incubation, 157/162 (96.9%) strains displayed growth at a MIC lower than 8 mg/L, 1 strain at 8 mg/L, 1 strain at 12 mg/L, and 2 (1.2%) (both *S. epidermidis*) strains failed to grow. The strains displaying growth at 8 mg/L were also tested for vancomycin resistance because according to the EUCAST criteria, the reading of teicoplanin at 8 mg/L is not enough in itself to assign a strain as vancomycin-resistant or as a heteroresistant strain. The two strains with high MET readings were confirmed using all other AST assays, except via AD-24 h and disc diffusion. The AST results are collated in Table 11.

Table 11. Comparison of strains with high MET values (≥8 mg/L) with other AST assays.

No.	MET	sBMD	hBMD	Vitek-2	sGDT	AD-24 h	AD-48 h	Screening McF 0.5	Disc Diffusion	Material	Strain ID
71	8	R	R	R	R	S	R	pos	S	BC	*S. haemolyticus*
72	12	R	R	R	R	R	R	pos	I	BC	*S. hominis*

sBMD, standard broth microdilution; hBMD, high-broth microdilution; sGDT, standard gradient strip; BC, blood culture; AD, agar dilution; S, susceptible; R, resistant; I, intermediate according to CLSI; pos, positive.

2.6. Disc Diffusion and Screening Agar

By means of disc diffusion, all the samples except two were susceptible, according to the CLSI criteria, thus confirming that this method does not reliably detect teicoplanin resistance.

By means of screening agar (5 mg/L teicoplanin) using a standard 0.5 McF inoculum, 113/162 (69.8%) strains were positive, suggesting a teicoplanin MIC of over 5 mg/L and thus resistant. The remaining 44 (27.2%) were negative, 4 could not be evaluated, and 1 was not performed. Notably, 147/162 (90.7%) strains were positive when an McF 2 inoculum was used, 12 were negative, 2 could not be evaluated, and in 1, this was not performed. The most positive strains in a screening method, 99 (61.1%) strains with McF 0.5 and 131 (80.9%) using McF 2, were among the strains tested susceptible with sBMD and therefore would be falsely assigned as teicoplanin-resistant, which would correspond to the highest ME among all the employed AST methods. A summary of the results comparing the AST methods is depicted in Table 12.

Table 12. Teicoplanin susceptibility tested via AST and the EA, CA, vME, and ME yielded when compared with sBMD.

Method	Strain	No. % Isolates		EA	CA	vME	ME
		Susceptible	Resistant				
		≤4	>4				
sBMD	All strains	146 (90.1)	16 (9.9)				
	S. epidermidis	145 (89.5)	12 (7.4)				
	S. haemolyticus	-	2 (1.2)				
	S. hominis	1 (0.6)	2 (1.2)				
hBMD	All strains	109 (67.3)	53 (32.7)	137 (84.6)	121 (74.7)	2 (1.2)	39 (24.1)
	S. epidermidis	107 (66)	50 (30.9)	132 (81.5)	117 (72.2)	1 (0.6)	39 (24.1)
	S. haemolyticus	-	2 (1.2)	2 (1.2)	2 (1.2)	-	-
	S. hominis	2 (1.2)	1 (0.6)	3 (1.8)	2 (1.2)	1 (0.6)	-
Vitek-2	All strains	88 (54.3)	74 (45.7)	103 (63.6)	94 (58.0)	5 (3.1)	63 (38.9)
	S. epidermidis	88 (54.3)	69 (42.6)	99 (61.1)	90 (55.6)	5 (3.1)	62 (38.2)
	S. haemolyticus	-	2 (1.2)	2 (1.2)	2 (1.2)	-	-
	S. hominis	-	3 (1.8)	2 (1.2)	2 (1.2)	-	1 (0.6)
AD-24 h [1]	All strains	128 (79)	33 (20.4)	146 (90.1)	138 (85.2)	3 (1.8)	20 (12.4)
	S. epidermidis	125 (77.2)	31 (19.1)	142 (87.7)	134 (82.7)	2 (1.2)	20 (12.4)
	S. haemolyticus	1 (0.6)	1 (0.6)	1 (0.6)	1 (0.6)	1 (0.6)	-
	S. hominis	1 (0.6)	2 (1.2)	3 (1.8)	3 (1.8)	-	-
AD-48 h [1]	All strains	97 (59.9)	64 (39.5)	132 (81.5)	109 (67.3)	2 (1.2)	50 (30.9)
	S. epidermidis	97 (59.9)	59 (36.4)	127 (78.4)	105 (64.8)	2 (1.2)	49 (30.2)
	S. haemolyticus	-	2 (1.2)	2 (1.2)	2 (1.2)	-	-
	S. hominis	-	3 (1.8)	3 (1.8)	2 (1.2)	-	1 (0.6)
sGDT [2]	All strains	156 (96.3)	3 (1.8)	118 (72.8)	146 (90.1)	13 (8.0)	-
	S. epidermidis	154 (95.1)	-	114 (70.4)	142 (87.7)	12 (7.4)	-
	S. haemolyticus	-	2 (1.2)	2 (1.2)	2 (1.2)	-	-
	S. hominis	2 (1.2)	1 (0.6)	2 (1.2)	2 (1.2)	1 (0.6)	-

Strains without growth: [1] one strain and [2] three strains.

3. Discussion

The AST results and the institutional yearly resistance statistics confirmed the previously published data [4,15–17] that CoNS strains are highly resistant to most commonly used beta-lactam antibiotic agents, leaving glycopeptides, linezolid, and daptomycin as the most important therapeutic options. A number of aspects should be considered when choosing the appropriate treatment, including side effects, risk of developing resistance during therapy, therapeutic drug monitoring, cost, and availability. Teicoplanin has been considered an alternative to vancomycin due to its lower nephrotoxicity, reduced drug interactions, and once-daily administration.

Teicoplanin resistance has been reported in the USA and the UK since the early 1980s, but the published data since then [18] do not reflect the actual incidence and its impact on therapeutical use. Teicoplanin resistance is an increasing and emerging challenge, but published data are inconclusive due to a number of factors. These include the different methods employed (e.g., broth microdilution vs. disk diffusion [4]); settings, diagnostic vs. research (e.g., broth microdilution vs. population analysis); the standards employed (e.g., CLSI vs. EUCAST defined breakpoints); the inclusion of diverse cohorts (e.g., catheter-related bacteremia vs. healthy volunteers [17,19]; the bacterial species studied (most studies have focused on *S. aureus* and fewer on CoNS [20]); clonal dissemination [21]; data generated at different time points [22]; or that teicoplanin was not tested. Thus, to date, reports have probably underestimated the true incidence of teicoplanin resistance and are still insufficient to identify its underlying mechanisms with certainty.

It is still unclear if increasing teicoplanin resistance should be attributed to one or several possible underlying mechanisms. The mechanism is neither well defined nor adequately studied. Several mechanisms have been proposed such as cellular aggregates and antibiotic retention [23] or cell-wall alteration through reorganization or thickening [24,25]. Perhaps even more worrying is that teicoplanin resistance has been shown to develop under therapy [26,27]. Biavasco et al. pointed out that the AST employed for teicoplanin can be easily influenced by technical factors such as methods, media, inoculum, and incubation time [28]. Furthermore, it has been shown that the physical properties of teicoplanin—a large, lipophilic, and negatively charged molecule—have an impact upon testing by generating a lower diffusion coefficient on agar compared with vancomycin [29].

Broth microdilution is generally regarded as the gold standard method for antibiotic susceptibility testing; however, few laboratories use it for routine purposes. To optimize laboratory workflow with a high sample throughput, semi-automated devices such as Vitek-2 are employed routinely for the AST of fast-growing bacteria. Generally speaking, Vitek-2 performs well: It is fast and robust, with minimum hands-on time, is cost-effective, and requires little technical expertise. In our laboratory, using Vitek-2, a rapid rise in teicoplanin-resistant CoNS strains was observed in 2015. Baris et al. also reported an increased number of teicoplanin-resistant strains with BD Phoenix [16]. As in our study, most of the samples tested as teicoplanin-resistant using Vitek-2 were not confirmed via sBMD, leading to the highest rate of ME among the AST methods. The majority of teicoplanin MIC, either 4 or 8 mg/L (56.8%), determined using Vitek-2 were close to the EUCAST epidemiological cut-off (ECOFF) value for CoNS (MIC 4 mg/L), thus having an impact upon the generated EA and CA. Meanwhile, most of the MIC in sBMD (54.7%) were concentrated at the upper limit of the range (2 or 4 mg/L), thus conforming to the published EUCAST MIC distribution determining the teicoplanin breakpoints for CoNS. Vaudaux et al. found a similar MIC distribution using macrodilution but not microdilution. Moreover, MIC distribution was different when performed using macrodilution or microdilution [30].

According to these results, the AST performance for teicoplanin does not fulfill the CLSI criteria of 90% agreement for both EA and CA [31]. It is difficult to establish a diagnostic workflow that reliably confirms teicoplanin resistance among routinely tested strains. Firstly, EA and CA differ in test, antibiotic, and methodology, confirming the results of Campana et al. Moreover, their results showed that EA and CA vary with species (e.g., CA for strip test for *S. aureus* (100%) vs. 75% for CoNS according to CLSI) [32]. Secondly, most of the routinely employed AST assays use a low bacterial inoculum and are fast, whereas the strains that might bear heteroresistance are first detected at CFU above 10^6 CFU/mL and after a longer incubation time (48 h). With a final inoculum of 5×10^5 CFU/well, the microdilution assay, the current gold standard method, is unable to reliably detect heteroresistance [30]. Routinely employed methods probably do not detect heteroresistant strains, which may have a negative impact on therapeutic outcomes.

Teicoplanin AST is easily influenced by inoculum, incubation time, media, and method and is more variable than vancomycin AST. All these suggest that a re-evaluation of diagnostic methods, breakpoints, and their capacity to accurately identify teicoplanin resistance and heteroresistance among clinically relevant CoNS strains is needed. A possible diagnostic algorithm should encompass different steps that can be carried out in a routine setting: a rapid automated AST to identify possible resistance, followed by a high inoculum and a longer incubation period method to confirm resistance or susceptibility. The second method should preferably be fast, commercially available for routine settings, have a low cost, and be reliable and reproducible. A possible option would be MET. MET is a method with low hands-on time, but adjustments are needed for it to be as reliable for use with CoNS as it is for *S. aureus*. The strains with suspected teicoplanin resistance could be further tested in reference laboratories by means of population analysis profiles (PAPs). PAP is the gold standard method to detect heteroresistance. This is a demanding time-consuming method, difficult to implement in a routine setting, and poses the risk of selecting resistance instead of finding it [33]. Using a different method in the second step is challenging because not all laboratories have the option to produce the necessary in-house plates.

These results do not confirm an increased vancomycin resistance as previously thought or predicted. This may be due to an underlying mechanism that involves only teicoplanin or that the teicoplanin molecule presents technical difficulties causing an unreliable result [3]. A similar situation applies to colistin [34].

In conclusion, extensive teicoplanin susceptibility testing showed that the results obtained using a single method could not be fully confirmed by employing various other methods. Due to a high discrepancy among the methods tested, no algorithm can be proposed to reliably detect teicoplanin resistance. The fact that the results were so diverse suggests that all the aspects involved in teicoplanin testing should be re-evaluated so that improvements can be made not only in the laboratory but also in establishing reliable breakpoints. Given the relevance that these results pose for antibiotic therapy, further clinical studies looking into the clinical efficacy of teicoplanin and in vitro teicoplanin testing are of great importance.

4. Materials and Methods

In accordance with the Declaration of Helsinki (2013), ethical approval for this study was granted by the Ethics Committee of the Medical Faculty of Heinrich Heine University, Dusseldorf (Study No. 5694/26.9.2016).

4.1. Bacterial Strains

For this study, 162 consecutive CoNS strains were collected from August 2015 to August 2016 at the Institute of Medical Microbiology and Hospital Hygiene, Heinrich Heine University Hospital, Düsseldorf. The strains were selected based on non-susceptibility against teicoplanin and were recovered from different samples such as blood culture, soft-tissue infections, or central lines. Routinely, putative clinically relevant isolates were subjected to identification and susceptibility testing. Identification was performed with Vitek® MS (bioMérieux, Marcy l'Etoile, France), a matrix-assisted laser desorption ionization–time of flight mass spectrometry method (MALDI-TOF MS). Antibiotic susceptibility testing was performed with Vitek 2 (bioMérieux, Marcy l'Etoile, France) AST- P654 cards.

The strains were stored in 80% glycerol in a Mueller–Hinton Broth (MHB) (commercially dehydrated base from Oxoid, Thermo Scientific, Basingstoke, United Kingdom) (v/v) at −80 °C until additional testing was performed. To perform further testing, the strains were subcultured on Columbia agar supplemented with 5% sheep blood (COS Agar) (bioMérieux, Marcy l'Etoile, France), incubated at 36 ± 1 °C in an atmosphere enriched with 5–10% CO_2 for 18–24 h. Subsequently, a single colony was picked, subcultured on COS Agar, and incubated for another 18–24 h under the same conditions.

4.2. Antimicrobial Susceptibility Testing (AST)

The minimal inhibitory concentration (MIC) was determined on a standard 0.5 McFarland bacterial suspension in a 0.85% saline solution using different susceptibility testing methods. The MIC is reported either in mg/L or µg/mL, and strains were classified as susceptible or resistant according to EUCAST breakpoints.

4.2.1. Broth Microdilution

Broth microdilution (BMD) was performed according to the method recommended by the European Committee on Antimicrobial Susceptibility Testing (EUCAST) (ISO 20776-1) and used as the reference method for antimicrobial susceptibility testing (AST) of rapidly growing aerobic bacteria. Both antibiotics used in this assay, vancomycin (V2002-100MG) and teicoplanin (T0578-100MG) (Sigma-Aldrich, Darmstadt, Germany), were resuspended in water at a concentration of 5120 mg/L (stock solution) and kept in aliquots at −80 °C until use. Ready-to-use antimicrobial solutions were freshly prepared from the stock solutions on the day of the assay using the Mueller–Hinton Broth (MHB). For the assay, 100 zµL MHB was added in each well of a 96-well flat bottom plate. Then, 100 µL antibiotic with the highest concentration (32 mg/L) was added to the wells of the first column using a multichannel pipette, and mixed (pipetted 5 times), thus achieving a final concentration of 16 mg/L in the first dilution (wells A1-H1). Afterward, 100 µL suspension was transferred to the corresponding well in the second column. This process was repeated up to the 10th column, from which 100 µL were discarded. As a result, a serial twofold dilution was generated to a final concentration of 0.03 mg/L. In addition to the ten antibiotic concentrations columns, growth/positive control (column 11—MHB and bacterial inoculum without antibiotic) and negative control (column 12—only MHB) were tested. To all the wells other than the negative control column, 10 µL of the standard bacterial inoculum (5×10^5 colony forming units/mL (CFU/mL)) was added. To obtain a standard inoculum, each strain was resuspended in 0.85% saline to a 0.5 on the McFarland scale (McF) ($1–2 \times 10^8$ CFU/mL), followed by a 1:20 dilution (5×10^6 CFU/mL). The 96-well-plate was sealed and incubated for 24 h at 36 ± 1 °C air (according to ISO 20776-1), and the OD was then measured at 620 nm with a Sunrise TW absorbance reader (Tecan Trading AG, Männedorf, Switzerland). An absorbance of >0.5 was considered positive for bacterial growth.

A second BMD assay was performed under similar conditions but with a higher bacterial inoculum (hBMD). For the bacterial inoculum, the strains were resuspended in 0.85% saline to a 0.5 McF, diluted 1:2 (5×10^7 CFU/mL), and 10 µL added to the well to a final concentration of 5×10^6 CFU/mL. The plates were sealed and incubated for 18 ± 2 h, and OD was measured.

4.2.2. Agar Dilution

For agar dilution (AD) assay, the Mueller–Hinton agar (dehydrated base from Oxoid, Thermo Scientific, Basingstoke, United Kingdom) was autoclaved and cooled to 45–50 °C and adjusted to a 7.3 pH, and teicoplanin from the stock solution was added to final concentrations of 0.25, 0.5, 1, 2, 4, and 8 mg/L. Additionally, drug-free plates were prepared and used for growth control. The prepared plates were kept wrapped at 4 °C and brought to room temperature before being subjected to previously described procedures [33,35,36]. Briefly, a 0.5 McF ($1–2 \times 10^8$ CFU/mL) standard bacterial suspension was serially diluted 1:10 to 10^3 CFU/mL, and 10 µL from each dilution was transferred to the plates and incubated for 20–24 h and 48 h, after which the colonies were counted.

4.2.3. Glycopeptide Antibiotic Susceptibility Testing (EUCAST)

EUCAST endorses the use of standard gradient diffusion test (sGDT), macrodilution gradient test (MET), and screening agar as detection methods of glycopeptide non-susceptible *S. aureus* strains [37]. These assays have been recommended by EUCAST for *S. aureus* for research use only but have neither been suggested nor validated for CoNS. The obtained results are therefore not suitable for clinical interpretation.

The teicoplanin standard gradient diffusion strip test (sGDT) was performed according to the manufacturer's instruction using teicoplanin MIC test strips (range 0.016–256 μg/mL) (MTS; Liofilchem, Italy) [38] on a 0.5 McF standard bacterial inoculum on Mueller–Hinton agar (MHE) plates (BioMérieux, France). The MIC in mg/L was read after 16–20 h incubation, representing the point where the formed symmetrical ellipse met the strip.

MET was performed according to EUCAST and the manufacturer's instructions. Briefly, colonies from a 24 h old culture were resuspended in 2 mL 0.85% saline to McF 2 (heavier inoculum), streaked evenly on a brain–heart infusion (BHI) agar (Graso Biotech, Poland), and left to dry. Teicoplanin gradient strips were applied to the surface, incubated at 37 °C air, and read after 24 and 48 h. Not only was the value documented but also the presence of hazes, microcolonies, and isolated colonies.

4.2.4. Screening Agar

For the agar screening method, in-house Mueller–Hinton agar plates with and without 5 mg/L teicoplanin were produced and used based on the previously described protocol [39]. Briefly, colonies were suspended in 0.85% saline to an McF 0.5 and McF 2.0, and 10 μL of each inoculum were evenly distributed on the surface of the agar, incubated at 37 °C in air, and the growth was assessed after 24 and 48 h.

4.2.5. Disc Diffusion

Disc diffusion was performed, even though this approach is no longer EUCAST-recommended. CLSI version 2012 released breakpoints for disc diffusion warning indicating that it is unknown if the method can discriminate between susceptible and resistant strains to teicoplanin. For disc diffusion, the bacterial inoculum was evenly distributed on MHE plates (bioMérieux, Marcy l'Etoile, France), teicoplanin 30 mg discs (Liofilchem, Italy) placed on the surface, and incubated at 36 ± 1 °C in air. The inhibition zone was read after 24 h and interpreted according to the Clinical and Laboratory Standards Institute (CLSI).

4.2.6. Quality Controls

All the performed tests included negative and positive controls. *S. aureus* ATCC 29213 (teicoplanin reference range 0.25–1 mg/L, vancomycin reference range 0.5–2 mg/L) was included as a positive control (quality controls; QC strains) in all the assays under the same conditions as the CoNS strains [40]. The test results were considered valid only when the QC strain was tested within the EUCAST-given ranges. The AD assay included three additional strains as controls: *Enterococcus faecalis* ATCC 29212 (teicoplanin reference range 0.25–1 mg/L, vancomycin reference range 1–4 mg/L); the vancomycin-resistant *S. aureus* (VRSA) strain Mu50 (ATCC 700699); and Mu3 (ATCC 700698), a methicillin-resistant *S. aureus* (MRSA) strain with heterogeneous resistance to vancomycin.

4.3. *EUCAST Rules, Results Interpretation, and Data Analysis*

All the AST results, except disc diffusion, were interpreted according to EUCAST breakpoints [41] and assigned to susceptible (MIC ≤ 4 mg/L) or resistant (MIC > 4 mg/L). The MIC values were reported in serial 1:2 dilutions and intermediate values as the next higher MIC. CLSI criteria were used to assess the results of disc diffusion and sBMD. According to CLSI, the strains were susceptible at MIC ≤ 8 mg/L, with zone diameter ≥14 mm; intermediate at MIC 16 mg/L, with zone diameter 11–13 mm; or resistant at MIC ≥ 32 mg/L, with zone diameter ≤ 10 mm [42].

Data were analyzed by comparing the measured MIC values and the corresponding interpretation generated using Vitek-2, hBMD, AD-24 h, AD-48 h, sGDT, MET, and screening agar with those from sBMD, the EUCAST recommended reference method. A very major error (vME) was defined as a false-susceptible result, whereas a major error (ME) was considered a false-resistant result compared with the results of sBMD. An essential agreement (EA) was considered when the MICs fell within the 1 $\log_2$ dilution of the MIC determined using sBMD, while categorical agreement (CA) was assigned to the isolate rated with the same interpretation category results (S/R) as sBMD. Acceptable performance for a method was defined as a percentage ≥90% for EA, CA, and ≤3% for vME or ME [31].

Author Contributions: Conceptualization, C.R.M. and A.D.B.; methodology, A.D.B.; writing—original draft preparation, A.D.B.; writing—review and editing, C.R.M. All authors have read and agreed to the published version of the manuscript.

Funding: This research received no external funding.

Institutional Review Board Statement: This study was conducted in accordance with the Declaration of Helsinki and approved by the Ethics Committee) of The Medical Faculty of Heinrich Heine University, Dusseldorf (Protocol Code 5694, on 26.09.2016).

Informed Consent Statement: Not applicable.

Data Availability Statement: The data presented in this study are presented in the tables, no additional data was generated.

Conflicts of Interest: The authors declare no conflict of interest.

References

1. Van Epps, J.S.; Younger, J.G. Implantable Device-Related Infection. *Shock* **2016**, *46*, 597–608. [CrossRef] [PubMed]
2. Plachouras, D.; Savey, A.; Palomar, M.; Moro, M.; Lebre, A.; McCoubrey, J. Incidence and microbiology of central line-associated bloodstream infections in European intensive care units: Results from the European Healthcare-Associated Infections surveillance Network (HAI-Net). In Proceedings of the European Conference of Clinical Microbiology and Infectious Diseases (ECCMID), Madrid, Spain, 24 April 2018.
3. Amat-Santos, I.J.; Messika-Zeitoun, D.; Eltchaninoff, H.; Kapadia, S.; Lerakis, S.; Cheema, A.N.; Gutiérrez-Ibanes, E.; Munoz-Garcia, A.J.; Pan, M.; Webb, J.G.; et al. Infective endocarditis after transcatheter aortic valve implantation: Results from a large multicenter registry. *Circulation* **2015**, *131*, 1566–1574. [CrossRef]
4. Bora, P.; Datta, P.; Gupta, V.; Singhal, L.; Chander, J. Characterization and antimicrobial susceptibility of coagulase-negative staphylococci isolated from clinical samples. *J. Lab. Physicians* **2018**, *10*, 414–419. [CrossRef]
5. ECDC. Healthcare-associated infections: Surgical site infections. In *Annual Epidemiological Report for 2017*; ECDC: Solna, Sweden, 2019.
6. Kloos, W.E.; Bannerman, T.L. Update on clinical significance of coagulase-negative staphylococci. *Clin. Microbiol. Rev.* **1994**, *7*, 117–140. [CrossRef]
7. Klingenberg, C.; Aarag, E.; Rønnestad, A.; Sollid, J.E.; Abrahamsen, T.G.; Kjeldsen, G.; Flaegstad, T. Coagulase-negative staphylococcal sepsis in neonates. Association between antibiotic resistance, biofilm formation and the host inflammatory response. *Pediatr. Infect. Dis. J.* **2005**, *24*, 817–822. [CrossRef]
8. ECDC. Healthcare-Associated Infections Acquired in Intensive Care Units. 2019. Available online: https://www.ecdc.europa.eu/sites/default/files/documents/AER_for_2017-HAI.pdf (accessed on 6 December 2021).
9. Kusumoto, F.M.; Schoenfeld, M.H.; Wilkoff, B.L.; Berul, C.I.; Birgersdotter-Green, U.M.; Carrillo, R.; Cha, Y.-M.; Clancy, J.; Deharo, J.-C.; Ellenbogen, K.A.; et al. 2017 HRS expert consensus statement on cardiovascular implantable electronic device lead management and extraction. *Heart Rhythm* **2017**, *14*, e503–e551. [CrossRef] [PubMed]
10. Butler, M.S.; Hansford, K.A.; Blaskovich, M.A.; Halai, R.; Cooper, M.A. Glycopeptide antibiotics: Back to the future. *J. Antibiot.* **2014**, *67*, 631–644. [CrossRef] [PubMed]
11. Yim, G.; Thaker, M.N.; Koteva, K.; Wright, G. Glycopeptide antibiotic biosynthesis. *J. Antibiot.* **2014**, *67*, 31–41. [CrossRef] [PubMed]
12. Svetitsky, S.; Leibovici, L.; Paul, M. Comparative efficacy and safety of vancomycin versus teicoplanin: Systematic review and meta-analysis. *Antimicrob. Agents Chemother.* **2009**, *53*, 4069–4079. [CrossRef]
13. Wood, M.J. The comparative efficacy and safety of teicoplanin and vancomycin. *J. Antimicrob. Chemother.* **1996**, *37*, 209–222. [CrossRef]
14. Kato-Hayashi, H.; Niwa, T.; Ohata, K.; Harada, S.; Matsumoto, T.; Kitagawa, J.; Tsurumi, H.; Suzuki, A. Comparative efficacy and safety of vancomycin versus teicoplanin in febrile neutropenic patients receiving hematopoietic stem cell transplantation. *J. Clin. Pharm. Ther.* **2019**, *44*, 888–894. [CrossRef]
15. Lee, J.Y.H.; Monk, I.R.; Gonçalves da Silva, A.; Seemann, T.; Chua, K.Y.L.; Kearns, A.; Hill, R.; Woodford, N.; Bartels, M.D.; Strommenger, B.; et al. Global spread of three multidrug-resistant lineages of Staphylococcus epidermidis. *Nat. Microbiol.* **2018**, *3*, 1175–1185. [CrossRef] [PubMed]
16. Baris, A.; Malkocoglu, G.; Buyukyanbolu, E.; Aslan, F.M.; Bayraktar, B.; Aktas, E. Evaluation of Teicoplanin Resistance Detected by Automated System in Coagulase Negative Staphylococci: A Comparison with Gradient Test and Broth Microdilution Methods. *Curr. Microbiol.* **2020**, *77*, 3355–3360. [CrossRef] [PubMed]

17. Marincola, G.; Liong, O.; Schoen, C.; Abouelfetouh, A.; Hamdy, A.; Wencker, F.D.R.; Marciniak, T.; Becker, K.; Köck, R.; Ziebuhr, W. Antimicrobial Resistance Profiles of Coagulase-Negative Staphylococci in Community-Based Healthy Individuals in Germany. *Front. Public Health.* **2021**, *9*, 684456. [CrossRef] [PubMed]
18. Del Bene, V.E.; John, J.F.; Twitty, J.A., Jr.; Lewis, J.W. Anti-staphylococcal activity of teicoplanin, vancomycin, and other antimicrobial agents: The significance of methicillin resistance. *J. Infect. Dis.* **1986**, *154*, 349–352. [CrossRef] [PubMed]
19. Cherifi, S.; Byl, B.; Deplano, A.; Nonhoff, C.; Denis, O.; Hallin, M. Comparative epidemiology of Staphylococcus epidermidis isolates from patients with catheter-related bacteremia and from healthy volunteers. *J. Clin. Microbiol.* **2013**, *51*, 1541–1547. [CrossRef] [PubMed]
20. Satola, S.W.; Farley, M.M.; Anderson, K.F.; Patel, J.B. Comparison of detection methods for heteroresistant vancomycin-intermediate Staphylococcus aureus, with the population analysis profile method as the reference method. *J. Clin. Microbiol.* **2011**, *49*, 177–183. [CrossRef] [PubMed]
21. Miragaia, M.; Couto, I.; Pereira, S.F.F.; Kristinsson, K.G.; Westh, H.; Jarløv, J.O.; Carriço, J.; Almeida, J.; Santos-Sanches, I.; de Lencastre, H. Molecular characterization of methicillin-resistant Staphylococcus epidermidis clones: Evidence of geographic dissemination. *J. Clin. Microbiol.* **2002**, *40*, 430–438. [CrossRef]
22. Camargo, C.H.; Mondelli, A.L.; Boas, P.J. Comparison of teicoplanin disk diffusion and broth microdilution methods against clinical isolates of Staphylococcus aureus and S. epidermidis. *Braz. J. Microbiol.* **2011**, *42*, 1265–1268. [CrossRef]
23. Sieradzki, K.; Villari, P.; Tomasz, A. Decreased susceptibilities to teicoplanin and vancomycin among coagulase-negative methicillin-resistant clinical isolates of staphylococci. *Antimicrob. Agents Chemother.* **1998**, *42*, 100–107. [CrossRef]
24. Howden, B.P.; Davies, J.K.; Johnson, P.D.; Stinear, T.P.; Grayson, M.L. Reduced vancomycin susceptibility in Staphylococcus aureus, including vancomycin-intermediate and heterogeneous vancomycin-intermediate strains: Resistance mechanisms, laboratory detection, and clinical implications. *Clin. Microbiol. Rev.* **2010**, *23*, 99–139. [CrossRef]
25. Becker, K.; Heilmann, C.; Peters, G. Coagulase-negative staphylococci. *Clin. Microbiol. Rev.* **2014**, *27*, 870–926. [CrossRef]
26. Schwalbe, R.S.; Stapleton, J.T.; Gilligan, P.H. Emergence of vancomycin resistance in coagulase-negative staphylococci. *N. Engl. J. Med.* **1987**, *316*, 927–931. [CrossRef] [PubMed]
27. Arioli, V.; Pallanza, R. Teicoplanin-resistant coagulase-negative staphylococci. *Lancet* **1987**, *1*, 39. [CrossRef] [PubMed]
28. Biavasco, F.; Vignaroli, C.; Varaldo, P.E. Glycopeptide resistance in coagulase-negative staphylococci. *Eur. J. Clin. Microbiol. Infect. Dis.* **2000**, *19*, 403–417. [CrossRef]
29. Cavenaghi, L.A.; Biganzoli, E.; Danese, A.; Parenti, F. Diffusion of teicoplanin and vancomycin in agar. *Diagn. Microbiol. Infect. Dis.* **1992**, *15*, 253–258. [CrossRef] [PubMed]
30. Vaudaux, P.; Huggler, E.; Bernard, L.; Ferry, T.; Renzoni, A.; Lew, D.P. Underestimation of vancomycin and teicoplanin MICs by broth microdilution leads to underdetection of glycopeptide-intermediate isolates of Staphylococcus aureus. *Antimicrob. Agents Chemother.* **2010**, *54*, 3861–3870. [CrossRef]
31. Humphries, R.M.; Ambler, J.; Mitchell, S.L.; Castanheira, M.; Dingle, T.; Hindler, J.A.; Koeth, L.; Sei, K. CLSI Methods Development and Standardization Working Group Best Practices for Evaluation of Antimicrobial Susceptibility Tests. *J. Clin. Microbiol.* **2018**, *56*, e01934-17. [CrossRef]
32. Campana, E.H.; Carvalhaes, C.G.; Nonato, B.; Machado, A.M.; Gales, A.C. Comparison of M.I.C.E. and Etest with CLSI agar dilution for antimicrobial susceptibility testing against oxacillin-resistant *Staphylococcus* spp. *PLoS ONE* **2014**, *9*, e94627. [CrossRef]
33. Wootton, M.; Howe, R.A.; Hillman, R.; Walsh, T.R.; Bennett, P.M.; MacGowan, A.P. A modified population analysis profile (PAP) method to detect hetero-resistance to vancomycin in Staphylococcus aureus in a UK hospital. *J. Antimicrob. Chemother.* **2001**, *47*, 399–403. [CrossRef]
34. Matuschek, E.; Åhman, J.; Webster, C.; Kahlmeter, G. Antimicrobial susceptibility testing of colistin—Evaluation of seven commercial MIC products against standard broth microdilution for *Escherichia coli*, *Klebsiella pneumoniae*, *Pseudomonas aeruginosa*, and *Acinetobacter* spp. *Clin. Microbiol. Infect.* **2018**, *24*, 865–870. [CrossRef] [PubMed]
35. EUCAST. Media Preparation for EUCAST Disk Diffusion Testing and for Determination of MIC Values by the Broth Microdilution Method. 2020. Available online: https://www.eucast.org/fileadmin/src/media/PDFs/EUCAST_files/Disk_test_documents/2020_manuals/Media_preparation_v_6.0_EUCAST_AST.pdf (accessed on 6 December 2021).
36. CLSI. Methods for Dilution Antimicrobial Susceptibility Tests for Bacteria That Grow Aerobically; Approved Standard—Ninth Edition. 2012. Available online: https://clsi.org/standards/products/microbiology/documents/m07/ (accessed on 6 December 2021).
37. EUCAST. EUCAST Guidelines for Detection of Resistance Mechanisms and Specific Resistances of Clinical and/or Epidemiological Importance, Version 2.0. 2017. Available online: https://www.eucast.org/fileadmin/src/media/PDFs/EUCAST_files/Resistance_mechanisms/EUCAST_detection_of_resistance_mechanisms_170711.pdf (accessed on 6 December 2021).
38. Liofilchem. MTSTM Technical Sheet Staphylococci-Rev.5/ 31 May 2021. 2021. Available online: https://www.liofilchem.com/images/brochure/mic_test_strip_patent/MTS20.pdf (accessed on 10 June 2022).
39. Hiramatsu, K.; Aritaka, N.; Hanaki, H.; Kawasaki, S.; Hosoda, Y.; Hori, S.; Fukuchi, Y.; Kobayashi, I. Dissemination in Japanese hospitals of strains of Staphylococcus aureus heterogeneously resistant to vancomycin. *Lancet* **1997**, *350*, 1670–1673. [CrossRef] [PubMed]

40. EUCAST. Routine and Extended Internal Quality Control for MIC Determination and Disk Diffusion as Recommended by EUCAST. 2016. Available online: https://www.eucast.org/fileadmin/src/media/PDFs/EUCAST_files/QC/v_6.1_EUCAST_QC_tables_routine_and_extended_QC.pdf (accessed on 10 June 2022).
41. EUCAST. Breakpoint Tables for Interpretation of MICs and Zone Diameters Version 7.1, valid from 10 March 2017. 2017. Available online: https://www.eucast.org/fileadmin/src/media/PDFs/EUCAST_files/Breakpoint_tables/v_7.1_Breakpoint_Tables.pdf (accessed on 10 June 2022).
42. CLSI. *Performance Standards for Antimicrobial Susceptibility Testing*; Twenty-Second Informational Supplement 2012; Clinical and Laboratory Standards Institute: Wayne, PA, USA, 2012.

antibiotics

MDPI

Article

Molecular Characterization of Community- and Hospital-Acquired Methicillin-Resistant *Staphylococcus aureus* Isolates during COVID-19 Pandemic

Muhammad Sohail [1,2,*], Moazza Muzzammil [3], Moaz Ahmad [3], Sabahat Rehman [4], Mohammed Garout [5], Taghreed M. Khojah [6], Kholoud M. Al-Eisa [6], Samar A. Breagesh [6], Rola M. Al Hamdan [6], Halimah I. Alibrahim [6], Zainab A. Alsoliabi [7], Ali A. Rabaan [8,9,10] and Naveed Ahmed [11,*]

1 Department of Microbiology, Chughtai Lab, Lahore 54000, Pakistan
2 Department of Medical Lab Technology, Faculty of Rehabilitation & Allied Health Sciences, Riphah International University Islamabad (QIE Campus), Lahore 54000, Pakistan
3 Department of Medical Education, King Edward Medical University, Lahore 54000, Pakistan
4 Department of Pathology, HITEC Institute of Medical Sciences, Heavy Industries Taxila Cantt, Taxila 47070, Pakistan
5 Department of Community Medicine and Health Care for Pilgrims, Faculty of Medicine, Umm Al-Qura University, Makkah 21955, Saudi Arabia
6 Molecular Biology Department, Riyadh Regional Laboratory, Riyadh 11425, Saudi Arabia
7 Pharmacy Department, Qatif Central Hospital, Qatif 32654, Saudi Arabia
8 Molecular Diagnostic Laboratory, Johns Hopkins Aramco Healthcare, Dhahran 31311, Saudi Arabia
9 College of Medicine, Alfaisal University, Riyadh 11533, Saudi Arabia
10 Department of Public Health and Nutrition, The University of Haripur, Haripur 22610, Pakistan
11 Department of Medical Microbiology and Parasitology, School of Medical Sciences, Universiti Sains Malaysia, Kubang Kerian 16150, Malaysia
* Correspondence: drsohailmmg@gmail.com (M.S.); namalik288@gmail.com (N.A.)

Abstract: Methicillin-resistant *Staphylococcus aureus* (MRSA) is a drug-resistant superbug that causes various types of community- and hospital-acquired infectious diseases. The current study was aimed to see the genetic characteristics and gene expression of MRSA isolates of nosocomial origin. A total of 221 MRSA isolates were identified from 2965 clinical samples. To identify the bacterial isolates, the clinical samples were inoculated on blood agar media plates first and incubated at 37 °C for 18–24 h. For further identification, the Gram staining and various biochemical tests were performed once the colonies appeared on the inoculated agar plates. The phenotypic identification of antibiotic susceptibility patterns was carried out using Kirby–Bauer disk diffusion method by following the Clinical and Laboratory Standards Institute (CLSI) 2019 guidelines. The biofilm-producing potentials of MRSA were checked quantitatively using a spectrophotometric assay. All strains were characterized genotypically by *SCCmec* and *agr* typing using the specific gene primers. Furthermore, a total of twelve adhesion genes were amplified in all MRSA isolates. MRSA was a frequently isolated pathogen (44% community acquired (CA)-MRSA and 56% hospital acquired (HA)-MRSA), respectively. Most of the MRSA isolates were weak biofilm producers (78%), followed by moderate (25%) and strong (7%) biofilm producers, respectively. Prominent adhesion genes were *clfB* (100%), *icaAD* (91%), *fib* (91%), *sdrC* (91%) followed by *eno* (89%), *fnbA* (77%), *sdrE* (67%), *icaBC* (65%), *clfA* (65%), *fnbB* (57%), *sdrD* (57%), and *cna* (48%), respectively. The results of the current study will help to understand and manage the spectrum of biofilm-producing MRSA-associated hospital-acquired infections and to provide potential molecular candidates for the identification of biofilm-producing MRSA.

Keywords: hospital-acquired infections; biofilm formation; adhesion genes; MRSA; antimicrobial resistance; AMR

1. Introduction

Methicillin-resistant *Staphylococcus aureus* (MRSA) is a notorious multidrug resistant (MDR) superbug, commonly associated with hospital-acquired infections (HAIs) [1,2], and these infections are always challenging to treat through a single intervention [3]. Previous studies have revealed that severity and diversity of infection caused by MRSA depend upon the combination of a set of adhesion genes and molecular typing of the pathogen [4,5]. It is believed that MRSA evolved gradually and acquired the antibacterial resistance inducing mobile genetic elements and biofilm production, which makes the MRSA a superbug of hospital-acquired (HA-MRSA) and community-acquired (CA-MRSA) infections [6].

Molecular typing involving SCC*mec*, *agr* typing, and microbial surface components identifying adhesive matrix molecules (MSCRAMM) are vital factors to depict the relatedness of MRSA and predict the severity of infection [7]. The most common MSCRAMM are clumping factor A and B (*clfA*, *clfB*), intercellular adhesin (*icaAD*, *icaBC*), laminin-binding protein (*eno*), fibronectin-binding proteins A and B (*fnbA*, *fnbB*), fibrinogen-binding protein (*fib*), serine–aspartate repeat proteins C, D and E (*sdrC*, *sdrD*, *sdrE*), and collagen-binding adhesin (*cna*), respectively [8]. Sequence type ST239 was the most frequently isolated HA-MRSA, whereas ST-59 was the most frequently isolated clone among CA-MRSA worldwide [9]. Similarly, clonal complex CC8, CC1, and sequence type ST8 and ST30, respectively, are most frequently reported in Pakistan [10].

Biofilm is a complex mechanism of bacterial communities that poses extraordinary resistance to antibiotics, persistent infections, and enhances the survival of pathogens inside the human body over extended periods. Biofilm formation is a multistep process that depends upon bacterial phenotypical characteristics, gene expression, and adhesion proteins present on the surface of bacteria [11,12]. Some MRSA strains having specific genotyping characteristics can form biofilm various infection sites [13].

The attachment, maturation, and dispersion are the three phases of biofilm development. The attachment phase may also be divided into the first reversible attachment and the irreversible attachment stages [12]. Stronger physical or chemical shear pressures may be tolerated by the irreversibly attached biofilm. The first stage of biofilm development for the human infections by *Staphylococcus aureus* (*S. aureus*) is adhesion to human matrix proteins including fibronectin (Fn), fibrinogen (Fg), and vitronectin (Vn), etc. [14]. The peptidoglycan on the cell wall is covalently linked to microbial surface components that recognized adhesive matrix molecules dependent adhesions. Cells of the same species or cells from other species are attracted to the biofilm from the bulk fluid after the initial layer of the biofilm has been formed. A thin layer of biofilm develops into a mushroom or tower-shaped structure. Bacteria are arranged in a thick biofilm (>100 layers) in accordance with their metabolism and aero tolerance [15]. The imprisoned bacteria produce additional biofilm scaffolds as the biofilm develops, including proteins, DNA, polysaccharides, etc. The dispersion phase, which is similarly crucial for the biofilm life cycle, comes after biofilm maturation. There are several factors for biofilms dispersion, including nutritional deficiency, stiff competition, population growth, etc. The whole biofilm may experience dispersal, or just a part of it. The emergence of new biofilms at different places is promoted by the release of planktonic bacteria [16,17].

While MRSA is one of the most important pathogens worldwide, some strains are restricted to one geographic area. This makes molecular characterization of great importance for investigating the epidemiology of MRSA both locally and globally. Biofilm formation is considered a virulence property of *S. aureus* and results in recurrent infections that are difficult to treat and lead to higher treatment costs. Therefore, the current study was conducted with the primary objective to see the overall prevalence of bacterial infections among possible community- and hospital-acquired infection during the COVID-19 pandemic. Simultaneously, the secondary objective was to investigate the phenotypic and genotyping characterization of MRSA isolates.

2. Materials and Methods

2.1. Study Design and Study Setting

Before starting the study, an ethical approval was obtained from the Institutional Review Board of Chughtai Lab, Lahore. The current study was conducted by the Department of Microbiology, Chughtai Lab, Lahore in collaboration with the Department of Medical Lab Technology, Faculty of Rehabilitation & Allied Health Sciences, Riphah International University Islamabad, QIE Campus Lahore, Pakistan, from the duration of 16 November 2021 to 4 March 2022. A total of 2965 clinical samples (blood, urine, sputum, tracheal aspirate, bronchoalveolar lavage, abscess, and wound swabs, respectively) were received from the different public and private sector hospitals for microbiological diagnosis. After the collection of samples, these were transported to the microbiology section of Chughtai Lab for further processing. Chughtai Lab is a private sector laboratory which has collection centers and stat labs throughout the country, providing the facilities for radiological and laboratory diagnosis.

At first, the clinical samples suspected of bacterial infection were processed in the Department of Microbiology, Chughtai Lab, for microbiological diagnosis, and then after the final diagnosis, only the MRSA isolates were collected and transported to the Department of Medical Lab Technology, Riphah International University Islamabad, QIE Campus Lahore for further molecular-based processing.

2.2. Data Collection

The data of patients was collected from the hospitals retrospectively using a pre-structured questionnaire. The data includes the coronavirus diseases-2019 (COVID-19) status of patients, and their comorbidities. The data about COVID-19 includes the current or previous history of acquiring COVID-19 infections and states if during the current study, the patients were tested for COVID-19 or not.

2.3. Pathogen Isolation and Identification

Following the standard microbiological techniques, samples were analyzed in the microbiology laboratory to isolate and identify the bacterial pathogens. After receiving the clinical samples in the microbiology laboratory, the samples were inoculated on blood and chocolate agar media plates with the sterile disposable wire loop and incubated at 37 °C for the period of 18–24 h. After the incubation period, the inoculated plates were taken out from the incubator and observed for the presence of bacterial colonies. The bacterial identification and confirmation was carried out by bacterial colony morphology, Gram staining, and biochemical (catalase, citrate, and coagulase) tests [18]. Once the bacterial colonies were identified, the isolated colonies were proceeded for antibiotic susceptibility testing.

2.4. Antimicrobial Sensitivity Testing

The antibiotic susceptibility testing was performed by following the guidelines from Clinical and Laboratory Standards Institute (CLSI) 2019. The antibiotic sensitivity testing of bacterial isolates was performed on Mueller-Hinton agar (Oxoid, UK) using the Kirby–Bauer disk diffusion method [19]. The panel of antibiotics against each of the isolated bacterium was selected based on the CLSI-2019 guidelines with certain extra drugs (amikacin and tobramycin). The tested isolates were declared sensitive or resistant by following the established criteria of the zone of inhibition (ZOIs) sizes around tested disk of antibiotic. The ZOIs of each tested antibiotic were given in CLSI guideline 2019 [19].

The tested antibiotics were amikacin (30 μg), chloramphenicol (30 μg), cefoxitin (30 μg), ciprofloxacin (5 μg), doxycycline (30 μg), ofloxacin (5 μg), trimethoprim/Sulfamethoxazole (23.75 μg), clindamycin (2 μg), azithromycin (15 μg), gentamicin (10 μg), linezolid (30 μg), tobramycin (10 μg), and vancomycin, respectively. The susceptibility pattern of vancomycin was checked by minimum inhibitory concentration (MIC) testing.

2.5. Biofilm Formation

The biofilm formation potential of MRSA isolates was assayed quantitatively using a spectrophotometric assay. Positive control included *Staphylococcus epidermidis* ATCC 12228. Standard bacterial inoculum (0.5 McFarland) was prepared in Blood Head Infusion (BHI) (Oxoid, UK) supplemented with 1% glucose (Sigma-Aldrich, Burlington, USA) and inoculated in 96 well plate. This polystyrene microtiter plate was incubated for 48 h at 37 °C without agitation. The cells were washed with physiological saline (Thermo Fisher Scientific, Waltham, MA, USA) three times and stained with 0.1% CV (crystal violet) (Sigma-Aldrich, Germany). After washing with physiological saline, the stain was dissolved in 200 µL of 95% ethanol and measured at 595 nm by an ELISA plate reader (Rayto, Shenzhen, China). This assay was performed in triplicate. Based on OD^{595}, biofilm formation capability was categorized as non-biofilm producers OD < 0.05, weak-biofilm producers OD > 0.5 and ≤1, moderate-biofilm producers OD > 1 and ≤2, and strong-biofilm producers having OD > 2, respectively [20].

2.6. Genotyping of MRSA

All strains were characterized genotypically for SCC*mec* [21] and *agr* typing [22] by using the primers and PCR conditions previously reported [23]. Twelve adhesion genes including *clfA*, *clfB*, *icaAD*, *icaBC*, *eno*, *fnbA*, *fnbB*, fib, *sdrC*, *sdrD*, *sdrE* and *cna* were amplified in all MRSA isolates following previously reported sets of primers and PCR conditions [10]. Sixteen strains of MRSA selected based on the antibiogram, SCC*mec*, and *agr* typing for MLST (Multilocus sequence typing) followed by gene expression studies [24].

2.7. Data Analysis

Statistical analysis was performed by employing the data in SPSS software version 22.0 (Chicago, IL, USA). The mean, standard deviation (SD) and percentages (%) were calculated.

3. Results

3.1. Isolation of MRSA

A total of 2965 clinical samples from 2692 patients for possible bacterial infection were analyzed to identify the pathogens and their antibiotic susceptibility patterns. From these 2965 samples, in total, 2637 were found to be positive for different bacterial infections. In total, 336 samples were found positive for two or three types of bacterial infections. The growth of *S. aureus* was seen in 24.82% (736/2965) of samples; among them, 30.02% (221/736) were MRSA. The prevalence of other pathogens is shown in Table 1.

Table 1. Overall prevalence of pathogens isolated from different clinical samples (n = 2973).

The Pathogen Identified following CLSI Guidelines		Number (n)	% Prevalence
S. aureus (n = 736)	MRSA	221	7.43
	MSSA	501	16.85
	CONS	14	0.47
Enterococcus faecalis		39	1.31
Klebsiella pneumoniae		778	26.16
Acinetobacter baumannii		336	11.30
Escherichia coli		801	26.94
Pseudomonas aeruginosa		257	8.64
Stenotrophomonas maltophilia		26	0.87

MRSA: Methicillin Resistance *Staphylococcus aureus*. CONS: Coagulase Negative *Staphylococcus aureus*. MSSA: Methicillin Sensitive *Staphylococcus aureus*.

From the 2692 patients included in the current study, a total of 876 were found positive for COVID-19 infection during the study time. A total of 1789 patients were hospitalized, whereas the remaining 903 were checked clinically by the clinical/physician, given medicines, and not hospitalized. Among the 876 COVID-19 patients, a total of 241 (27.51%) were found positive for bacterial infections. The prevalence of MRSA among COVID-19 patients who were coinfected with different bacterial infections was 17.4% (n = 42).

3.2. Antimicrobial Sensitivity Testing (AST)

The antimicrobial sensitivity profile of 221 strains of MRSA was determined following the standard concentration of antibiotics recommended by CLSI. All strains of MRSA were sensitive to vancomycin and linezolid. MRSA strains were susceptible to chloramphenicol (79%) and doxycycline (59%). On the other hand, MRSA was highly resistant to amikacin (81%), gentamycin (93%), tobramycin (96%), azithromycin (97%), ciprofloxacin (96%), ofloxacin (96%), Trimethoprim/Sulfamethoxazole (89%), and clindamycin (85%). The prevalence of CA-MRSA and HA-MRSA is 44% (98/221) and 56% (123/221) based upon genotypic characteristics. HA-MRSA are more resistant to tested antibiotics, especially amikacin, doxycycline, trimethoprim/sulfamethoxazole, and clindamycin compared to CA-MRSA except for chloramphenicol (Figure 1).

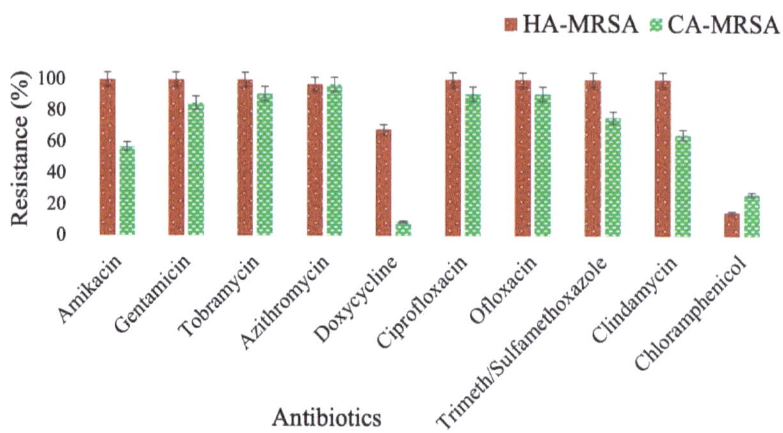

Figure 1. Antibiogram of HA-MRSA and CA-MRSA against recommended antibiotics.

3.3. Biofilm Assay

3.3.1. Congo Red Agar

MRSA showed varying degrees of slime production from very black colonies (7%), black (15%), and light back or pink colonies (78%), respectively. ATCC 35556 *S. aureus* produced very black colonies after 48 h of incubations, used as a positive reference strain, and ATCC 1228 was used as a negative control.

3.3.2. Quantitative Microtiter Plate Method

All MRSA, 100% strains, showed biofilm production with varying degree of adhesion from strong 7% (Optical density at 595 >1.0 nm), moderate 15% (Optical density at 595 < 1.0 nm and >0.6 nm), to weak adhesion 78% (Optical density at 595 < 0.6 nm). Reference strain ATCC 35556 *S. aureus* firmly adhered to the microtiter plate, whereas ATCC 1228 did not adhere to the plate. Based upon an antibiogram, most resistant strains of MRSA (n = 16) were selected for molecular studies. Biofilm production by CA-MRSA and HA-MRSA is illustrated in Figure 2.

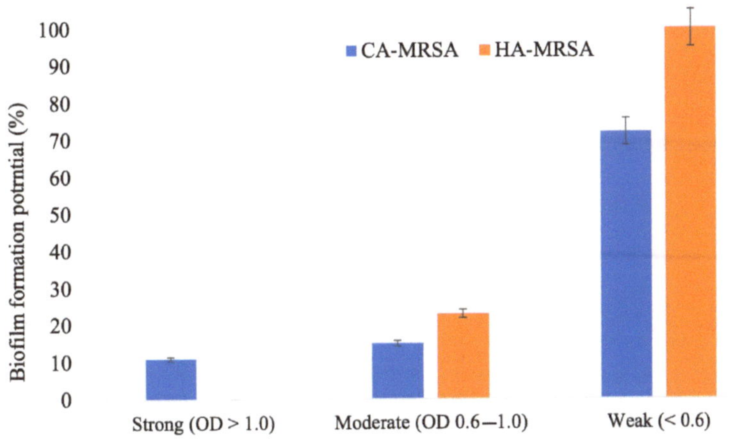

Figure 2. Biofilm production of CA-MRSA and HA-MRSA isolates using the Microtiter Plate method.

3.4. Detection of Biofilm Genes

Twelve biofilm-associated adhesion genes were detected among 221 strains of MRSA isolates. The prevalence of 12 gene involved in biofilm production is: *clfA* (65%), *clfB* (100%), *icaAD* (91%), *icaBC* (65%), *eno* (89%), *fnbA* (77%), *fnbB* (57%), *fib* (91%), *sdrC* (91%), *sdrD* (57%), *sdrE* (67%), and *cna* (48%), respectively.

3.5. Molecular Characterization of MRSA

SCC*mec* and *agr* typing elucidated that MRSA strains isolates are genetically diverse. Most of the strains were classified as SCC*mec* type II (20%), type III (17%), type IV (35%), type V (6%), and type VI (2%), and *agr* type I (40%), type II (8%), type III (5%), and type IV (4%). Some strains were not typed by SCC*mec* or/and *agr* typing. No statistically significant difference was found between biofilm formation protentional and SCC*mec* or *agr* typing. The correlation of biofilm formation protentional, adhesion genes, SCC*mec*, and *agr* typing is illustrated in Table 2.

Table 2. Molecular characterization of biofilm-producing MRSA isolates.

Biofilm	Biofilm Formation Genes	*SCCmec* Typing	*agr* Typing
Strong (OD > 1.0)	*clfA* (21%), *clfB* (24%), *icaAD* (23%), *icaBC* (24%), *eno* (24%), *fnbA* (22%), *fnbB* (20%), *fib* (23%), *sdrC* (25%), *sdrD* (24%), *sdrE* (27%), and *cna* (14%)	*SCCmec* II (25%), *SCCmec* III (15%), *SCCmec* IV (21%), *SCCmec* V (6%)	*agr* I (54%), *agr* II (15%), *agr* III (17%), *agr* IV (14%)
Moderate (OD 0.6–1.0)	*clfA* (27%), *clfB* (28%), *icaAD* (28%), *icaBC* (24%), *eno* (28%), *fnbA* (27%), *fnbB* (23%), *fib* (28%), *sdrC* (24%), *sdrD* (21%), *sdrE* (24%), and *cna* (20%)	*SCCmec* II (15%), *SCCmec* III (13%), *SCCmec* IV (33%), *SCCmec* V (12%)	*agr* I (15%), *agr* II (13%), *agr* III (33%), *agr* IV (13%)
Weak (< 0.6)	*clfA* (52%), *clfB* (49%), *icaAD* (49%), *icaBC* (53%), *eno* (48%), *fnbA* (50%), *fnbB* (57%), *fib* (49%), *sdrC* (51%), *sdrD* (54%), *sdrE* (48%), and *cna* (38%)	*SCCmec* II (14%), *SCCmec* III (21%), *SCCmec* IV (39%), *SCCmec* V (7%)	*agr* I (27%), *agr* II (5%), *agr* III (2%), *agr* IV (67%)

Genetic characterization of 16 MRSA strains was conducted to elucidate the relationship between biofilm formation potential and clonal lineages. Molecular analysis revealed that ST2490 (5/16) was the most frequent type, followed by ST8 (3/16), ST5 (2/16), and ST72 (2/16) that are responsible for biofilm formation.

4. Discussion

The infections by superbugs such as MRSA have always been a substantial threat to public health services and the healthcare system [25,26]. The HAIs are notoriously challenging to treat because of emerging and inherited antimicrobial resistance (AMR), biofilm formation, and low penetration at the site of infection [27]. The current study found that *S. aureus* isolates, primarily the MRSA, are predominantly associated with HAIs. The majority of MRSA (>50%) were resistant to available recommended antibiotics except for linezolid and vancomycin, to which MRSA was 100% sensitive. HA-MRSA was more resistant to tested antibiotics than CA-MRSA, especially amikacin, doxycycline, trimethoprim/sulfamethoxazole, and clindamycin. On the other hand, CA-MRSA was a strong biofilm producer as compared to HA-MRSA.

A previous study from the United Kingdom (UK) and Europe concluded that MRSA was a significant (>40%) pathogen for different infections [28]. A study from Lahore, Pakistan showed that the MRSA was 14.9% prevalent during the COVID-19 pandemic [29]. Some studies [30,31] also reported CONS as a significant pathogen for different infections. However, results of the current study showed the majority of the infections by *S. aureus* were caused MSSA strains (16.85%). This difference in prevalence is justifiable based on surgical practices and the tendency to review cases of surgical procedures and health facilities in general.

A study conducted on 105 strains of *S. aureus* isolated from various clinical specimens at Kabul University, Afghanistan concluded that all MRSA were sensitive to vancomycin, and 8.5% of MRSA were resistant to clindamycin; similar investigations were reported in the current study [32]. There are studies conducted in China, United States, South Africa, Japan, and Australia which reported linezolid and vancomycin susceptivity is 100% towards MRSA isolated from clinical specimens such as pus, bone and joint infections, and prosthetic device [33,34]. A study from India reported 1% resistance to linezolid and vancomycin [15]. Another study reported that the MRSA isolated from skin and soft tissue infection were 15% susceptible to ciprofloxacin, 53% erythromycin, 77% clindamycin, 86% doxycycline, 67% gentamicin, 48% cotrimoxazole, and 94% chloramphenicol [35], respectively; the results of this study were opposite to the current results - possibly due to the nature of specimens, empirical therapy, and guidelines for the usage of antibiotics for MRSA infections. On the other hand, it also reported that CA-MRSA was more sensitive to antibiotics compared with HA-MRSA, in agreement with the current study [35]. A study published recently on the antibiogram of *S. aureus* in Pakistan reported 1% resistance to linezolid, 2% to vancomycin, 16% to chloramphenicol, 42% to doxycycline, 56 % to gentamycin, 62% to azithromycin, 55% to ciprofloxacin, 56% to ofloxacin, 43% to Trimethoprim/Sulfamethoxazole, and 41% to clindamycin [36], respectively. These results are opposite to the results of the current study except for chloramphenicol and doxycycline. The difference in the antibiogram was justifiable in terms of the nature of specimens such as poultry, and animal-related infections were the most difficult to treat infections and demand more extended antibiotic therapy and a hospital stay which contributes to the emergence of antibiotic resistance; other factors may include prescription of antibiotics, availability of antibiotics, and biofilm-forming potential of the causative agent.

Biofilm formation involves a complex community of pathogens on the biotic and non-biotic surfaces and are labeled as major contributing factors in the AMR in biofilms [37,38]. A previous study has isolated 305 MRSA and revealed that all strains (100%) were biofilm producers; among them, only 13% were strong biofilm producers [39]. These investigations are in line with the current study outcomes.

A previous study on the biofilm formation potential of MRSA concluded that prevalence of *fib* is 90%, *cna* 93%, and *fnbB* 53% [20], respectively. These results are the same as reported by our study except *cna,* in which we reported 48%. Previous studies showed variable results of biofilm-related genes such as *icaA and icaD* (34%) [17], *clfA* (100%), *eno* (78%) *clfB* (100%), *fib* (74%), *fnbB* (46%), *fnbA* (56%), and *cna* (54%) [40], respectively. Another study reported the *fnbA* (78%), *fnbB* (81%), *clfA* (59%), and *cna* (73%) in MRSA isolated from pharyngitis patients and have the potential of biofilm production [15]; a recent study reported *clfA* and *clfB* completely dominant followed by *fnbA* (80%), *fnbB* (77%), *sdrC* (68%), *icaA* (63%), *icaD* (58%), *sdrD* (54%), *can* (25%), and *fib* (20%) in biofilm producer MRSA [41], respectively. These dissimilarities in the prevalence of virulence genes might be influenced by genetic factors, epidemiological variations, specimen source, transmission routes, and environmental factors, respectively.

This study revealed that most of the potent biofilm producer strains of MRSA belonged to SCC*mec* II and *agr*I, moderate biofilm producers belonged to SCC*mec* IV, and *agr* III and weak biofilm producers belonged to SCC*mec* IV and *agr* IV, respectively. A previous study on MRSA isolated from bacteremia patients reported that SCC*mec* IV is predominant in biofilm development [42]. Previous studies involving biofilm formation potential of MRSA and SCC*mec* and *agr* typing demonstrated the same results as indicated in this study with some differences which are justifiable in terms of specimen nature and geographical privileges [14,16,43]. Biofilm formation potential was not significantly attributed to the SCC*mec* and *agr* typing, but both schemes typed all MRSA strains that produced biofilm.

Study Limitations: This current study successfully justified the outcomes and explained the variations among various studies conducted in different parts of the world. However, it has certain limitations. Besides the possibility of bacterial infections, the possibility of other respiratory viral infections, or infections by atypical bacteria, were not investigated, which might be important to rule out the possibility of other lethal bacterial infections. The current study did not report the data on clinical and subclinical conditions of patients because of the restriction in ethical approval from the target institutions. Furthermore, because of the financial limitations, the sequencing analysis of PCR products could not be carried out in order to further identify the isolates as well as to check their phylogenetics. Hence, some factors might be missing that contribute to HAIs. Further large-scale and multi-institutional studies are recommended.

5. Conclusions

This genetic characterization of MRSA isolates revealed that most of the potent biofilm producer strains belonged to SCC*mec* II and *agr*I, moderate biofilm producers belonged to SCC*mec* IV, and *agr* III, and weak biofilm producers belonged to SCC*mec* IV and *agr* IV, respectively. Furthermore, the results of the current study significantly contributed to understanding and managing the spectrum of biofilm-producing MRSA-associated infectious diseases probably in hospital settings. The results also provide the potential molecular candidates for biofilm-producing MRSA.

Author Contributions: Conceptualization, M.S. and A.A.R.; methodology, M.M., M.A., S.R. and M.S.; software, M.M. and M.A.; validation, M.G., T.M.K., K.M.A.-E., S.A.B. and R.M.A.H.; formal analysis, M.M., M.A., S.R. and M.S.; investigation, M.M., M.A., S.R. and M.S.; resources, S.A.B., R.M.A.H., H.I.A. and M.S.; data curation, S.A.B., R.M.A.H., H.I.A., M.S. and A.A.R.; writing—original draft preparation, M.S., M.M., M.A., H.I.A. and M.G.; writing—review and editing, N.A., T.M.K., K.M.A.-E., S.A.B., Z.A.A. and A.A.R.; visualization, M.S.; supervision, A.A.R.; project administration, N.A. and M.S. All authors have read and agreed to the published version of the manuscript.

Funding: This research received no external funding.

Institutional Review Board Statement: The study was conducted in accordance with the Declaration of Helsinki, and approved by the Institutional Review Board of Chughtai Lab (protocol code: CLL-IRB/AL/2021-11/002 and was approved on 11 November 2021.

Informed Consent Statement: Informed consent was obtained from all subjects involved in the study. Written informed consent has been obtained from the patient(s) to publish this paper.

Data Availability Statement: Data related to the current study can be accessed upon a reasonable request to drsohailmmg@gmail.com.

Acknowledgments: The authors would like to acknowledge the University of the Punjab Lahore, Pakistan, and the Chughtai Lab for providing workplace and excellent research opportunities.

Conflicts of Interest: The authors declare no conflict of interest.

References

1. Schwotzer, N.; Wahl, P.; Fracheboud, D.; Gautier, E.; Chuard, C. Optimal culture incubation time in orthopedic device-associated infections: A retrospective analysis of prolonged 14-day incubation. *J. Clin. Microbiol.* **2014**, *52*, 61–66. [CrossRef] [PubMed]
2. Parveen, S.; Saqib, S.; Ahmed, A.; Shahzad, A.; Ahmed, N. Prevalence of MRSA colonization among healthcare-workers and effectiveness of decolonization regimen in ICU of a Tertiary care Hospital, Lahore, Pakistan. *Adv. Life Sci.* **2020**, *8*, 38–41.
3. Crosby, H.A.; Tiwari, N.; Kwiecinski, J.M.; Xu, Z.; Dykstra, A.; Jenul, C.; Fuentes, E.J.; Horswill, A.R. The Staphylococcus aureus ArlRS two-component system regulates virulence factor expression through MgrA. *Mol. Microbiol.* **2020**, *113*, 103–122. [CrossRef] [PubMed]
4. Crandall, H.; Kapusta, A.; Killpack, J.; Heyrend, C.; Nilsson, K.; Dickey, M.; Daly, J.A.; Ampofo, K.; Pavia, A.T.; Mulvey, M.A. Clinical and molecular epidemiology of invasive Staphylococcus aureus infection in Utah children; continued dominance of MSSA over MRSA. *PLoS ONE* **2020**, *15*, e0238991. [CrossRef]
5. Okorie-Kanu, O.J.; Anyanwu, M.U.; Ezenduka, E.V.; Mgbeahuruike, A.C.; Thapaliya, D.; Gerbig, G.; Ugwuijem, E.E.; Okorie-Kanu, C.O.; Agbowo, P.; Olorunleke, S. Molecular epidemiology, genetic diversity and antimicrobial resistance of Staphylococcus aureus isolated from chicken and pig carcasses, and carcass handlers. *PLoS ONE* **2020**, *15*, e0232913. [CrossRef]
6. Qayoom, I.; Verma, R.; Murugan, P.A.; Raina, D.B.; Teotia, A.K.; Matheshwaran, S.; Nair, N.N.; Tägil, M.; Lidgren, L.; Kumar, A. A biphasic nanohydroxyapatite/calcium sulphate carrier containing Rifampicin and Isoniazid for local delivery gives sustained and effective antibiotic release and prevents biofilm formation. *Sci. Rep.* **2020**, *10*, 1–14. [CrossRef]
7. Su, M.; Satola, S.W.; Read, T.D. Genome-based prediction of bacterial antibiotic resistance. *J. Clin. Microbiol.* **2019**, *57*, e01405–e01418. [CrossRef]
8. Hussain, S.; Joo, J.; Kang, J.; Kim, B.; Braun, G.B.; She, Z.-G.; Kim, D.; Mann, A.P.; Mölder, T.; Teesalu, T. Antibiotic-loaded nanoparticles targeted to the site of infection enhance antibacterial efficacy. *Nat. Biomed. Eng.* **2018**, *2*, 95–103. [CrossRef]
9. Peng, H.; Liu, D.; Ma, Y.; Gao, W. Comparison of community-and healthcare-associated methicillin-resistant Staphylococcus aureus isolates at a Chinese tertiary hospital, 2012–2017. *Sci. Rep.* **2018**, *8*, 1–8. [CrossRef]
10. Sohail, M.; Latif, Z. Molecular analysis, biofilm formation, and susceptibility of methicillin-resistant Staphylococcus aureus strains causing community-and health care-associated infections in central venous catheters. *Rev. Soc. Bras. Med. Trop.* **2018**, *51*, 603–609. [CrossRef]
11. Tursi, S.A.; Puligedda, R.D.; Szabo, P.; Nicastro, L.K.; Miller, A.L.; Qiu, C.; Gallucci, S.; Relkin, N.R.; Buttaro, B.A.; Dessain, S.K. Salmonella Typhimurium biofilm disruption by a human antibody that binds a pan-amyloid epitope on curli. *Nat. Commun.* **2020**, *11*, 1–13. [CrossRef] [PubMed]
12. Vestby, L.K.; Grønseth, T.; Simm, R.; Nesse, L.L. Bacterial biofilm and its role in the pathogenesis of disease. *Antibiotics* **2020**, *9*, 59. [CrossRef] [PubMed]
13. Baldan, R.; Sendi, P. Precision medicine in the diagnosis and management of orthopedic biofilm infections. *Front. Med.* **2020**, *7*, 580671. [CrossRef] [PubMed]
14. Vanhommerig, E.; Moons, P.; Pirici, D.; Lammens, C.; Hernalsteens, J.-P.; De Greve, H.; Kumar-Singh, S.; Goossens, H.; Malhotra-Kumar, S. Comparison of biofilm formation between major clonal lineages of methicillin resistant Staphylococcus aureus. *PloS ONE* **2014**, *9*, e104561. [CrossRef] [PubMed]
15. Gowrishankar, S.; Kamaladevi, A.; Balamurugan, K.; Pandian, S.K. In vitro and in vivo biofilm characterization of methicillin-resistant Staphylococcus aureus from patients associated with pharyngitis infection. *BioMed Res. Int.* **2016**, *2016*, 1289157. [CrossRef] [PubMed]
16. Lade, H.; Park, J.H.; Chung, S.H.; Kim, I.H.; Kim, J.-M.; Joo, H.-S.; Kim, J.-S. Biofilm formation by Staphylococcus aureus clinical isolates is differentially affected by glucose and sodium chloride supplemented culture media. *J. Clin. Med.* **2019**, *8*, 1853. [CrossRef] [PubMed]
17. Nasr, R.A.; AbuShady, H.M.; Hussein, H.S. Biofilm formation and presence of icaAD gene in clinical isolates of staphylococci. *Egypt. J. Med. Hum. Genet.* **2012**, *13*, 269–274. [CrossRef]
18. Carnicer-Pont, D.; Bailey, K.A.; Mason, B.; Walker, A.; Evans, M.R.; Salmon, R. Risk factors for hospital-acquired methicillin-resistant Staphylococcus aureus bacteraemia: A case-control study. *Epidemiol. Infect.* **2006**, *134*, 1167–1173. [CrossRef]
19. Wayne, P. *Performance Standards for Antimicrobial Susceptibility Testing*; Clinical and Laboratory Standards Institute: Wayne, PA, USA, 2019.
20. Atshan, S.S.; Shamsudin, M.N.; Thian Lung, L.T.; Sekawi, Z.; Ghaznavi-Rad, E.; Pei Pei, C. Comparative characterisation of genotypically different clones of MRSA in the production of biofilms. *J. Biomed. Biotechnol.* **2012**, *2012*, 417247. [CrossRef]
21. Boye, K.; Bartels, M.; Andersen, I.; Møller, J.; Westh, H. A new multiplex PCR for easy screening of methicillin-resistant Staphylococcus aureus SCCmec types I–V. *Clin. Microbiol. Infect.* **2007**, *13*, 725–727. [CrossRef]
22. Shopsin, B.; Mathema, B.; Alcabes, P.; Said-Salim, B.; Lina, G.; Matsuka, A.; Martinez, J.; Kreiswirth, B. Prevalence of agr specificity groups among Staphylococcus aureus strains colonizing children and their guardians. *J. Clin. Microbiol.* **2003**, *41*, 456–459. [CrossRef] [PubMed]

23. Abbasian, S.; Farahani, N.N.; Mir, Z.; Alinejad, F.; Haeili, M.; Dahmardehei, M.; Mirzaii, M.; Khoramrooz, S.S.; Nasiri, M.J.; Darban-Sarokhalil, D. Genotypic characterization of Staphylococcus aureus isolated from a burn centre by using agr, spa and SCCmec typing methods. *New Microbes New Infect.* **2018**, *26*, 15–19. [CrossRef] [PubMed]
24. Maiden, M.C.; Bygraves, J.A.; Feil, E.; Morelli, G.; Russell, J.E.; Urwin, R.; Zhang, Q.; Zhou, J.; Zurth, K.; Caugant, D.A. Multilocus sequence typing: A portable approach to the identification of clones within populations of pathogenic microorganisms. *Proc. Natl. Acad. Sci. USA* **1998**, *95*, 3140–3145. [CrossRef]
25. Nishitani, K.; Sutipornpalangkul, W.; de Mesy Bentley, K.L.; Varrone, J.J.; Bello-Irizarry, S.N.; Ito, H.; Matsuda, S.; Kates, S.L.; Daiss, J.L.; Schwarz, E.M. Quantifying the natural history of biofilm formation in vivo during the establishment of chronic implant-associated Staphylococcus aureus osteomyelitis in mice to identify critical pathogen and host factors. *J. Orthop. Res.* **2015**, *33*, 1311–1319. [CrossRef]
26. Schwarz, E.M.; Parvizi, J.; Gehrke, T.; Aiyer, A.; Battenberg, A.; Brown, S.A.; Callaghan, J.J.; Citak, M.; Egol, K.; Garrigues, G.E. 2018 international consensus meeting on musculoskeletal infection: Research priorities from the general assembly questions. *J. Orthop. Res.* **2019**, *37*, 997–1006. [CrossRef]
27. Urish, K.L.; Cassat, J.E. Staphylococcus aureus osteomyelitis: Bone, bugs, and surgery. *Infect. Immun.* **2020**, *88*, e00932-19. [CrossRef] [PubMed]
28. Walls, R.; Roche, S.; O'Rourke, A.; McCabe, J. Surgical site infection with methicillin-resistant Staphylococcus aureus after primary total hip replacement. *J. Bone Jt. Surgery. Br. Vol.* **2008**, *90*, 292–298. [CrossRef]
29. Rizvi, A.; Saeed, M.U.; Nadeem, A.; Yaqoob, A.; Rabaan, A.A.; Bakhrebah, M.A.; Al Mutair, A.; Alhumaid, S.; Aljeldah, M.; Al Shammari, B.R.; et al. Evaluation of Bi-Lateral Co-Infections and Antibiotic Resistance Rates among COVID-19 Patients in Lahore, Pakistan. *Medicina* **2022**, *58*, 904. [CrossRef]
30. Jahanmard, F.; Croes, M.; Castilho, M.; Majed, A.; Steenbergen, M.; Lietaert, K.; Vogely, H.; Van Der Wal, B.; Stapels, D.; Malda, J. Bactericidal coating to prevent early and delayed implant-related infections. *J. Control. Release* **2020**, *326*, 38–52. [CrossRef]
31. Zimmerli, W.; Widmer, A.F.; Blatter, M.; Frei, R.; Ochsner, P.E. Role of rifampin for treatment of orthopedic implant–related staphylococcal infections: A randomized controlled trial. *Jama* **1998**, *279*, 1537–1541. [CrossRef]
32. Naimi, H.M.; Rasekh, H.; Noori, A.Z.; Bahaduri, M.A. Determination of antimicrobial susceptibility patterns in Staphylococcus aureus strains recovered from patients at two main health facilities in Kabul, Afghanistan. *BMC Infect. Dis.* **2017**, *17*, 1–7. [CrossRef]
33. Li, X.; Huang, T.; Xu, K.; Li, C.; Li, Y. Molecular characteristics and virulence gene profiles of Staphylococcus aureus isolates in Hainan, China. *BMC Infect. Dis.* **2019**, *19*, 1–12. [CrossRef]
34. Yu, Y.; Yao, Y.; Weng, Q.; Li, J.; Huang, J.; Liao, Y.; Zhu, F.; Zhao, Q.; Shen, X.; Niu, J. Dissemination and molecular characterization of Staphylococcus aureus at a Tertiary Referral Hospital in Xiamen City, China. *BioMed Res. Int.* **2017**, *2017*, 1367179. [CrossRef]
35. Alvarez-Uria, G.; Reddy, R. Prevalence and antibiotic susceptibility of community-associated methicillin-resistant Staphylococcus aureus in a rural area of India: Is MRSA replacing methicillin-susceptible Staphylococcus aureus in the community? *Int. Sch. Res. Not.* **2012**, *2012*, 248951. [CrossRef] [PubMed]
36. Bilal, H.; Khan, M.N.; Rehman, T.; Hameed, M.F.; Yang, X. Antibiotic resistance in Pakistan: A systematic review of past decade. *BMC Infect. Dis.* **2021**, *21*, 1–19. [CrossRef]
37. Chait, R.; Craney, A.; Kishony, R. Antibiotic interactions that select against resistance. *Nature* **2007**, *446*, 668–671. [CrossRef]
38. Rabaan, A.A.; Alhumaid, S.; Mutair, A.A.; Garout, M.; Abulhamayel, Y.; Halwani, M.A.; Alestad, J.H.; Bshabshe, A.A.; Sulaiman, T.; AlFonaisan, M.K.; et al. Application of Artificial Intelligence in Combating High Antimicrobial Resistance Rates. *Antibiotics* **2022**, *11*, 784. [CrossRef]
39. Post, V.; Wahl, P.; Uçkay, I.; Ochsner, P.; Zimmerli, W.; Corvec, S.; Loiez, C.; Richards, R.G.; Moriarty, T.F. Phenotypic and genotypic characterisation of Staphylococcus aureus causing musculoskeletal infections. *Int. J. Med. Microbiol.* **2014**, *304*, 565–576. [CrossRef]
40. Ghasemian, A.; Peerayeh, S.N.; Bakhshi, B.; Mirzaee, M. Comparison of biofilm formation between methicillin-resistant and methicillin-susceptible isolates of Staphylococcus aureus. *Iran. Biomed. J.* **2016**, *20*, 175.
41. Navidinia, M.; Mohammadi, A.; Arjmand, R.; Dadashi, M.; Goudarzi, M. Molecular typing, biofilm formation, and analysis of adhesion factors in Staphylococcus aureus strains isolated from urinary tract infections. *Gene Rep.* **2021**, *22*, 101008. [CrossRef]
42. Hamada, M.; Yamaguchi, T.; Sato, A.; Ono, D.; Aoki, K.; Kajiwara, C.; Kimura, S.; Maeda, T.; Sasaki, M.; Murakami, H. Increased incidence and plasma-biofilm formation ability of SCC mec type IV methicillin-resistant Staphylococcus aureus (MRSA) isolated from patients with bacteremia. *Front. Cell. Infect. Microbiol.* **2021**, *11*, 602833. [CrossRef] [PubMed]
43. Lee, A.S.; De Lencastre, H.; Garau, J.; Kluytmans, J.; Malhotra-Kumar, S.; Peschel, A.; Harbarth, S. Methicillin-resistant Staphylococcus aureus. *Nat. Rev. Dis. Prim.* **2018**, *4*, 1–23. [CrossRef] [PubMed]

antibiotics

MDPI

Article

Molecular Epidemiology of Methicillin-Resistant *Staphylococcus aureus* among Patients Diagnosed with Surgical Site Infection at Four Hospitals in Ethiopia

Seble Worku [1,2,3,*], Tamrat Abebe [1], Berhanu Seyoum [3], Ashenafi Alemu [3], Yidenek Shimelash [4], Marechign Yimer [3], Alemseged Abdissa [3], Getachew Tesfaye Beyene [3], Göte Swedberg [5] and Adane Mihret [1,3]

1 Department of Microbiology, Immunology and Parasitology, School of Medicine, College of Health Sciences, Addis Ababa University, Addis Ababa 1165, Ethiopia; tamrat.abebe@aau.edu.et (T.A.); adane.mihret@ahri.gov.et (A.M.)
2 Department of Medical Laboratory Science, College of Health Sciences, Debre Tabor University, Debre Tabor P.O. Box 272, Ethiopia
3 Bacterial and Viral Diseases Research Directorate, Armauer Hansen Research Institute, Addis Ababa 1165, Ethiopia; berhanu.seyoum@ahri.gov.et (B.S.); ashenafialemu07@gmail.com (A.A.); marechign01@yahoo.com (M.Y.); alemseged.abdissa@ahri.gov.et (A.A.); getachew.tesfaye@ahri.gov.et (G.T.B.)
4 Debre Tabor Comprehensive Specialized Hospital, Debre Tabor P.O. Box 272, Ethiopia; yidenek.shim2008@gmail.com
5 Department of Medical Biochemistry and Microbiology, Uppsala University, 750 08 Uppsala, Sweden; gote.swedberg@imbim.uu.se
* Correspondence: workuseble@ymail.com; Tel.: +251-911782487

Abstract: Methicillin-resistant *Staphylococcus aureus* (MRSA) is a common cause of severe surgical site infections (SSI). The molecular epidemiology of MRSA is poorly documented in Ethiopia. This study is designed to determine the prevalence of MRSA and associated factors among patients diagnosed with SSI. A multicenter study was conducted at four hospitals in Ethiopia. A wound culture was performed among 752 SSI patients. This study isolated *S. aureus* and identified MRSA using standard bacteriology, Matrix-Assisted Laser Desorption/Ionization Time-of-Flight Mass Spectrometry (MALDI-TOF MS), and cefoxitin disk diffusion test. The genes *mecA*, *femA*, *vanA*, and *vanB* were detected through PCR tests. *S. aureus* was identified in 21.6% of participants, with 24.5% of these being methicillin-resistant Staphylococci and 0.6% showing vancomycin resistance. Using MALDI-TOF MS for the 40 methicillin-resistant Staphylococci, we confirmed that 31 (77.5%) were *S. aureus*, 6 (15%) were *Mammaliicoccus sciuri*, and the other 3 (2.5%) were *Staphylococcus warneri*, *Staphylococcus epidermidis*, and *Staphylococcus haemolyticus*. The gene *mecA* was detected from 27.5% (11/40) of Staphylococci through PCR. Only 36.4% (4/11) were detected in *S. aureus*, and no *vanA* or *vanB* genes were identified. Out of 11 *mecA*-gene-positive Staphylococci, 8 (72.7%) were detected in Debre Tabor Comprehensive Specialized Hospital. Methicillin-resistant staphylococcal infections were associated with the following risk factors: age $\geq$ 61 years, prolonged duration of hospital stay, and history of previous antibiotic use, p-values < 0.05. Hospitals should strengthen infection prevention and control strategies and start antimicrobial stewardship programs.

Keywords: surgical site infection; methicillin-resistant *Staphylococci*; molecular epidemiology; antimicrobial resistance; Ethiopia

1. Introduction

Staphylococcus aureus (*S. aureus*) is a Gram-positive coccus that causes significant infections worldwide, including bacteremia, endocarditis, osteomyelitis, and skin and soft tissue infections, due to its easy transmission and commensal nature [1]. Not only *S. aureus* but also coagulase-negative Staphylococci (CoNS), which currently are defined as more than 40 species, are frequently associated with opportunistic human infections.

S. epidermidis and *S. haemolyticus* are the major species of CoNS frequently isolated from clinical specimens [2]. Furthermore, *Mammaliicoccus sciuri* (previously called *S. sciuri*) is part of the normal flora of goats and camels, and it is a rare opportunistic pathogen in humans [3]. *S. aureus* possesses a unique set of virulence factors, including toxins, enzymes, and metallophores, which enable it to survive extreme conditions, promote tissue colonization, cause systemic infection, *and* evade the host's immunity [4]. By utilizing metallophores, this bacterium can sequester metal ions from its environment [5]. *S. aureus* infections have previously been treated with beta-lactams, including penicillin and, later, methicillin, as well as sulfonamides, tetracyclines, and macrolides [6]. However, antibiotic-resistant strains of *S. aureus* have developed due to repeated exposure to antibiotics, leading to an increase in methicillin-resistant *S. aureus* (MRSA) infections globally. MRSA is one of the most causative pathogens of surgical site infections (SSIs), and it is a prevalent bacterium that frequently colonizes hospital environments and causes hospital-acquired infections [6] and community-acquired infections [7].

MRSA is characterized by resistance to penicillins, cephalosporins, and carbapenems, with the exception of the new anti-MRSA cephalosporins ceftaroline and ceftobiprole antimicrobial agents [6,8]. The main mechanism of resistance is an altered penicillin-binding protein (PBP2a/c) encoded by the *mecA* gene [1]. The *mecA* gene is regarded as the gold standard for identifying isolates of MRSA, and it is a helpful marker. It is highly conserved in staphylococcal strains and acquired through horizontal gene transfer. *mecA* is carried/located on the mobile genetic element staphylococcal cassette chromosome (SCC) mec, and it codes for the low-affinity PBP2a [9]. Other chromosomal factors, such as the high-level expression of *femA* and *femB*, also seem to be essential for high-level methicillin resistance [10]. The current treatment options for more serious MRSA infections requiring hospitalization include parenteral antimicrobials, such as teicoplanin, tigecycline, linezolid [8], trimethoprim–sulfamethoxazole, doxycycline, daptomycin [6], and vancomycin [8]. However, the majority of MRSA strains are capable of evolution and have acquired resistance to a variety of antibiotics, including those listed above [9,11].

Vancomycin resistance in MRSA was first discovered in 1996 in Japan following a few years of commercializing the antibiotic [12]. Vancomycin resistance is acquired through mutations and cell wall modification [12,13] mediated by a *vanA* gene cluster that can be acquired from vancomycin-resistant enterococcus (VRE) [11] through mobile genetic elements like transposonTn1546 [14]. Vancomycin-resistant *S. aureus* (VRSA) infections are treated with antibiotics like tigecycline, quinupristin, daptomycin, ceftobiprole, iclaprim, linezolid, and new glycopeptides (telavancin, oritavancin, and dabavancin) [15].

Globally, the prevalence of VRSA was 16% in Africa, 1% in Europe, 3% in South America, 4% in North America, and 5% in Asia [16]. A systematic review and meta-analysis revealed a highly variable prevalence of VRSA and MRSA in Ethiopian *S. aureus* isolates. The MRSA prevalence ranged from 8.3% to 77.3% (with a pooled prevalence of 32.5%) [17]. In the same way, there was a 5.1% to 44.3% variation in VRSA prevalence. [18]. These days, MRSA is considered a serious threat to public health, and it is one of the pathogens that needs to be treated with high priority. However, the molecular epidemiology of MRSA and VRSA is less well documented in Ethiopia, and published reports on MRSA- and VRSA-causing SSIs are scarce. Furthermore, almost all earlier reports depend on phenotypic laboratory methods. Therefore, in this study, we used the Matrix-Assisted Laser Desorption/Ionization Time-of-Flight Mass Spectrometry (MALDI-TOF MS) technique for the confirmation of bacterial isolates and multiplex polymerase chain reaction (PCR) for the detection of *mecA*, *femA*, *vanA*, and *vanB*.

2. Results

In this study, a total of 752 participants were included. *S. aureus* was isolated from 21.7% (163) of these participants. Following that, a cefoxitin disc diffusion test was used as a substitute marker for oxacillin and other penicillinase-resistant penicillins to ascertain the percentage of MRSA.

Of all participants, 5.3% (40/752) carried bacteria characterized as MRSA, while among isolates of *S. aureus*, the frequency of MRSA was 24.5% (Table 1). All methicillin-resistant isolates were also tested for vancomycin susceptibility. Except for one isolate (2.5%), all tested isolates for vancomycin were sensitive.

Table 1. Antibiotic resistance pattern of *S. aureus* isolates from patients diagnosed with surgical site infection at four different hospitals in Ethiopia between July 2020 and August 2021.

Antibiotics	Methods	N (%), % = N/752	N (%), % = N/163	AST Results	Strain
Cefoxitin	I-Z ≥ 22 mm	123 (16.4)	123 (75.5)	S	MSSA
	≤21 mm	40 (5.3)	40 (24.5)	R	MRSA

Abbreviations: S, susceptible; R, resistant; I, intermediate; I-Z, inhibition zone; MSSA, methicillin-sensitive *S. aureus*; MRSA, methicillin-resistant *S. aureus*; AST, antimicrobial susceptibility test.

The age of study participants with MRSA ranged from 8 days to 85 years, with a mean age (±standard deviation) of 35 ± 28.3 years and a median of 30 years, and 58.3% (95) were males. Fifty-nine (36.2%) of the participants received antimicrobial prophylaxis before the procedure, and 47.2% (63) underwent surgeries lasting longer than an hour (Table 2).

Table 2. Socio-demographic characteristics and clinical data of *S. aureus* among patients diagnosed with SSI at four different hospitals in Ethiopia from July 2020 to August 2021.

Variables/Characteristics		Frequency of *S. aureus* (%)
Gender	Male	95 (58.3)
	Female	68 (41.7)
Age in (years)	≤18	25 (20.9)
	19–40	77 (54)
	41–60	19 (13.5)
	≥61	42 (11.7)
Surgical site infection	Superficial	79 (48.5)
	Deep	84 (51.5)
Preoperative hospital stays	<7	77 (47.2)
	≥7	86 (52.8)
Previous use of antibiotics	Yes	79 (48.5)
	No	84 (51.5)
Smoking	Yes	16 (9.8)
	No	147 (90.1)
Alcoholic	Yes	48 (29.4)
	No	115 (70.6)
Nature of surgery	Emergency	55 (68.1)
	Elective	108 (31.9)
Type of surgery	Clean/Clean contaminated surgery	148 (90.8)
	Contaminated surgery	15 (9.2)
Timing of surgical antimicrobial prophylaxis	Before the operation	59 (36.2)
	During the operation	104 (63.8)
Duration of operation	<1 h	100 (52.8)
	≥1 h	63 (47.2)

The likelihood of MRSA SSI occurrence was about 3.7 times higher among patients aged ≥ 61 (AOR = 3.729 (1.179–11.791)) compared to those aged ≤ 60. Similarly, the relative risk of MRSA SSI occurrence was about 1.9 times more likely among patients who had a hospital stay ≥ 7 days (AOR = 1.856 (0.688–5.311)). Also, those who had a history of antibiotic use had a 3.7 times higher risk of developing methicillin-resistant Staphylococci infections (AOR = 3.692 (1.059–2.800)) than methicillin-sensitive *S. aureus* (MSSA) SSI. The likelihood of SSI occurrence was about 3.16 times more likely among patients who had antimicrobial prophylaxis during the operation (AOR = 3.066 (1.101–9.392)) than those who had antimicrobial prophylaxis before the operation. All p-value < 0.05 (Table 3).

Table 3. Bivariate and multivariate analysis for identification of methicillin resistance Staphylococci predictors among patients diagnosed with surgical site infection at four different hospitals in Ethiopia from July 2020 to August 2021.

Characteristics		Bacterial Growth		*p*-Value	Crude-OR (95%CI)	Adjusted-OR (95%CI)	*p*-Value
		MRSA	MSSA				
Gender	Male	29 (17.8)	66 (40.5)	0.039	2.276 (1.0444–4.9633)	1.638 (0.597–4.489)	0.337
	Female	11 (6.7)	57 (35)			1	
Age in (years)	≤18	2 (1.2)	23 (14.1)	0.000	2.788 (1.8716–4.154)	0.556 (0.1014–3.046)	0.499
	19–40	11 (6.7)	66 (40.5)			1	
	41–60	2 (1.2)	17 (10.4)			1.556 (0.259–9.328)	0.628
	≥61	25 (15.3)	17 (10.4)			3.729 (1.179–11.791)	**0.025**
Preoperative hospital stays	≤7	13 (8)	64 (39.3)	0.034	2.253 (1.064–4.771)	1	
	>7	27 (16.7)	59 (36.2)			1.856 (0.688–5.311)	**0.000**
Previous use of antibiotics	Yes	26 (16)	53 (32.5)			3.692 (1.059–2.800)	**0.025**
	No	14 (8.9)	70 (42.9)	0.001	3.256 (1.724–7.634)	1	
History of alcohol intake	Yes	18 (11)	30 (18.4)			1.075 (0.1331–8.6925)	0.945
	No	22 (13.5)	93 (57.1)	0.015	2.536 (1.202–5.351)	1	
Nature of surgery	Elective	16 (9.8)	92 (56.4)			1	
	Emergency	24 (14.7)	31 (19)	0.000	4.452 (2.098–9.445)	1.962 (0.0619–6.224)	0.000
Timing of surgical antimicrobial prophylaxis	Before the operation	7 (4.3)	57 (35)			1	
	After the operation	33 (20.2)	71(43.6)	0.006	3.453 (2.098–9.445)	3.066 (1.001–9.392)	0.05
Duration of operation	≤1 h	19 (11.7)	81 (49.7)			1	
	>1 h	21 (12.9)	42 (25.8)	0.004	2.132 (1.034–4.396)	1.890 (0.6321–5.652)	0.235

Bold: *p*-value statistically significant association.

2.1. MALDI-TOF MS Identification of Methicillin-Resistant Staphylococcus Isolates

Of the 40 phenotypic MRSA bacterial isolates, MALDI-TOF MS only identified 77.5% (31/40) as *S. aureus*, while 6 were identified as *M. sciuri*, and the other three as *S. warneri*, *S. epidermidis*, and *S. haemolyticus* (Figure 1). The majority (70%) of methicillin-resistant Staphylococcus isolates were identified from Debre Tabor Comprehensive Specialized Hospital (Figure 2) with 47.5% as *S. aureus*, 15% as *M. sciuri*, 2.5% as *S. warneri*, 2.5% as *S. epidermidis* and 2.5% as *S. haemolyticus*.

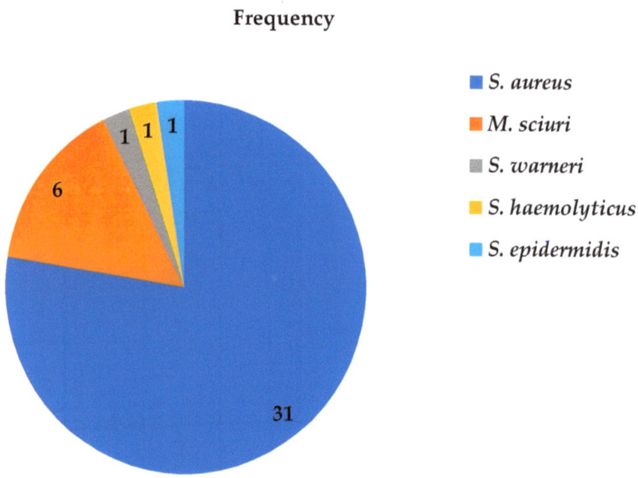

Figure 1. Frequency of methicillin-resistant Staphylococci isolates from patients diagnosed with surgical site infection at four different hospitals in Ethiopia using MALDI-TOF MS between July 2020 and August 2021.

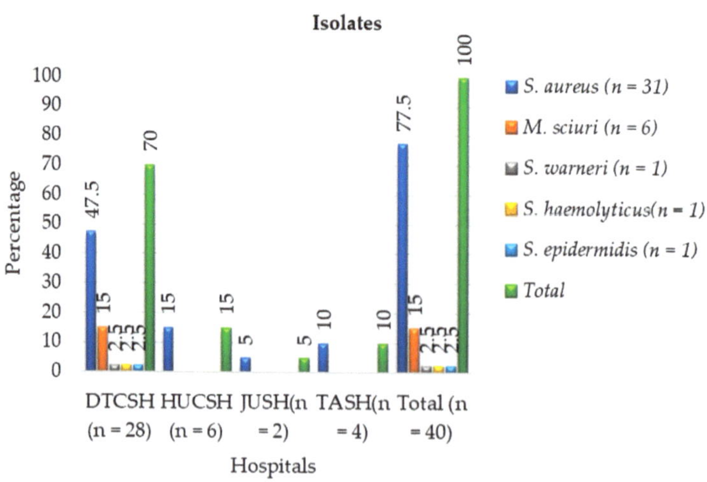

Figure 2. Frequencies of MALDI-TOF MS identification and distribution of phenotypic methicillin-resistant Staphylococci and *M. sciuri* isolates at each hospital between July 2020 and August 2021. DTCSH: Debre Tabor Comprehensive Specialized Hospital; HUCSH: Hawassa University Comprehensive Specialized Hospital; JUTSH: Jimma University Teaching Specialized Hospital; TASH: Tikur Anbessa Specialized Hospital.

2.2. *PCR amplification of mecA, femA, van A, and vanB*

Detection of *mecA, femA, van A*, and *vanB* was performed for all MRSA and methicillin-resistant Staphylococci other than *S. aureus* (MRSOSA). The PCR tests revealed that 27.5% (11/40) contained the *mecA* gene, 25% (10/40) were both *mecA*- and *femA*- positive, and 92.5% (37/40) showed the *femA* gene (Figure 3). On the other hand, from all eleven isolates that contained the *mecA* gene only, four were *S. aureus*, whereas five were *M. sciuri*, one was *S. warneri*, and one was *S. haemolyticus*, respectively (Figure 4 and Table 4). Among *S. aureus* isolates, only 12.9% (4/31) carried the *mecA* gene (MRSA), whereas 83.3% (5/6) of *M. sciuri* and both *S. warneri* and *S. haemolyticus* isolates carried *mecA* (Figure 3).

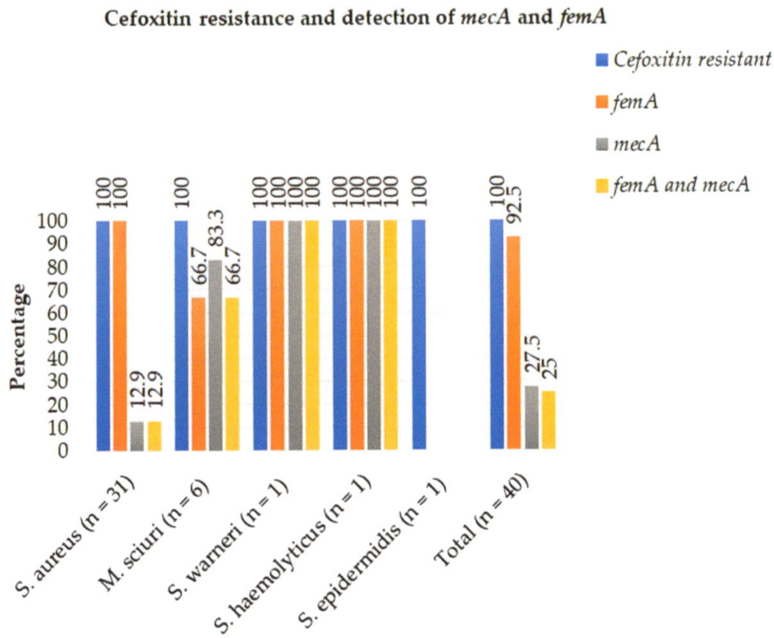

Figure 3. Frequency and distribution of cefoxitin resistance and PCR-confirmed gene among each Staphylococci isolate in patient diagnosed with surgical site infection between July 2020 and August 2021.

Table 4. Presentation of the cefoxitin and vancomycin resistance patterns of *mecA* carrying Staphylococci, and the distribution of *femA* and *van* genes among patients diagnosed with SSI at four different hospitals in Ethiopia from July 2020 to August 2021.

Lane (*mecA*pos)	MALDI-TOF MS	Study Site	Cefoxitin	Vancomycin	*femA*	*mecA*	*vanA* and *vanB*
1	*S. aureus*	DTCSH	R	S	Pos	Pos	Neg
5	*S. aureus*	JUTSH	R	S	Pos	Pos	Neg
7	*M. sciuri*	DTCSH	R	S	Pos	Pos	Neg
11	*S. haemolyticus*	DTCSH	R	R	Pos	Pos	Neg
12	*S. warneri*	DTCSH	R	S	Pos	Pos	Neg
18	*S. aureus*	TASH	R	S	Pos	Pos	Neg
19	*S. aureus*	HUCSH	R	S	Pos	Pos	Neg
23	*M. sciuri*	DTCSH	R	S	Pos	Pos	Neg
24	*M. sciuri*	DTCSH	R	S	Pos	Pos	Neg
29	*M. sciuri*	DTCSH	R	S	Pos	Pos	Neg
31	*M. sciuri*	DTCSH	R	S	Neg	Pos	Neg

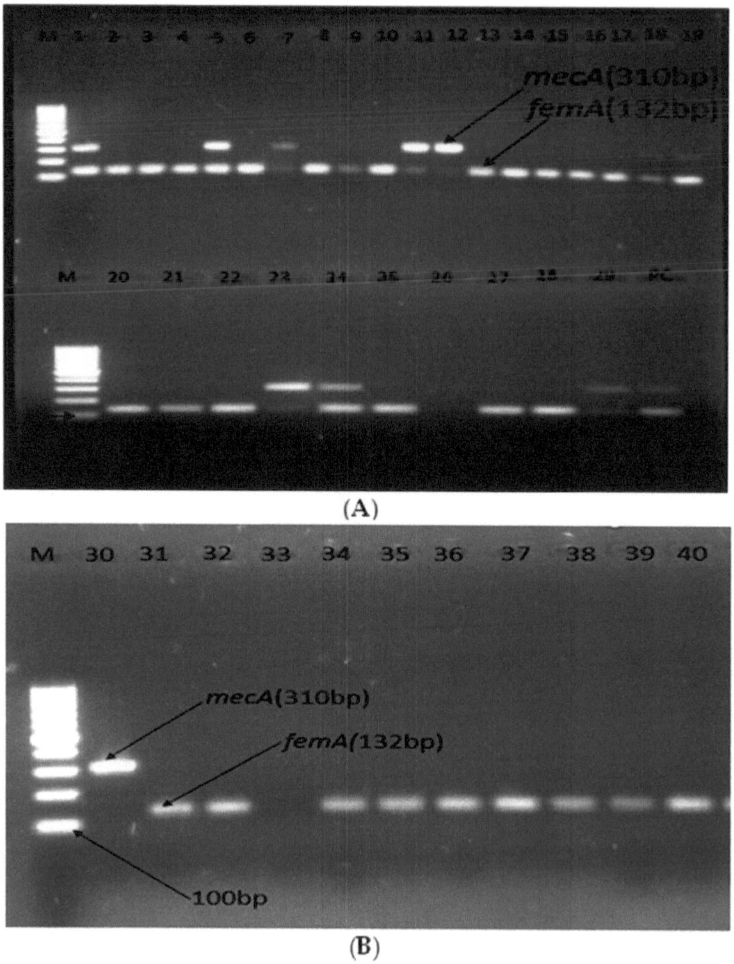

Figure 4. (**A**,**B**) Agarose gel electrophoresis showing bands of *femA* and *mecA* genes of methicillin-resistant Staphylococcic strains from patients diagnosed with surgical site infection at four different hospitals in Ethiopia; lane M1: 100 bp molecular weight ladder; lane PC: positive control; lanes 1–40 are tested isolates, and positive amplification of *femA* and *mecA* is indicated by 132 bp and 310 bp PCR amplicons, respectively.

The 11 isolates that contained the *mecA* gene, as shown in Figure 4A,B, were *S. aureus* (lanes 1, 5, 18, and 19), *M. sciuri* (lanes 7, 23, 24, 29, and 31), *S. hemolyticus* (lane 11), and *S. warneri* (lane 12). These were analyzed for *vanA* and *vanB*, and none of these isolates showed *vanA* or *vanB* in the gel electrophoresis (Table 5).

Most of the isolates carrying both the *mecA* and *femA* gene were reported from Debre Tabor (Figure 5). At Debre Tabor Comprehensive Specialized Hospital, 72.7% of *mecA*-positive, 70% of cefoxitin-resistant, and 67.7% of *femA*-positive Staphylococci were discovered (Figure 5).

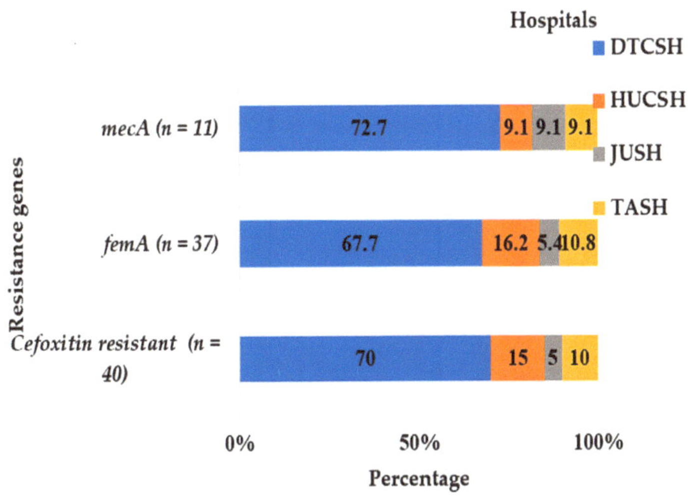

Figure 5. Frequency and distribution of cefoxitin-resistant isolates and *mecA* and *femA* genes from the total number of Staphylococci and *M. sciuri* isolates at each hospital between July 2020 and August 2021. DTCSH: Debre Tabor Comprehensive Specialized Hospital; HUCSH: Hawassa University Comprehensive Specialized Hospital; JUTSH: Jimma University Teaching Specialized Hospital; TASH: Tikur Anbessa Specialized Hospital.

3. Discussion

MRSA is one of the primary bacteria responsible for surgical site infections [19]. The bacteria are human commensals [20], and they can cause a variety of infections, including simple skin and wound infections; they can also infect visceral organs. If not diagnosed and treated properly, many of these illnesses can quickly become life-threatening diseases [1].

In our study, among the 752 wound swab samples processed, we detected 21.7% (163/752) of *S. aureus* phenotypically. The present finding is similar to previous studies reported from Jimma Ethiopia (23.6%) [21] and Brazil (20%) [22]. On the other hand, this finding is lower than those of studies conducted in other parts of Ethiopia, such as Dessie (34.5%) [23] and Debre Markos (39.7%) [24]. The variation in prevalence between studies might be due to variations in the study subjects, the conducted time, and the method employed for the detection of *S. aureus* [25].

The proportion of MRSA among the isolates based on disc diffusions was 24.5% (40/163). This study's findings were similar to a previous study conducted in an Indian Hospital (21.7%) [26]. On the other hand, the finding showed higher frequency than earlier studies in Ethiopia from Dessie (9.8%) [23] and Debre Markos (13.2%) [24], but it was below the national pooled prevalence estimate of Ethiopia (32.5%) [17], Addis Ababa (68.4%) [27], Arba Minch (82.3%) [28], and Nigeria (44%) [29]. Variations in MRSA prevalence across countries are influenced by demographics, antibiotic prescription policies, infection prevention and control programs, staff and elderly hygiene education, healthcare system structure, and MRSA diagnostic facilities [30,31].

From those tested for vancomycin resistance, one isolate had a minimum inhibitory concentration for vancomycin greater than 8 μg/mL, and it was identified as a vancomycin-resistant Staphylococcus. This result was consistent with Pournajaf et al.'s [32] finding that vancomycin resistance was 2.5%, and this figure was lower than that from a review from Ethiopia, where the pooled prevalence of VRSA was 5.3% [17], as well as the findings from Debre Markos (14.1%) [33] and elsewhere (29.4%) [34].

From all methicillin-resistant Staphylococci, the *mecA* gene was carried by 27.5% of the isolates. This finding was comparable with a study from Nigeria, where 30.5% of the isolates carried the *mecA* gene [29]. In the present study, 12.9% of *S. aureus* carried the *mecA* gene, which is lower than studies reported in Ethiopia (20%) [35], Bangladesh (25%) [36], Nigeria (38%) [29], and Iran (45.1%) [32]. It should be noted that the majority of isolates exhibiting the *mecA* gene were discovered in Debre Tabor. Eight (72.7%) of the ten *mecA*-positive isolates were detected at Debre Tabor Hospital. The reason might be poor socioeconomic status, personal demographics, antibiotic prescription practice, and infection control practices, which are associated with increased MRSA infection rates [30,31].

In our present study, the *femA* gene was detected in all *S. aureus* isolates, except two cefoxitin-resistant strains (6.7%). This finding was comparable with a study from China [37]. Additionally, in the present study, the *femA* gene was found in *S. haemolyticus*, *S. warneri*, and 83.3% of *M. sciuri* cefoxitin-resistant strains. On the other hand, neither *mecA* nor *femA* were detected in *S. epidermidis* [37]. The primers used should be specific to *S. aureus*; therefore, it is somewhat surprising that the two *S. aureus* lack the gene and that several other non-aureus isolates carry the gene. The explanation could be mutational changes in *S. aureus* and gene transfer to other species. All of these isolates have been sent for whole genome sequencing, and this matter will be analyzed further when the results are ready.

It is interesting that a significantly higher proportion of CoNS isolates harbour methicillin resistance genes, where 83.3% of *M. sciuri* and 50% of *S. haemolyticus* carried the *mecA* gene. This is in agreement with early reports that CoNS were the most common species in nosocomial infections and exhibit higher antibiotic resistance rates than *S. aureus*. This may be explained by the high prevalence of methicillin resistance linked with staphylococcal cassette chromosome (SCCmec) elements in CoNS [38], and they are considered a major reservoir of SCCmec [39]. For instance, Berglund et al. described the likely transfer of a type V SCCmec from methicillin-resistant *S. haemolyticus* to MSSA, thus transforming into MRSA [9,40]. Another study revealed that the *mecA* homologue in *M. sciuri* may be an evolutionary precursor to MRSA pathogenic strains, highlighting the main routes of antibiotic resistance gene transfer [41]. Furthermore, the report demonstrated that MSSA become MRSA by acquiring SCCmec from *S. epidermidis* through horizontal transfer [42]. These accounts suggest that horizontal interspecies transfer of mobile genetic elements could be a crucial element for MRSA global dissemination [40,41,43].

The absence of the *mecA* and *vanA* genes in the MRSA and VRSA samples does not imply the absence of resistance, as resistance may be due to other mutations or cassette-containing resistance genes [44]. Globally, resistant staphylococcal isolates lacking the *mecA* gene show the possibility for additional mechanisms to compete with *mecA* in the establishment of MRSA [45,46]. MRSA's resistance against beta-lactams and methicillin is further complicated by its ability to develop resistance to vancomycin through accidental transmission of the *vanA* gene from Enterococcal strains [47]. Vancomycin is a glycopeptide antibiotic that prevents the formation of the peptidoglycan layer by binding to the peptide precursor. Antibiotic overuse leads to bacterial resistance, thus prompting the search for new antimicrobial strategies [48]. Genomics can identify antibiotic targets, and live non-multiplying bacteria can be targeted for new antibacterials, potentially resulting in new antibacterials that shorten therapy microorganisms, reduce adverse effects, and potentially reduce antibacterial resistance [49]. Preclinical research explores metal uptake via bacterial metallophores [48]. Bacteriophages have been demonstrated to be antibacterial in animals that are susceptible to certain infectious diseases [49].

In the present study, the likelihood of methicillin-resistant staphylococcus SSI increased among patients aged $\geq$ 61 years (p = 0.025, AOR = 3.729 (1.179–11.791)). Similar findings have been reported in Brazil [22]. Previous studies have suggested that patients on antibiotics (p = 0. 017), who had a previous wound infection (p = 0.006), and with a hospital stay > 72 h showed an association with MRSA infection [33]. Similarly to our finding, previous use of antibiotics (p = 0.025, AOR = 3.066 (1.101–9.392)) and preoperative hospital stays > 7 days (p = 0.000, AOR = 1.856 (0.688–5.311)) demonstrated an association with methicillin-resistant Staphylococci for SSI. Unlike our study, a report by X. Yang et al. [50] showed that long, invasive procedures used in the ICU, such as tracheal intubation and ventilator usage, along with patients with cerebral infarction and other embolisms increase the likelihood of developing MRSA colonization and further infections.

4. Materials and Methods

4.1. Study Area and Design

A cross-sectional study was conducted in four purposively selected University Teaching Hospitals, including Debre Tabor Comprehensive Specialized Hospital (DTCSH), Tikur Anbessa Specialized Hospital (TASH), Hawassa University Teaching Hospital (HUTH), and Jimma University Teaching Hospital (JUTH) in Ethiopia. These hospitals provide a range of services in both outpatient and inpatient units under different wards, such as general surgical, gynecology, obstetric/maternity, and orthopedics, and they all have microbiology laboratories for culture and antimicrobial sensitivity testing. This study was conducted between July 2020 and August 2021.

4.2. Variables

The variables in this study were MRSA and VRSA infections, socio-demographic characteristics, clinical data, and risk factor variables, such as age, sex, surgical site, length of hospital stay, history of hospital admission, previous use of antibiotics, smoking history, alcohol consumption, type and nature of the surgery, type of antimicrobial prophylaxis, history of previous antibiotic use, surgical procedure performed, and duration of the operation.

4.3. Study Population and Sampling

The study population consisted of patients admitted for elective and emergency surgery in general surgery, gynecology/obstetric, and orthopedics wards. All surgical patients, regardless of their age, who underwent surgery during the study period and developed signs and symptoms of surgical site infection (SSI) within 30 days were included in this study. Consent and/or assent was secured from each participant before the commencement of data collection. Patients who developed SSIs after 30 days following the operation, those who refused to participate, patients with infected burn wounds, and those on treatment were excluded from this study.

4.4. Sample Size and Sampling Technique

A total of 752 clinically diagnosed cases of SSI from different wards were enrolled in this study. The sample size was calculated based on a single population proportion sample size estimation formula ($n = Z^2 P (1 - P)/d^2$) using a proportion (P) of 20% [51]. As this was a multicenter study, to increase the sample size, a precision (d) of 0.03 was used, where Z stands for Z statistic with a confidence level of 95% and a Z value of 1.96. Considering a 10% non-response rate, the final total sample size was estimated at 752. Enrollments continued until the necessary sample size was achieved, with proportional allocation among the different hospitals based on patient flow.

4.5. Specimen Collection, Isolation, and Identification of S. aureus

Wound swabs or aspirates were collected based on standard operation procedure (SOP). Conventional bacteriological techniques, such as morphological, cultural, and biochemical characterization, were used to identify strains of *S. aureus* [52]. The specimens were inoculated on blood agar plates (BAP) (Oxoid, UK), and mannitol salt agar (MSA) (Oxoid) and then incubated at 35 °C for 24 h. The *S. aureus* isolates were identified through Gram staining, catalase and coagulase tests, including golden yellow colonies on MSA, which were considered phenotypic identification tests.

4.6. Identification Confirmation of the Species of Bacteria Strain Using Matrix-Assisted Laser Desorption/Ionization Time-of-Flight Mass Spectrometry (MALDI-TOF MS)

All phenotypically cefoxitin-resistant *S. aureus* isolates were re-identified using MALDI-TOF MS [53] at the Clinical Microbiology Department, Uppsala University Hospital in Sweden.

A single colony of bacteria from fresh cultures was smeared onto a MALDI-TOF plate, air-dried, treated with formic acid and MALDI matrix solution, and again air-dried before reading. MALDI-TOF identification scores were automatically generated by the system software [54], and isolates with scores of two and above were accepted, while those with scores below 1.7 and flagged red were rejected. Samples with scores between 1.7 and 2 and flagged yellow were re-analyzed.

4.7. *Antimicrobial Susceptibility Testing*

Antimicrobial susceptibility testing (AST) was carried out using the cefoxitin disc diffusion test, which is a surrogate marker test for oxacillin resistance following the clinical and laboratory standard Institute (CLSI) protocol [55], and the minimum inhibitory concentration (MIC) of the vancomycin strip was determined using the E-test method on MHA. The reference strain *S. aureus* (ATCC® 25923, Seattle, DC, USA) was used as a quality control. Evidence showed that MRSA is a requisite for VISA [56]. Hence, we screened VRSA/VISA from MRSA isolates.

4.8. *Identification of Methicillin-Resistant S. aureus Strains*

4.8.1. DNA Extraction

DNA was extracted from all cefoxitin-resistant *S. aureus* isolates through the boiling method, as described previously [57]. Briefly, each isolate was grown overnight on nutrient agar (Oxoid, UK), and 3 to 5 colonies of that culture were suspended in 300 µL of 1× Tris EDTA buffer. The suspension was subjected to 10 min of boiling at 94 °C in a water bath (Thermo Fisher Scientific, CA, USA), followed by 10 minutes of freezing at −20 °C, 1 min at room temperature, and 5 min of centrifugation at 14,000× *g*. Finally, 150 µL of the supernatant was transferred into a nuclease-free Eppendorf tube and measured using Nanodrop (Thermo Scientific) for the quality and quantity of DNA prior to storage at −20 °C until analysis.

4.8.2. Standardization of Multiplex PCR for the Detection of *Staphylococci mecA*, *femA*, *vanA*, and *vanB*

Multiplex PCR was used to amplify different genes that are associated with methicillin resistance from Staphylococci. The primers and annealing temperatures were standardized for the detection of *S. aureus mecA*, *vanA*, *vanB*, and *femA*. The PCR products were analyzed through gel electrophoresis. Positive and negative control strains were included in all amplification reactions to ensure accuracy of the test results.

First, PCR was standardized using a range of annealing temperatures to establish the optimum annealing reaction condition for all of the primers. All PCR primers are described in Table 5. The reaction mixture contained 12.5 µL of hot star master mix (Qiagen, Hilden, Germany), 0.5 µL each of the forward and reverse primers, 9 µL of molecular-grade water, and 2.5 µL of the template, with a final volume of 25 µL. Amplification for *mecA* and *femA* was carried out over 40 cycles of initial heat activation at 95 °C for 15 min, denaturation at 94 °C for 30 s, followed by annealing at 52 °C for 45 s, extension at 72 °C for 1 min, and final extension at 72 °C for 10 min. Amplification for *vanA* and *vanB* was carried out over 40 cycles of initial heat activation at 95 °C for 15 min, denaturation at 94 °C for 30 s, followed by annealing at 56 °C for 90 s, extension at 72 °C for 90 min, and final extension at 72 °C for 10 min. The PCR products were analyzed through electrophoresis on a 2% agarose gel and detected through staining in ethidium bromide with the aid of a gel imaging system, GelDoc (Bio-Rad). The following controls were included in all amplification reactions: ATCC 33591 (*mecA*-positive *S. aureus*) and ATCC 25923 (*mecA*-negative *S. aureus*).

Table 5. Primers used in multiplex PCR for the detection of the *mecA*, *vanA*, *vanB*, and *femA* genes.

Target Gene	Primer Name	Primer Sequence (5′-3′)	Size bp	References
mecA	MF MR	GTAGAAATGACTGAACGTCCGATAA CCAATTCCACATTGTTTCGGTCTAA	310	[58]
vanA	VF VR	GGGAAAACGACAATTGC GTACAATGCGGCCGTTA	732	[59]
vanB	VF VR	ACCTACCCTGTCTTTGTGAA AATGTCTGCTGGAACGATA	300	
femA	FF FR	AAAAAAGCACATAACAAGCG GATAAAGAAGAAACCAGCAG	132	[60]

4.9. Quality Assurance

Specimens were collected according to the recommended standard operating procedures (SOPs). The performance of all prepared culture media (BAP and MSA) was also checked by inoculating control strains, *S. aureus* (ATCC® 25923), for each new batch of agar plates prepared. In addition, the sterility of culture media was checked by incubating 5% of the prepared media at 37 °C for 24–48 h. In addition, reagents for Gram stain and biochemical tests were checked against control strains of *S. aureus.* The 0.5 McFarland standard was used to standardize the bacterial suspension inoculum density for the susceptibility test. Each MALDI-TOF run also included *S. aureus* (ATCC® 25923) as a quality control strain. Furthermore, the performance of the antibiotic disks was evaluated using American-type cell culture (ATCC) controls. As such, *S. aureus* ATCC® 25923 (cefoxitin zone 21–29 mm) and *S. aureus* ATCC® 43300 (zone $\leq$ 21 mm) were used as control strains to determine the performance of the cefoxitin disc diffusion test. *S. aureus* ATCC® 29213 MIC of vancomycin broth value 0.5–2.0 µg/mL was used as a control strain to measure the performance of vancomycin [55].

4.10. Data Entry and Analysis

The data were checked for completeness, missing values, and coding of questionnaires entered into the Research Electronic Data Capture (RED-Cap). A double data entry method was used to ensure the accuracy of the data, and data were analyzed using STATA version 25. Descriptive statistics were used to present antimicrobial susceptibility patterns. Frequencies and cross-tabulations were used to summarize descriptive statistics. Logistic regression was used to study the effect of independent variables on the dependent variables. *p*-values less than 0.05 were considered statistically significant.

4.11. Ethical Considerations

The Department of Medical Microbiology, Immunology, and Parasitology (DMIP) and the AHRI/ALERT Research Ethics Committee (AAREC) reviewed and approved this study. Institutional review board (IRB) approval was also obtained from Addis Ababa University's College of Health Sciences, AAUMF03-008/2020. Selected hospitals received a formal letter from the AHRI and DMIP, and each hospital's medical directors gave their consent. Written consent/assent was taken from each study participant before initiation of the actual data collection.

Patient information was kept confidential by sharing the laboratory results of research participants only with the designated accountable clinicians. Patients who experienced SSIs were managed according to hospital policy. In general, this study was conducted in accordance with the Declaration of Helsinki.

5. Conclusions

A multicenter study identified 11 mecA-positive Staphylococci species, with 36.4% being MRSA, but no VRSA was found among these MRSA. What is more captivating in this study is a significantly high prevalence of *mecA* carriage among CoNS, suggesting difficulties in the treatment of patients with CoNS infections. Furthermore, this signifies a huge potential of MSSA conversion to MRSA through horizontal gene transfer, which would make things more complicated. In terms of geographic distribution, out of 11 *mecA*-gene-positive Staphylococci, 8 (72.7%) were detected in DTCSH, with significant variations between hospitals, suggesting that strategies to control methicillin-resistant Staphylococci should be tailored to specific hospitals. The presence of staphylococcal isolates was linked to factors like older age, hospital stay, antibiotic history, and prophylaxis. Prompt prevention and control measures for MRSA-high-risk populations, including strict adherence to infection prevention methods, periodic surveillance, and antibiotic stewardship programs, are crucial for effective treatment and prevention strategies.

Author Contributions: S.W., conceptualization, data collection, laboratory investigation, drafting the manuscript, and write-up. Y.S., M.Y. and A.A. (Ashenafi Alemu) substantially participated in laboratory investigation. T.A., B.S., A.A. (Alemseged Abdissa), G.T.B., G.S. and A.M. substantially participated in the design of this study, reviewed the manuscript and provided critical intellectual content. S.W. and G.S. participated in the analysis and interpretation of the findings. All authors have read and agreed to the published version of the manuscript.

Funding: Collection of samples and laboratory work in Ethiopia was covered by grants from Addis Ababa University, and Armauer Hansen Research Institute. For work in Sweden, grants from SIDA/SAREC bilateral research cooperation with Ethiopia (5108013506) and the Centre for Innovative Drug Development and Therapeutic Trials for Africa (CDT-Africa) were used to cover travel and subsistence for SW during a three-month visit. Laboratory work in Sweden was done without external grants.

Institutional Review Board Statement: Ethical clearance and approval were obtained from Addis Ababa University's College of Health Sciences and AAREC. Institutional Review Board Statement: The research was granted approval by the Institutional Review Board (AAUMF 03-008-2020) of Addis Abeba University's College of Health Sciences and the Department of Microbiology, Immunology, and Parasitology Ethical Review Committee (DEREC002/2020). This study was also approved by the AHRI/ALERT Ethics Review Committee (protocol number: PO/29/19) of the Armauer Hansen Research Institute.

Informed Consent Statement: Data were collected only when written consent or assent was obtained from each participant or guardian.

Data Availability Statement: The data sets generated during and/or analyzed during the current study are available from the corresponding authors upon reasonable request. The data are not publicly available due to privacy restrictions.

Acknowledgments: The authors would like to extend our gratitude to The Centre for Innovative Drug Development and Therapeutic Trials for Africa (CDT-Africa) for supporting this study (cover Sweden travel cost and subsistence for SW). We would like to thank the Debre Tabor Compressive Specialized Hospital, Tikur Anbessa Specialized Hospital, Hawassa University Teaching Hospital, and Jimma University Teaching Specialized Hospital for allowing us to conduct this study. We appreciate all of the research sites' doctors, nurses, and microbiologists who assisted us in conducting this investigation. And, lastly, we would like to thank everyone who took part in this study.

Conflicts of Interest: The authors declare no competing interest. The funders had no role in the design of this study; in the collection, analyses, or interpretation of data; in the writing of the manuscript; or in the decision to publish the results.

Abbreviations

CLSI, Clinical and Laboratory Standards Institute; CoNS, Coagulase Negative Staphylococcous; MALDI-TOF MS, Matrix-Assisted Laser Desorption/Ionization Time-of-Flight Mass Spectrometry; MHA, Muller–Hinton agar; MIC, Minimum Inhibitory Concentration; MRSA, Methicillin-Resistant *S. aureus*; MSSA, Methicillin-Sensitive *S. aureus*; SSI, Surgical Site Infection; SCC*mec*, Staphylococcal Cassette Chromosome mec; VISA, Vancomycin-Intermediate *S. aureus*; VRSA, Vancomycin-Resistant

S. aureus; VSSA, Vancomycin-Sensitive *S. aureus*.

References

1. Turner, N.A.; Sharma-Kuinkel, B.K.; Maskarinec, S.A.; Eichenberger, E.M.; Shah, P.P.; Carugati, M.; Holland, T.L.; Fowler, V.G. Methicillin-resistant *Staphylococcus aureus*: An overview of basic and clinical research. *Nat. Rev. Microbiol.* **2019**, *17*, 203–218. [CrossRef]
2. Martínez-Meléndez, A.; Morfín-Otero, R.; Villarreal-Treviño, L.; González-González, G.; Llaca-Díaz, J.; Rodríguez-Noriega, E.; Camacho-Ortíz, A.; Garza-González, E. Staphylococcal cassette chromosome mec (SCCmec) in coagulase negative staphylococci. *Med. Univ.* **2015**, *17*, 229–233. [CrossRef]
3. Beims, H.; Overmann, A.; Fulde, M.; Steinert, M.; Bergmann, S. Isolation of Staphylococcus sciuri from horse skin infection. *Open Vet. J.* **2016**, *6*, 242–246. [CrossRef]
4. Cheung, G.Y.C.; Bae, J.S.; Otto, M. Pathogenicity and virulence of *Staphylococcus aureus*. *Virulence* **2021**, *12*, 547–569. [CrossRef] [PubMed]
5. Ghssein, G.; Ezzeddine, Z. The Key Element Role of Metallophores in the Pathogenicity and Virulence of *Staphylococcus aureus*: A Review. *Biology* **2022**, *11*, 1525. [CrossRef] [PubMed]
6. David, M.Z.; Daum, R.S. Treatment of *Staphylococcus aureus* Infections. *Curr. Top. Microbiol. Immunol.* **2017**, *409*, 325–383.
7. Hiramatsu, K.; Katayama, Y.; Matsuo, M.; Sasaki, T.; Morimoto, Y.; Sekiguchi, A.; Baba, T. Multi-drug-resistant *Staphylococcus aureus* and future chemotherapy. *J. Infect. Chemother.* **2014**, *20*, 593–601. [CrossRef] [PubMed]
8. Mahjabeen, F.; Saha, U.; Mostafa, M.N.; Siddique, F.; Ahsan, E.; Fathma, S.; Tasnim, A.; Rahman, T.; Faruq, R.; Sakibuzzaman, M.; et al. An Update on Treatment Options for Methicillin-Resistant *Staphylococcus aureus* (MRSA) Bacteremia: A Systematic Review. *Cureus* **2022**, *14*, e31486. [CrossRef]
9. Vestergaard, M.; Frees, D.; Ingmer, H. Antibiotic Resistance and the MRSA Problem. *Microbiol. Spectr.* **2019**, *7*. [CrossRef]
10. Li, X.; Xiong, Y.; Fan, X.; Feng, P.; Tang, H.; Zhou, T. The role of femA regulating gene on methicillin-resistant *Staphylococcus aureus* clinical isolates. *Med. Mal. Infect.* **2012**, *42*, 218–225. [CrossRef]
11. Cong, Y.; Yang, S.; Rao, X. Vancomycin resistant *Staphylococcus aureus* infections: A review of case updating and clinical features. *J. Adv. Res.* **2020**, *21*, 169–176. [CrossRef]
12. Stogios, P.J.; Savchenko, A. Molecular mechanisms of vancomycin resistance. *Protein Sci.* **2020**, *29*, 654–669. [CrossRef]
13. Li, G.; Walker, M.J.; De Oliveira, D.M.P. Vancomycin Resistance in Enterococcus and *Staphylococcus aureus*. *Microorganisms* **2022**, *11*, 24. [CrossRef]
14. Arredondo-Alonso, S.; Top, J.; Corander, J.; Willems, R.J.L.; Schürch, A.C. Mode and dynamics of vanA-type vancomycin resistance dissemination in Dutch hospitals. *Genome Med.* **2021**, *13*, 9.
15. Micek, S.T. Alternatives to vancomycin for the treatment of methicillin-resistant *Staphylococcus aureus* infections. *Clin. Infect. Dis.* **2007**, *45* (Suppl. S3), S184–S190. [CrossRef] [PubMed]
16. Wu, Q.; Sabokroo, N.; Wang, Y.; Hashemian, M.; Karamollahi, S.; Kouhsari, E. Systematic review and meta-analysis of the epidemiology of vancomycin-resistance *Staphylococcus aureus* isolates. *Antimicrob. Resist. Infect. Control* **2021**, *10*, 101. [CrossRef] [PubMed]
17. Eshetie, S.; Tarekegn, F.; Moges, F.; Amsalu, A.; Birhan, W.; Huruy, K. Methicillin resistant *Staphylococcus aureus* in Ethiopia: A meta-analysis. *BMC Infect. Dis.* **2016**, *16*, 689. [CrossRef]
18. Anagaw, B.; Shiferaw, Y.; Anagaw, B.; Biadglegne, F.; Moges, F.; Kassu, A.; Unakal, C.; Mulu, A. Frequency of methicillin-resistant *Staphylococcus aureus* isolates from clinical specimens in Gondar University Hospital, Northwest Ethiopia. *Asian J. Med. Sci.* **2013**, *5*, 59–64. [CrossRef]
19. Pal, S.; Sayana, A.; Joshi, A.; Juyal, D. *Staphylococcus aureus*: A predominant cause of surgical site infections in a rural healthcare setup of Uttarakhand. *J. Fam. Med. Prim. Care* **2019**, *8*, 3600–3606.
20. Haaber, J.; Penadés, J.R.; Ingmer, H. Transfer of antibiotic resistance in *Staphylococcus aureus*. *Trends Microbiol.* **2017**, *25*, 893–905. [CrossRef] [PubMed]
21. Godebo, G.; Kibru, G.; Tassew, H. Multidrug-resistant bacterial isolates in infected wounds at Jimma University Specialized Hospital, Ethiopia. *Ann. Clin. Microbiol. Antimicrob.* **2013**, *12*, 17. [CrossRef]
22. Almeida, G.C.; dos Santos, M.M.; Lima, N.G.; Cidral, T.A.; Melo, M.C.; Lima, K.C. Prevalence and factors associated with wound colonization by Staphylococcus spp. and *Staphylococcus aureus* in hospitalized patients in inland northeastern Brazil: A cross-sectional study. *BMC Infect. Dis.* **2014**, *14*, 328. [CrossRef]
23. Tsige, Y.; Tadesse, S.; G/Eyesus, T.; Tefera, M.M.; Amsalu, A.; Menberu, M.A.; Gelaw, B. Prevalence of Methicillin-Resistant *Staphylococcus aureus* and Associated Risk Factors among Patients with Wound Infection at Referral Hospital, Northeast Ethiopia. *J. Pathog.* **2020**, *2020*, 3168325. [CrossRef] [PubMed]
24. Kahsay, A.; Mihret, A.; Abebe, T.; Andualem, T. Isolation and antimicrobial susceptibility pattern of *Staphylococcus aureus* in patients with surgical site infection at Debre Markos Referral Hospital, Amhara Region, Ethiopia. *Arch. Public Health* **2014**, *72*, 16. [CrossRef] [PubMed]

25. Congdon, S.T.; Guaglione, J.A.; Ricketts, O.M.A.; Murphy, K.V.; Anderson, M.G.; Trowbridge, D.A.; Abduladheem, Y.A.; Phillips, A.M.; Beausoleil, A.M.; Stanley, A.J.; et al. Prevalence and antibiotic resistance of *Staphylococcus aureus* associated with a college-aged cohort: Life-style factors that contribute to nasal carriage. *Front. Cell. Infect. Microbiol.* **2023**, *13*, 1195758. [CrossRef] [PubMed]
26. Kownhar, H.; Shankar, E.M.; Vignesh, R.; Sekar, R.; Velu, V.; Rao, U.A. High isolation rate of *Staphylococcus aureus* from surgical site infections in an Indian hospital. *J. Antimicrob. Chemother.* **2008**, *61*, 758–760. [CrossRef] [PubMed]
27. Tadesse, S.; Alemayehu, H.; Tenna, A.; Tadesse, G.; Tessema, T.S.; Shibeshi, W.; Eguale, T. Antimicrobial resistance profile of *Staphylococcus aureus* isolated from patients with infection at Tikur Anbessa Specialized Hospital, Addis Ababa, Ethiopia. *BMC Pharmacol. Toxicol.* **2018**, *19*, 24. [CrossRef] [PubMed]
28. Mama, M.; Aklilu, A.; Misgna, K.; Tadesse, M.; Alemayehu, E. Methicillin-and inducible clindamycin-resistant *Staphylococcus aureus* among patients with wound infection attending Arba Minch Hospital, South Ethiopia. *Int. J. Microbiol.* **2019**, *2019*, 2965190. [CrossRef] [PubMed]
29. Ibadin, E.E.; Enabulele, I.O.; Muinah, F. Prevalence of mecA gene among staphylococci from clinical samples of a tertiary hospital in Benin City, Nigeria. *Afr. Health Sci.* **2017**, *17*, 1000–1010. [CrossRef]
30. Kavanagh, K.T. Control of MSSA and MRSA in the United States: Protocols, policies, risk adjustment and excuses. *Antimicrob. Resist. Infect. Control* **2019**, *8*, 103. [CrossRef]
31. Al-Orphaly, M.; Hadi, H.A.; Eltayeb, F.K.; Al-Hail, H.; Samuel, B.G.; Sultan, A.A.; Skariah, S. Epidemiology of multidrug-resistant Pseudomonas aeruginosa in the Middle East and North Africa Region. *mSphere* **2021**, *6*, e00202–e00221. [CrossRef]
32. Pournajaf, A.; Ardebili, A.; Goudarzi, L.; Khodabandeh, M.; Narimani, T.; Abbaszadeh, H. PCR-based identification of methicillin–resistant *Staphylococcus aureus* strains and their antibiotic resistance profiles. *Asian Pac. J. Trop. Biomed.* **2014**, *4*, S293–S297. [CrossRef] [PubMed]
33. Tefera, S.; Awoke, T.; Mekonnen, D. Methicillin and vancomycin resistant *Staphylococcus aureus* and associated factors from surgical ward inpatients at Debre Markos Referral Hospital, Northwest Ethiopia. *Infect. Drug Resist.* **2021**, *14*, 3053–3062. [CrossRef] [PubMed]
34. Alani, H.A.; Hassawi, D.S.; Flayih, M.T. Patterns of antibiotic resistance in *Staphylococcus aureus* isolates and detection the heteroresistance to vancomycin by population analysis method. *JUAPS* **2017**, *11*, 26–33. [CrossRef]
35. Moges, F.; Tamiru, T.; Amare, A.; Mengistu, G.; Eshetie, S.; Dagnew, M.; Feleke, T.; Gizachew, M.; Abebe, W. Prevalence of Methicillin-Resistant *Staphylococcus aureus* and Multidrug-Resistant Strains from Patients Attending the Referral Hospitals of Amhara Regional State, Ethiopia. *Int. J. Microbiol.* **2023**, *2023*, 3848073. [CrossRef]
36. Zahan, N.A.; Hossain, M.A.; Musa, A.K.; Shamsuzzaman, A.K.; Mahamud, M.C.; Mamun, A.A.; Paul, S.K.; Ahmed, S.; Sumona, A.A.; Begum, Z.; et al. PCR for mecA gene of methicillin resistant *Staphylococcus aureus*. *Mymensingh Med. J.* **2009**, *18*, 21–26. [PubMed]
37. Kobayashi, N.; Wu, H.; Kojima, K.; Taniguchi, K.; Urasawa, S.; Uehara, N.; Omizu, Y.; Kishi, Y.; Yagihashi, A.; Kurokawa, I. Detection of *mecA*, *femA*, and *femB* genes in clinical strains of using polymerase chain reaction. *Epidemiol. Infect.* **1994**, *113*, 259–266. [CrossRef]
38. Garza-Gonzalez, E.; Morfin-Otero, R.; Llaca-Diaz, J.M.; Rodriguez-Noriega, E. Staphylococcal cassette chromosome mec (SCCmec) in methicillin-resistant coagulase-negative staphylococci. *Epidemiol. Infect.* **2010**, *138*, 645–654. [CrossRef]
39. Zong, Z.; Peng, C.; Lü, X. Diversity of SCC mec elements in methicillin-resistant coagulase-negative staphylococci clinical isolates. *PLoS ONE* **2011**, *6*, e20191. [CrossRef]
40. Berglund, C.; Söderquist, B. The origin of a methicillin-resistant *Staphylococcus aureus* isolate at a neonatal ward in Sweden—Possible horizontal transfer of a staphylococcal cassette chromosome mec between methicillin-resistant Staphylococcus haemolyticus and *Staphylococcus aureus*. *Clin. Microbiol. Infect.* **2008**, *14*, 1048–1056. [CrossRef]
41. Wu, S.W.; de Lencastre, H.; Tomasz, A. Recruitment of the mecA gene homologue of Staphylococcus sciuri into a resistance determinant and expression of the resistant phenotype in *Staphylococcus aureus*. *J. Bacteriol.* **2001**, *183*, 2417–2424. [CrossRef]
42. Bloemendaal, A.L.A.; Brouwer, E.C.; Fluit, A.C. Methicillin resistance transfer from Staphylocccus epidermidis to methicillin-susceptible *Staphylococcus aureus* in a patient during antibiotic therapy. *PLoS ONE* **2010**, *5*, e11841. [CrossRef] [PubMed]
43. Miragaia, M. Factors Contributing to the Evolution of mecA-Mediated β-lactam Resistance in Staphylococci: Update and New Insights From Whole Genome Sequencing (WGS). *Front. Microbiol.* **2018**, *9*, 2723. [CrossRef]
44. Sun, J.; Deng, Z.; Yan, A. Bacterial multidrug efflux pumps: Mechanisms, physiology and pharmacological exploitations. *Biochem. Biophys. Res.* **2014**, *453*, 254–267. [CrossRef] [PubMed]
45. Zong, Z. Characterization of a complex context containing mecA but lacking genes encoding cassette chromosome recombinases in Staphylococcus haemolyticus. *BMC Microbiol.* **2013**, *13*, 64. [CrossRef] [PubMed]
46. Elhassan, M.M.; Ozbak, H.A.; Hemeg, H.A.; Elmekki, M.A.; Ahmed, L.M. Absence of the *mecA* gene in methicillin resistant *Staphylococcus aureus* isolated from different clinical specimens in Shendi city, Sudan. *BioMed. Res. Int.* **2015**, *2015*, 895860. [CrossRef] [PubMed]
47. McGuinness, W.A.; Malachowa, N.; DeLeo, F.R. Vancomycin Resistance in *Staphylococcus aureus*. *Yale J. Biol. Med.* **2017**, *90*, 269–281.
48. Ezzeddine, Z.; Ghssein, G. Towards new antibiotics classes targeting bacterial metallophores. *Microb. Pathog.* **2023**, *182*, 106221. [CrossRef]

49. Coates, A.R.; Hu, Y. Novel approaches to developing new antibiotics for bacterial infections. *Br. J. Pharmacol.* **2007**, *152*, 1147–1154. [CrossRef]
50. Yang, X.; Zhao, J.; Wang, Y.; Wu, J.; Wang, X.; Wang, Y.; Zhang, Y.; Li, H. Molecular Epidemiology of Methicillin-Resistant *Staphylococcus aureus* in Hospitalized Patients in Eastern Heilongjiang Province, China. *Infect. Drug Resist.* **2021**, *14*, 1635–1643. [CrossRef]
51. Mengesha, R.E.; Kasa, B.G.; Saravanan, M.; Berhe, D.F.; Wasihun, A.G. Aerobic bacteria in post surgical wound infections and pattern of their antimicrobial susceptibility in Ayder Teaching and Referral Hospital, Mekelle, Ethiopia. *BMC Res. Notes* **2014**, *7*, 575. [CrossRef]
52. Bendary, M.M.; Solyman, S.M.; Azab, M.M.; Mahmoud, N.F.; Hanora, A.M. Genetic diversity of multidrug resistant *Staphylococcus aureus* isolated from clinical and non clinical samples in Egypt. *Cell. Mol. Biol.* **2016**, *62*, 55–61. [PubMed]
53. Torres-Sangiao, E.; Leal Rodriguez, C.; García-Riestra, C. Application and Perspectives of MALDI–TOF Mass Spectrometry in Clinical Microbiology Laboratories. *Microorganisms* **2021**, *9*, 1539. [CrossRef] [PubMed]
54. Hou, T.-Y.; Chiang-Ni, C.; Teng, S.-H. Current status of MALDI-TOF mass spectrometry in clinical microbiology. *J. Food Drug Anal.* **2019**, *27*, 404–414. [CrossRef] [PubMed]
55. Clinical Laboratory Standard Institute. *Performans Standards for Antimicrobial Susceptablity Testing, CLSI Supplement M100*, 30th ed.; Clinical and Laboratory Standared Institiute: Wayne, PA, USA, 2021.
56. Centers for Disease Control and Prevention. *Laboratory Detection of Vancomycin-Intermediate/Resistant Staphylococcus aureus (VISA/VRSA)*; Centers for Disease Control and Prevention: Atlanta, GA, USA, 2006.
57. Yamagishi, J.; Sato, Y.; Shinozaki, N.; Ye, B.; Tsuboi, A.; Nagasaki, M.; Yamashita, R. Comparison of boiling and robotics automation method in DNA extraction for metagenomic sequencing of human oral microbes. *PLoS ONE* **2016**, *11*, e0154389. [CrossRef]
58. Perez-Roth, E.; Claverie-Martın, F.; Villar, J.; Mendez-Alvarez, S. Multiplex PCR for simultaneous identification of *Staphylococcus aureus* and detection of methicillin and mupirocin resistance. *J. Clin. Microbiol.* **2001**, *39*, 4037–4041. [CrossRef]
59. Emamie, A.; Zolfaghari, P.; Zarei, A.; Ghorbani, M. Prevalence and antibiotic resistance of ESKAPE pathogens isolated from patients with bacteremia in Tehran, Iran. *Indian J. Med. Spec.* **2023**, *14*, 97.
60. Al-Marzoqi, A.; Azize, H.; Al Dulaimi, T.; Ahmed, N. Phenotypic detection of resistance in *Staphylococcus aureus* isolates: Detection of (mec A and fem A) gene in methicillin resistant *Staphylococcus aureus* (MRSA) by Polymerase Chain Reaction. *J. Nat. Sci. Res.* **2014**, *4*, 112–118.

antibiotics

MDPI

Article

Co-Colonization of Non-*difficile* Clostridial Species in Antibiotic-Associated Diarrhea Caused by *Clostridioides difficile*

Daniel Salas-Treviño, Samantha Flores-Treviño, Carlos Cisneros-Rendón, Cristian Valdemar Domínguez-Rivera and Adrián Camacho-Ortiz *

Department of Infectious Diseases, Hospital Universitario Dr. José Eleuterio González, Universidad Autónoma de Nuevo León, Monterrey 64460, Mexico; daniel.salastr@uanl.edu.mx (D.S.-T.); samantha.florestr@uanl.edu.mx (S.F.-T.); carlos.cisnerosrn@uanl.edu.mx (C.C.-R.); valdemar.dominguezrvr@uanl.edu.mx (C.V.D.-R.)

* Correspondence: adrian.camachoort@uanl.edu.mx

Abstract: Background/Objectives: Antibiotic-associated diarrhea (AAD) is a public health problem that develops in the hospital setting. The most common causative agent of AAD is *Clostridioides difficile* infection (CDI), although other non-*difficile* Clostridia (NDC) might also be present. NDC include members of the RIC group such as *Clostridium ramosum* [*T. ramosa*], *Clostridium innocuum* and *Clostridium clostridioforme* [*E. clostridioformis*]. The co-colonization of NDC and CDI in patients with AAD has not been fully analyzed. **Methods**: We compared clinical and laboratory data of patients with *C. difficile* infection (CDI) plus NDC against patients with only CDI. This study was a retrospective, case–control study. Hospitalized confirmed CDI cases were analyzed. CDI detection was performed using a 2-step diagnostic algorithm, including glutamate dehydrogenase (GDH) with toxin A/toxin B assays and molecular detection of the *tpi* gene. Stool samples were cultured and colonies morphologically compatible with any Clostridia were identified with matrix-assisted laser desorption/ionization time-of-flight mass spectrometry. Fisher's exact test and odds ratio (OR) were calculated to determine the degree of correlation between the variables and the study groups. **Results**: In the CDI + NDC group (n = 7), positive culture was observed for *C. ramosum* [*T. ramosa*] (n = 3), *C. innocuum* (n = 3), and *C. butyricum* (n = 1). According to our results, CDI + NDC patients received more days of antibiotic therapy, took more days to reduce diarrhea, had a significant increase in the number of days to suppress diarrhea, and previous hospitalizations were more frequently reported. **Conclusions**: In conclusion, the positive culture of NDC species such as *C. innocuum* or *C. ramosum* in patients with AAD caused by CDI correlates with treatment extension and/or failure.

Keywords: antibiotic-associated diarrhea; *Clostridioides difficile*; non-*difficile* Clostridia; *Clostridium innocuum*; *Clostridium ramosum*

1. Introduction

Antibiotic-associated diarrhea (AAD) is a public health problem that develops mostly in the hospital setting, with an incidence of between 5 and 25% of patients, depending on the treatment administered [1]. The most common infectious entity responsible for this condition is *Clostridioides difficile* [1], although the presence of other non-*difficile* Clostridia (NDC) has been not fully analyzed in this context. NDC can refer to the members of the RIC group, which includes *Clostridium ramosum* [*T. ramosa*], *Clostridium innocuum* and *Clostridium clostridioforme [E. clostridioformis]*. Under microscopic examination, strains belonging to the RIC group can appear as Gram-positively variable and in diverse shapes. *C. ramosum* [*T. ramosa*] can be straight or helicoidal curved rods [2] whereas *C. innocuum* is observed as straight bacilli in the shape of a double spoon or with oval ends due to its spores [3].

C. ramosum [*T. ramosa*], recently reclassified as *Thomasclavelia ramosa* based on 16S rRNA phylogeny [4], is part of the intestinal microbiota. However, it can also become pathogenic under certain circumstances such as bacteremia or extraintestinal infections caused by trauma or intestinal perforations [5,6], which can be treated with metronidazole, amoxicillin/clavulanate, piperacillin/tazobactam or meropenem [7]. *C. innocuum* is also a commensal bacteria colonizing the gut of up to 80% of adults [8,9], which can also cause several intestinal diseases including AAD and even bacteremia [8,10]. *C. clostridioforme,* now reclassified as *Enterocloster clostridioformis* based on 16S rRNA phylogeny [11], is also a causative agent of anaerobic bacteremia [12]. Accurate identification of the *Clostridium* strains within the RIC group is important due to their resistance to several antimicrobial agents [12].

Several methods can be used to identify anaerobes, including growth characteristics, colony morphology and susceptibility to specific antibiotics. Matrix-assisted laser desorption/ionization time-of-flight (MALDI-TOF) mass spectrometry is based on the spectra obtained from conserved ribosomal proteins, allowing rapid and accurate species identification [13].

This study aimed to identify the correlation between co-colonization with NDC and CDI in AAD. Therefore, we describe the clinical characteristics and outcomes of patients with AAD caused by *C. difficile* infection plus Clostridia non-*difficile* co-colonization (CDI + NDC) matching those of patients with AAD caused only by *C. difficile* infection (CDI).

2. Results

2.1. Characteristics of CDI Patients Included

We compared the clinical and prognostic characteristics effects of patients with CDI confirmed by 2-step algorithm testing and a positive culture and those from clinically similar patients with a virtually identical diagnostic test in whom positive culture of NDC was detected. At the time of diagnosis and culture collection, the patients had not been treated with any antibiotics during their current hospital stay.

The stool samples of twenty eight patients with positive GDH were analyzed. The CDI group (n = 21) included patients with CDI confirmed by PCR and growth of only C. difficile in cultures. In the CDI + NDC group (n = 7), significant growth (50,000 CFU/mL) was observed for *T. ramosa* (n = 3), *C. innocuum* (n = 3), and *C. butyricum* (n = 1). Molecular detection by PCR of *C. difficile* Topoisomerase I (*tpi*) gene was positive (Figure 1A) in all samples. The macroscopic and microscopic morphologies of the colonies and bacteria are shown in Figure 1B,C.

In both groups, patients on average were 45 years of age. Both weight and body mass index were higher in the CDI group (p = 0.003 and 0.017, respectively) (Table 1). Likewise, the length of hospital stay, days in intensive care, ATLAS score and Charlson score (comorbidity) were higher in the CDI group.

Table 1. Quantitative values of clinical and demographic characteristics of the study populations.

Patient Characteristic	CDI + NDC (n = 7)	CDI (n = 21)	p-Value
Age (years)	45.57 ± 23.02	45.33 ± 22.06	**0.9807** [a]
Patient weight (kg)	59.16 ± 9.55	72.38 ± 9.17	**0.0030** [a]

Table 1. *Cont.*

Patient Characteristic	CDI + NDC (n = 7)	CDI (n = 21)	p-Value
BMI	22.48 ± 4.21	26.41 ± 3.31	**0.0179** [a]
Hospital LOS	25.57 ± 17.90	32.00 ± 26.64	0.5592 [a]
Charlson Score	2.85 ± 2.03	3.23 ± 3.46	0.7584 [b]
LOS in ICU	0.83 ± 1.32	2.04 ± 6.02	0.7250 [b]
Bowel movements per day	6.71 ± 2.81	4.90 ± 2.24	0.0967 [a]
Total leukocytes (Cel/μL)	11.07 ± 4.05	12.11 ± 7.52	0.7324 [a]
Albumin (g/dL)	2.11 ± 0.51	2.35 ± 0.77	0.6497 [b]
Creatinine (mg/dL)	1.47 ± 1.51	3.47 ± 6.27	0.6027 [b]
ATLAS score	3.28 ± 0.75	4.14 ± 1.59	0.1846 [a]
Days of antibiotic treatment	11.14 ±1.95	8.65 ± 4.69	0.1888 [a]
Time to cessation of diarrhea	5.57 ± 1.81	3.13 ± 0.65	**0.0027** [a]

[a] Unpaired t-test; [b] Mann–Whitney U test. Bold p-values represent statistical significance. LOS: length of stay (days). Cel/μL: cells per microliter of whole blood, kg: kilograms, g/dL: grams per deciliter of serum, mg/dL: milligrams per deciliter of serum. ATLAS: age, treatment with systemic antibiotics, leukocyte count, albumin, and serum creatinine; CDI: *Clostridioides difficile* infection; NDC: non-*difficile* Clostridia; ICU: intensive care unit.

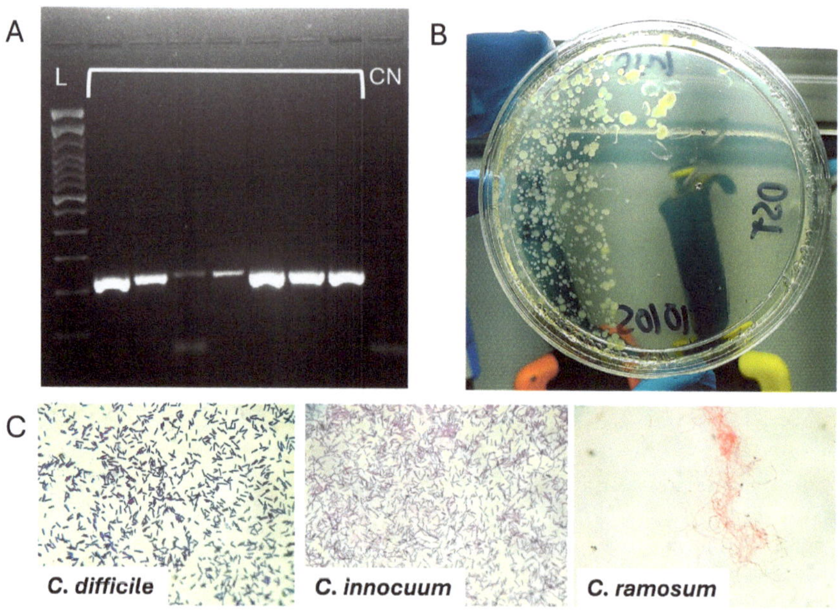

Figure 1. Molecular detection of *Clostridioides difficile* (CDI) and macro–microscopic morphology of non-*difficile* Clostridia (NDC). (**A**) Electrophoretic gel of the molecular detection by PCR of the *tpi* gene (230 bp) of *C. difficile*, L: DNA ladder (100 bp). Clasp lanes: samples, CN: negative control. (**B**) Macroscopic morphology of the colonies of an NDC (*Clostridium innocuum*), which are observed as small, whitish, irregular colonies with an opaque center and a light brown tone on CDA agar. (**C**) Microscopic morphology of CDI and NDC; the latter has a poor affinity to Gram and a very different morphology to CDI, being thinner bacilli, and in the case of *T. ramosa*, long and coiled. Total magnification: 400×.

2.2. Characteristics of CDI + NDC Group

The CDI + NDC group registered a greater number of stool movements compared to CDI patients, although it did not reach statistical significance (p = 0.096). Patients in the CDI + NDC group received more days of antibiotic treatment and significantly increased the time to cessation of diarrhea (p = 0.0027) compared to the CDI group (Table 1). Compared to

CDI patients, leukocytosis (>16 K Cells/µL) and the detection of toxin B were less reported in the CDI + NDC patients; furthermore, all the CDI + NDC patients were previously hospitalized (p = 0.0302) (Table 2).

Table 2. Frequency analysis of clinical variables of the study population.

Patient Characteristic	CDI + NDC (*n*, %)	CDI (*n*, %)	OR (CI)	*p*-Value
Toxin A detectable	6 (85.7)	18 (100)	0.11 (0.004–3.25)	0.2800
Toxin B detectable	4 (57.1)	16 (88.9)	0.16 (0.02–1.35)	0.1130
PPIs	6 (85.7)	12 (57.1)	4.50 (0.45–44.31)	0.3642
Leukocytes >16 K Cel/µL	0 (0.0)	8 (38.1)	0.10 (0.005–2.10)	0.0749
Treatment with Metronidazole/Vancomycin	3 (42.9)	4 (19.0)	3.18 (0.50–20.31)	0.3183
Treatment with Vancomycin	5 (71.4)	19 (90.5)	0.26 (0.02–2.36)	0.2530
Antibiotic switch treatment	2 (28.6)	1 (5.6)	6.80 (0.50–91.55)	0.1796
Previous Hospitalization	7 (100)	11 (52.4)	13.70 (0.69–270.5)	**0.0302**
Hospitalization in ICU	3 (42.9)	4 (19.0)	3.18 (0.50–20.31)	0.3183
Attributable mortality to CDI	0 (0.0)	3 (14.3)	0.35 (0.01–7.69)	0.5513

Fisher's exact test. OR: odds ratio. CI: confidence interval. PPIs: proton-pump inhibitors, Cel/µL: cells per microliter of whole blood, ICU: intensive care unit, CDI: *Clostridioides difficile* infection, NDC: non-*difficile* Clostridia. Bold letters indicate statistical significance $p < 0.05$.

3. Discussion

Traditionally, *Clostridium* are anaerobic Gram-positive spore-forming bacteria. However, some species are not spore-forming, such as *T. ramosa*, some can be oxygen-tolerant, and others can be microscopically visualized as Gram-negative (*T. ramose* and *E. clostridiformis*) [13]. Therefore, based on 16S rRNA phylogeny, several species were reclassified to other genera, such as *Thomasclavelia* and *Enterocloster,* among others. Particularly, [*T. ramosa*] and *C. clostridioforme* [*E. clostridioformis*] were renamed [4,11]. Toxin-producing *C. difficile* strains can cause AAD, pseudomembranous colitis and toxic megacolon [13]. Proper treatment of clostridial infections involves the administration of antibiotic therapy, including penicillin, vancomycin or metronidazole. Thus, precise identification of the *Clostridium* strains within NDC is important due to their discrepancies among treatment options [12].

T. ramosa often goes unnoticed in laboratory diagnosis due to its high heterogeneity in its affinity for Gram staining and its variable colony morphology. In the last decade, *T. ramosa* reporting increased due to more reliable identification methods such as MALDI-TOF-MS [14]. To our knowledge, the presence of viable *T. ramosa* co-colonization in patients with AAD caused by CDI has not been reported, such patients showed delay in gastrointestinal stabilization and recovery. *T. ramosa* can show susceptibility to metronidazole, amoxicillin/clavulanate, piperacillin/tazobactam or meropenem, whereas resistance to penicillin, ciprofloxacin, clindamycin, imipenem and ertapenem was also reported [7].

C. innocuum is a frequent colonizing microorganism of the gut microbiota in adults [9]. The relative abundance of *C. innocuum* can be higher in patients exposed to antibiotics and septic patients compared with healthy controls [15]. *C. innocuum* can show susceptibility to clindamycin, metronidazole, penicillin, piperacillin and ampicillin-sulbatam but is intrinsically resistant to vancomycin and can produce biofilm [15,16]. Moreover, *tcdAB*-like genes, associated with toxins, have been identified in *C. innocuum* strains [17], exhibiting its potential ability to cause AAD in hospitalized patients. Co-colonization of *C. difficile* and *C. innocuum* is frequent in patients with AAD [18,19]. Other studies rule out such an association [20]; however, all studies adopted a molecular detection approach and none of them analyzed the viability of these species in culture. In our study, we found that patients with positive *C. innocuum* cultures were associated with longer duration of treatments to cessation of diarrhea. Other studies have also reported *C. innocuum* in patients with intestinal diseases and diarrhea, which could explain the duration of treatments to stabilize diarrhea in our patients with co-colonization [18].

C. clostridioforme, also a member of the RIC group, generally show resistance to beta-lactams, glycopeptides, macrolides, chloramphenicol, lincosamides, rifampin, linezolid, bacitracin, aminoglycosides and tetracyclines. These gut commensals can act as a reservoir of antimicrobial resistances, mainly conferring vancomycin resistance [12,21]. In our study, we did not detect any strain identified as *C. clostridioforme*. Nevertheless, it is important to consider the colonization with this microorganism and its antimicrobial susceptibilities when investigating AAD causative agents.

C. butyricum, also part of the intestinal microbiota in almost 20% of adults, shows a protective gastrointestinal role against CDI [22,23]. *C. butyricum* can be used as a probiotic to stabilize dysbiosis due to the production of short-chain fatty acids (SCFAs) [24]; however, some strains and nutritional conditions (e.g., lactose consumption) are related to pathogenic properties of this species [23]. Therefore, strain typification must be evaluated for the determination of its role in AAD. Although not part of the RIC group, one strain was detected in our study.

Gut microbiota dysbiosis also plays a role in the pathogenesis of AAD, which can be either positive or negative, depending on the relative abundance of specific species [25]. In our study, prior hospitalizations were associated with co-colonization with NDC and C. difficile. Hospitalizations are often accompanied by antibiotic treatment, which causes intestinal dysbiosis, leading to the dominance of spore-forming bacteria such as these species [1]. Therefore, it is important to consider either *C. innocuum* or *T. ramosa* infection among probable causes when the suspicion of CDI does not subside even after proper treatment. Indeed, this reason could be attributable to our patients in the NDC group, who took more days to suppress diarrhea and needed longer antibiotic treatments.

Our study has some limitations, such as the sample size. While this is a small study and requires further analysis, the association between the factors we report and consider important is significant. Also, it would be interesting to analyze the minimum inhibitory concentration values for drugs such as vancomycin/metronidazole and the biofilm formation of the NDC strains to correlate them with persistence of symptoms or treatment failure. Likewise, a more in-depth analysis of the microbiome at baseline (pre-infection) and its comparison with the expression during infection could also help define the role of NDCs and their effect on the severity of clinical conditions.

It is plausible that the presence of RIC species in patients with *Clostridioides difficile* infection (CDI) plays an active role in contributing to pathological synergy, potentially exacerbating the infection and its symptoms. Alternatively, it may simply indicate a more profound dysbiosis, a disruption in the balance of gut microbiota, particularly in patients who have undergone prior antibiotic treatments. These treatments may have disrupted the normal microbial community, allowing the overgrowth of specific species like RIC, which may not necessarily be causally linked to the pathology but rather reflect an imbalance in the gut ecosystem.

4. Materials and Methods

4.1. Study Design

This study was a retrospective, case-control age-matched study.

4.2. CDI Case Selection

Hospitalized adult patients with a confirmed case of CDI from January 2023 to April 2024 were analyzed. The study was conducted at the University Hospital in Monterrey,

Mexico. Hospitalized patients who met the clinical case definition of 3 or more stool movements with a Bristol scale of 6 or 7 for 24 h were screened for CDI by a 2-step algorithm mentioned below and stool culture for *C. difficile*.

4.3. CDI Detection

The 2-step diagnostic algorithm included Glutamate dehydrogenase (GDH) and toxin A/toxin B assays (Meridian ImmunoCard Toxins A&B, Meridian Bioscience, Memphis, TN, USA) and molecular detection of *C. difficile* (*tpi* gene) through end-point PCR according to Lemee et al. [26].

All patients with a positive GDH test, at least one toxin assay positive and positive cultures with a significant growth (>50,000 CFU/mL) in anaerobic cultures of any Clostridial species were analyzed. We collect relevant clinical and laboratory data from medical records and our databases. For every CDI + NCD case, there were 3 CDI age-matched controls.

4.4. Ethics Statement

All procedures complied with relevant laws and institutional guidelines and have been approved by the appropriate institutional committee(s) with the approval code IF23-00001. Written informed consent was waived.

4.5. Stool Culture

Cultures were performed using samples used for diagnostic testing of GDH and toxins. These samples were obtained before any treatment targeting *C. difficile*. Stool samples were cultured on Reinforced Clostridial Medium EP/USP (Condalab, Madrid, Spain) and incubated under anaerobic conditions (CO_2 10%, H_2 5%, balanced with N_2) at 37 °C for 4 days. Colonies morphologically compatible with Clostridia, i.e., irregular, raised, convex grey colonies of 4–6 mm diameter, were further selected.

4.6. Clostridia Species Identification

Species identification was performed in 10 fully growing colonies of each morphotype compatible with clostridia. The identification was carried out with matrix-assisted laser desorption/ionization time-of-flight mass spectrometry (MALDI-TOF MS) (Microflex LT system, Bruker Daltonics, Bremen, Germany) according to the manufacturer's instructions, using the protein tube extraction method. The colonies were resuspended in 300 µL of water, 900 µL of ethanol was added, and the sample was centrifuged at 13,000 rpm for 2 min and the supernatant was discarded. Then, 20 µL of 70% formic acid (Fermont, Monterrey, Mexico) and 20 µL of acetonitrile (Fermont, NL, Mexico) were added. After centrifugation at 13,000 rpm for 2 min, 1 µL of the supernatant was transferred into a 96-well stainless-steel plate (Bruker Daltonics) and 1 µL of alpha-cyano-4-hydroxycinnamic acid matrix solution (Sigma Aldrich, Toluca, Mexico) was added. The plate was analyzed using the MALDI Biotyper 3.0 software to obtain spectral profiles and match them with the database. Scores above 2.00 were used as acceptable criteria for species-level identification.

4.7. Statistical Analysis

Quantifiable clinical data were described as means and standard deviation, and qualitative data were reported as frequencies and percentages. Comparative tests of means such as the Student's *t*-test and the Mann–Whitney U test were performed according to the data distribution. Fisher's exact test and odds ratio (OR) were calculated to determine the degree of correlation between the variables and the study groups. A p-value < 0.05 was considered as statistically significant and a value between 0.05 and 0.1 was considered as a statistical trend. The alpha error value was 5.0% in all the tests. The software SPSS® version 25.0 (IBM™, New York, NY, USA) was used.

5. Conclusions

In this study, we observed laboratory and clinical data related to CDI + NDC compared to CDI, such as history of previous hospitalization and longer treatment days to suppress diarrhea. These clinical and laboratory values are relevant when establishing the diagnosis and treatment of AAD, especially those that do not remit with conventional treatment, as it increases the possibility of NDC co-colonization. Although the number of samples and cultures performed in this short report are limited, we consider that cultivable co-colonization of NDC such as *C. innocuum* or *T. ramose* with *C. difficile* in AAD patients can be associated with treatment failure or prolongation.

Supplementary Materials: The following supporting information can be downloaded at: https://www.mdpi.com/article/10.3390/antibiotics14040397/s1, The results of the identification of the Non-*difficile* Clostridia isolates by MALDI-TOF used in this study.

Author Contributions: Conceptualization, A.C.-O. and D.S.-T.; methodology, A.C.-O. and D.S.-T.; software, D.S.-T.; validation, A.C.-O., D.S.-T. and S.F.-T.; formal analysis, D.S.-T., C.C.-R. and C.V.D.-R.; investigation, D.S.-T., C.C.-R. and C.V.D.-R.; resources, C.C.-R. and C.V.D.-R.; data curation, C.C.-R. and C.V.D.-R.; writing—original draft preparation, D.S.-T.; writing—review and editing, S.F.-T. and A.C.-O.; visualization, D.S.-T. and S.F.-T.; supervision, A.C.-O.; project administration, A.C.-O.; funding acquisition, A.C.-O. All authors have read and agreed to the published version of the manuscript.

Funding: This research received no external funding.

Institutional Review Board Statement: The study was conducted following the Declaration of Helsinki and approved by the Ethics Committee of the University Hospital Dr. José Eleuterio González (IF23-00001). Patient records were anonymized, de-identified, and access was restricted to the research team to ensure privacy.

Informed Consent Statement: Because this work was retrospective and used results obtained from conventional treatment and diagnosis (without experimental activities), informed consent was not necessary.

Data Availability Statement: Data used are contained within the article and Supplementary Material.

Conflicts of Interest: The authors declare no conflicts of interest.

References

1. Motamedi, H.; Fathollahi, M.; Abiri, R.; Kadivarian, S.; Rostamian, M.; Alvandi, A. A worldwide systematic review and meta-analysis of bacteria related to antibiotic-associated diarrhea in hospitalized patients. *PLoS ONE* **2021**, *16*, e0260667. [CrossRef] [PubMed]
2. Yutin, N.; Galperin, M.Y. A genomic update on clostridial phylogeny: Gram-negative spore formers and other misplaced clostridia. *Environ. Microbiol.* **2013**, *15*, 2631–2641. [CrossRef] [PubMed]
3. Cherny, K.E.; Muscat, E.B.; Reyna, M.E.; Kociolek, L.K. *Clostridium innocuum*: Microbiological and clinical characteristics of a potential emerging pathogen. *Anaerobe* **2021**, *71*, 102418. [CrossRef] [PubMed]
4. Lawson, P.A.; Perez, L.S.; Sankaranarayanan, K. Reclassification of *Clostridium cocleatum*, *Clostridium ramosum*, *Clostridium spiroforme* and *Clostridium saccharogumia* as *Thomasclavelia cocleata* gen. nov., comb. nov., *Thomasclavelia ramosa* comb. nov., gen. nov., *Thomasclavelia spiroformis* comb. nov. and *Thomasclavelia saccharogumia* comb. nov. *Int. J. Syst. Evol. Microbiol.* **2023**, *73*, 005694.
5. Yamairi, K.; Niki, M.; Imoto, W.; Kuwabara, G.; Shibata, W.; Oshima, K.; Yamada, K.; Kaneko, Y.; Kakeya, H. Two cases of Clostridium ramosum bacteremia with intestinal perforation: The antimicrobial susceptibility of clinical strains. *Anaerobe* **2023**, *80*, 102695. [CrossRef]
6. Shinzato, T.; Yonaha, T.; Oshiro, Y.; Ishiki, H. *Clostridium ramosum* bacteremia: A case series at a general acute care hospital. *J. Infect. Chemother.* **2023**, *29*, 78–81. [CrossRef]

7. Milosavljevic, M.N.; Kostic, M.; Milovanovic, J.; Zaric, R.Z.; Stojadinovic, M.; Jankovic, S.M.; Stefanovic, S.M. Antimicrobial treatment of Erysipelatoclostridium ramosum invasive infections: A systematic review. *Rev. Inst. Med. Trop. Sao Paulo* **2021**, *63*, e30. [CrossRef]
8. Chen, Y.-C.; Le, P.-H.; Wang, Y.-H.; Chuang, T.-C.; Yeh, Y.-M.; Chiu, C.-H. Gut Colonization and Antibiotic-Associated Diarrhea by *Clostridium innocuum* in Children and Adults. *Clin. Infect. Dis.* **2022**, *76*, 369–371. [CrossRef]
9. Luo, X.; Wang, X.; Wang, J.; Laranjo, M. *Clostridium innocuum*: More Important Than Ever. *Int. J. Clin. Pract.* **2024**, *2024*, 5797671. [CrossRef]
10. Cobo, F.; Pérez-Carrasco, V.; Tarriño-León, M.; Aguilera-Franco, M.; García-Salcedo, J.A.; Navarro-Marí, J.M. Bacteremia due to Clostridium innocuum: Analysis of four cases and literature review. *Anaerobe* **2023**, *83*, 102771. [CrossRef]
11. Haas, K.N.; Blanchard, J.L. Reclassification of the *Clostridium clostridioforme* and *Clostridium sphenoides* clades as *Enterocloster* gen. nov. and *Lacrimispora* gen. nov., including reclassification of 15 taxa. *Int. J. Syst. Evol. Microbiol.* **2020**, *70*, 23–34. [CrossRef] [PubMed]
12. Sermet, K.; Kipnis, E.; Duployez, C.; Wallet, F.; Dessein, R.; Le Guern, R. Answer to January 2022 Photo Quiz. *J. Clin. Microbiol.* **2022**, *60*, e00330-21. [CrossRef] [PubMed]
13. Gajdács, M.; Spengler, G.; Urbán, E. Identification and Antimicrobial Susceptibility Testing of Anaerobic Bacteria: Rubik's Cube of Clinical Microbiology? *Antibiotics* **2017**, *6*, 25. [CrossRef] [PubMed]
14. Legaria, M.; García, S.; Tudanca, V.; Barberis, C.; Cipolla, L.; Cornet, L.; Famiglietti, A.; Stecher, D.; Vay, C. *Clostridium ramosum* rapidly identified by MALDI-TOF MS. A rare gram-variable agent of bacteraemia. *Access Microbiol.* **2020**, *2*, acmi000137. [CrossRef]
15. Chiang-Ni, C.; Huang, J.-Y.; Hsu, C.-Y.; Lo, Y.-C.; Chen, Y.-Y.M.; Lai, C.-H.; Chiu, C.-H. Genetic diversity, biofilm formation, and Vancomycin resistance of clinical Clostridium innocuum isolates. *BMC Microbiol.* **2024**, *24*, 353. [CrossRef]
16. Chia, J.-H.; Feng, Y.; Su, L.-H.; Wu, T.-L.; Chen, C.-L.; Liang, Y.-H.; Chiu, C.-H. Clostridium innocuum is a significant vancomycin-resistant pathogen for extraintestinal clostridial infection. *Clin. Microbiol. Infect.* **2017**, *23*, 560–566. [CrossRef]
17. Bhattacharjee, D.; Flores, C.; Woelfel-Monsivais, C.; Seekatz, A.M. Diversity and Prevalence of *Clostridium innocuum* in the Human Gut Microbiota. *mSphere* **2022**, *8*, e00569-22. [CrossRef]
18. Skinner, A.M.; Petrella, L.; Spandoni, S.; Serna-Perez, F.; Johnson, S. Can *Clostridium innocuum* Masquerade as *Clostridioides difficile*? *Clin. Infect. Dis.* **2022**, *75*, 1268–1269. [CrossRef]
19. Cherny, K.; Balaji, A.; Mukherjee, J.; Goo, Y.; Hauser, A.; Ozer, E.; Satchell, K.; Bachta, K.; Kochan, T.; Mitra, S.; et al. Identification of Clostridium innocuum hypothetical protein that is cross-reactive with *C. difficile* anti-toxin antibodies. *Anaerobe* **2022**, *75*, 102555. [CrossRef]
20. Cherny, K.E.; Muscat, E.B.; Balaji, A.; Mukherjee, J.; Ozer, E.A.; Angarone, M.P.; Hauser, A.R.; Sichel, J.S.; Amponsah, E.; Kociolek, L.K. Association Between *Clostridium innocuum* and Antibiotic-Associated Diarrhea in Adults and Children: A Cross-sectional Study and Comparative Genomics Analysis. *Clin. Infect. Dis.* **2022**, *76*, e1244–e1251. [CrossRef]
21. Dehoux, P.; Marvaud, J.C.; Abouelleil, A.; Earl, A.M.; Lambert, T.; Dauga, C. Comparative genomics of Clostridium bolteae and *Clostridium clostridioforme* reveals species-specific genomic properties and numerous putative antibiotic resistance determinants. *BMC Genom.* **2016**, *17*, 819. [CrossRef] [PubMed]
22. Stoeva, M.K.; Garcia-So, J.; Justice, N.; Myers, J.; Tyagi, S.; Nemchek, M.; McMurdie, P.J.; Kolterman, O.; Eid, J. Butyrate-producing human gut symbiont, *Clostridium butyricum*, and its role in health and disease. *Gut Microbes* **2021**, *13*, 1907272. [CrossRef] [PubMed]
23. Cassir, N.; Benamar, S.; La Scola, B. *Clostridium butyricum*: From beneficial to a new emerging pathogen. *Clin. Microbiol. Infect.* **2016**, *22*, 37–45. [CrossRef] [PubMed]
24. Ariyoshi, T.; Hagihara, M.; Takahashi, M.; Mikamo, H. Effect of *Clostridium butyricum* on Gastrointestinal Infections. *Biomedicines* **2022**, *10*, 483. [CrossRef]
25. Choi, M.H.; Kim, D.; Lee, K.H.; Kim, H.J.; Sul, W.J.; Jeong, S.H. Dysbiosis of the gut microbiota is associated with in-hospital mortality in patients with antibiotic-associated diarrhoea: A metagenomic analysis. *Int. J. Antimicrob. Agents* **2024**, *64*, 107330. [CrossRef]
26. Lemee, L.; Dhalluin, A.; Testelin, S.; Mattrat, M.A.; Maillard, K.; Lemeland, J.F.; Pons, J.L. Multiplex PCR targeting tpi (triose phosphate isomerase), tcdA (Toxin A), and tcdB (Toxin B) genes for toxigenic culture of *Clostridium difficile*. *J. Clin. Microbiol.* **2004**, *42*, 5710–5714. [CrossRef]

MDPI AG
Grosspeteranlage 5
4052 Basel
Switzerland
Tel.: +41 61 683 77 34

Antibiotics Editorial Office
E-mail: antibiotics@mdpi.com
www.mdpi.com/journal/antibiotics